"In this lucid and innovative book, Mike Sheridan traces the remarkable global histories and cultural meanings of two powerfully charged boundary-marking plants – dracaena and cordyline – through five case studies of indigenous societies in Tanzania, Cameroon, New Guinea, French Polynesia, and the Caribbean. In critical dialogue with posthumanist approaches, his multi-sited ethnography of human-plant relations sets a model for political ecology. In the tradition of Eric Wolf, he masterfully combines history, culture, and power!"

Alf Hornborg, *Professor of Human Ecology, Lund University*

"In *Roots of Power*, it is fascinating to see tropical boundary plants being used to 'beat the bounds' of social science theorizing in so many directions. The author follows two sets of cultivars whose uncannily repetitive evocations offer a sense of definition in an entangled field!"

Marilyn Strathern, *Emeritus Professor of Social Anthropology, University of Cambridge*

"This brilliantly integrative ethnography explores the living boundaries of properties, and the properties people imbue in boundary plants themselves. Sheridan shows what else these mean in people's lives – material, social, and spiritual. *Roots of Power* marks a conjunction in the field of anthropology itself, conjoining deep work with far-reaching comparative study!"

Parker Shipton, *Professor of Anthropology, Boston University*

"Michael Sheridan's *Roots of Power: The Political Ecology of Boundary Plants* tells five stories of plants, people, property, peace, and protection. With assured prose and impressive scholarship Sheridan shows how two humble plants, used traditionally throughout the global tropics to set property boundaries, are also conduits for social harmony, resistance against oppression, and spiritual journeys. This is an important book, not only for the professional scholar, but also for those interested in the majesty of small things that so profoundly influence the gamut of human life, from the comity of neighbors to the deepest entanglements of social and political power!"

Judith Carney, *Distinguished Research Professor, Geography, UCLA*

"Historically, plants have been used as boundary markers in different communities around the world. Through the lens of dracaena and cordyline plants, Michael Sheridan delivers the first global comparative study of plants as social, political, and cultural boundary markers. A theoretical tour de force, this book demonstrates the power of anthropology to understand socio-cultural phenomena over time and space. Based on multi-sited ethnography in five countries, Sheridan analyses processes of socioecological change involving property relations, group identities, land use, and domains of meaning. This book is a new and refreshingly productive approach to the expanding field of political ecology, and to the discourse on property rights everywhere!"

N. Thomas Håkansson, *Professor Emeritus, Swedish University of Agricultural Sciences; Adjunct Full Professor of Anthropology, University of Kentucky*

Roots of Power

Roots of Power tells five stories of plants, people, property, politics, peace, and protection in tropical societies. In Cameroon, French Polynesia, Papua New Guinea, St. Vincent, and Tanzania, dracaena and cordyline plants are simultaneously property rights institutions, markers of social organization, and expressions of life-force and vitality.

In addition to their localized roles in forming landscapes and societies, these plants mark multiple boundaries and demonstrate deep historical connections across much of the planet's tropics. These plants' deep roots in society and culture have made them the routes through which postcolonial agrarian societies have negotiated both social and cultural continuity and change. This book is a multi-sited ethnographic political ecology of ethnobotanical institutions. It uses five parallel case studies to investigate the central phenomenon of "boundary plants" and establish the linkages among the case studies via both ancient and relatively recent demographic transformations such as the Bantu expansion across tropical Africa, the Austronesian expansion into the Pacific, and the colonial system of plantation slavery in the Black Atlantic. Each case study is a social-ecological system with distinctive characteristics stemming from the ways that power is organized by kinship and gender, social ranking, or racialized capitalism. This book contributes to the literature on property rights institutions and land management by arguing that tropical boundary plants' social entanglements and cultural legitimacy make them effective foundations for development policy. Formal recognition of these institutions could reduce contradiction, conflict, and ambiguity between resource managers and states in postcolonial societies and contribute to sustainable livelihoods and landscapes.

This book will appeal to scholars and students of environmental anthropology, political ecology, ethnobotany, landscape studies, colonial history, and development studies, and readers will benefit from its demonstration of the comparative method.

Michael Sheridan teaches anthropology and environmental studies at Middlebury College in Vermont.

Routledge Studies in Political Ecology

The *Routledge Studies in Political Ecology* series provides a forum for original, innovative and vibrant research surrounding the diverse field of political ecology. This series promotes interdisciplinary scholarly work drawing on a wide range of subject areas such as geography, anthropology, sociology, politics and environmental history. Titles within the series reflect the wealth of research being undertaken within this diverse and exciting field.

Roots of Power
The Political Ecology
of Boundary Plants

Michael Sheridan

Routledge
Taylor & Francis Group

LONDON AND NEW YORK

First published 2023
by Routledge
4 Park Square, Milton Park, Abingdon, Oxon OX14 4RN

and by Routledge
605 Third Avenue, New York, NY 10158

*Routledge is an imprint of the Taylor & Francis Group,
an informa business*

© 2023 Michael Sheridan

British Library Cataloguing-in-Publication Data
A catalogue record for this book is available from the British Library

ISBN: 9781032411408 (hbk)
ISBN: 9781032411422 (pbk)
ISBN: 9781003356462 (ebk)

DOI: 10.4324/9781003356462

Typeset in Times New Roman
by KnowledgeWorks Global Ltd.

For Kristina, Gaia, and Kieran

Contents

8 Conclusion: Beyond Boundaries 215

Figures

Acknowledgments

Parts of Chapter 2 were originally published in 2016 in a special issue of *Environment and Society: Advances in Research* (7(1): 29–49). I thank E&S editors Jerry Jacka and Amelia Moore for permission to revise and reprint "Boundary plants, the social production of space, and vegetative agency in agrarian societies."

This project would not have been possible without the guidance and insight of these research assistants:

Charles Ao (Papua New Guinea)
Hurlton Blackman (Tobago)
Eric Chu (Cameroon)
Caspar Dama (Papua New Guinea)
Errol Hassanali (Trinidad)
Chris Henry (St. Vincent)
Poerani Thia Kime-Ebb (French Polynesia)
Haruni Mbilinyi (Tanzania)
Livingstone Mboya (Tanzania)
Hope Nshiom (Cameroon)
Alfred Rolle (Dominica)
Bernard Tyrell (St. Vincent)

These friends and colleagues at Middlebury College provided support, encouragement, critical readings, and Heady Topper IPA:

Mez Baker-Médard, Kristy Bright, Svea Closser, James Fitzsimmons, Tracy Himmel-Isham, Jon Isham, Beth and Chris Keathley, Brad Koehler, Angela and Matt Landis, Rachel Manning, Marybeth Nevins, Ariane Ngabeu, Ellen Oxfeld, Laurie Patton, Mari Price, Jim Ralph, David Stoll, and Trinh Tran.

These students gave me invaluable feedback about what makes esoteric academic research interesting, and what a student-friendly text looks like:

Jaska Bradeen, Lila Buckley, Torrey Crim, Dan Krugman, Hannah Laga Abram, Massimo Sassi, Lizzie Stears, Katie van der Merwe, and Celia Woods-Smith.

These colleagues provided information, commentary, and excellent suggestions:

Shirley Ardener, John Njakoi Bah, John Beardsley, Jean Besson, Judith Carney, Emily Donaldson, Pamela Feldman-Savelsberg, Ian Fowler, Tom Håkansson, Päivi Hasu, Andreas Hemp, Barry Higman, Lennox Honychurch, Alf Hornborg, Jerry Jacka, Amelia Moore, Sally Falk Moore, Celia Nyamweru, Nancy Lee Peluso, John Rashford, Anthony Richards, David Schoenbrun, Parker Shipton, Glenn Stone, Marilyn Strathern, Mattias Tagseth, Amy Trubek, Jean-Pierre Warnier, Mats Widgren, Wallace Zane, and David Zeitlyn.

The stars of this show, however, are the people who invited me into their communities and patiently explained how and why boundary plants are so significant.

1 Introduction

Approaching the Boundary

Athumani's hardware store lay behind a thick hedge of dracaena shrubs, like many properties in Usangi, a town in Tanzania's North Pare Mountains. The masons were slinging cement and joking loudly behind the shop when I stopped by to check on the progress of his tourist hotel. I was informally advising a cultural tourism project sponsored by a European NGO, mostly by helping the guides write tour brochures. I was also teaching Athumani's wife Hawa how to make the pizza and pasta dishes that she expected to be tourist staples, and I needed to know if she'd found a reliable source of cheese. They weren't sure if they'd be feeding any tourists because of a quarrel within the project. Edgar, the headmaster of the town's secondary school, wanted to host tourists in the school's guest accommodations, and the emerging struggle between the two hoteliers threatened to break up the project. After a quick conference with my research assistant Haruni, I intervened in the conflict in the culturally appropriate way I had learned from elderly sacred grove caretakers. The same plant species that marks the boundaries of farms and the limits of sacred sites could, they said, settle disputes. I plucked a glossy green leaf of dracaena from Athumani's hedge, carried it to Edgar's office at the school, and announced that there should be peace in the tourism project. He received the leaf with both hands, laughing at the sight of a white American performing a traditional Pare reconciliation ritual. I picked a second dracaena leaf from a corner of the school grounds and presented it to Athumani, who grinned and observed that I had "really trapped him in the peace!" The project committee decided several days later to rotate the tourists between the two sites.

I thought that these uses and meanings of the *Dracaena fragrans* plant in North Pare were an arbitrary example of a cultural construct shaping a social landscape. In another place and time, I might have carried an olive branch instead. Then while working on a volume about African sacred groves (Sheridan and Nyamweru 2008), I was astonished to find references to the same plant imposing peace in West Africa. Why, I wondered, was this plant's social life so similar in places thousands of kilometers apart? This inspired me to track dracaena across tropical Africa and hypothesize

DOI: 10.4324/9781003356462-1

its long-term role in landscape formation (Sheridan 2008). A few sources for that project suggested that dracaena was doing similar work as a property rights institution, marker of sacred space, and peace symbol in the Caribbean. After I discovered that these references were actually about *Cordyline fruticosa*, the mystery deepened. Cordyline was only introduced to the Caribbean by British colonial botanists, and is famous in environmental anthropology as the symbolic linchpin of the New Guinean cultural and ecological system described in Roy Rappaport's groundbreaking *Pigs for the Ancestors* (1984). Rappaport describes how the plant marks territory, enacts social organization by gender and kinship, and symbolizes peace. Cordyline is also deeply meaningful throughout the Pacific, where it is commonly known as the ti plant. Why were Afro-Caribbean people using an Oceanic plant in ways that looked so similar to African uses of dracaena? Setting aside the simplistic idea that these webs of significance were fixed cultural traits that had diffused across the tropics, I followed the social roles, cultural significances, and historical contexts of dracaena and cordyline across the tropics in 2014–2015. I coined the term "boundary plants" to discuss the "vegetative manifestations of social institutions that assign resources to social groups and clothe them with culturally-defined legitimacy" (Sheridan 2016, 30) and to serve as the cord binding together a bundle of case studies.

This book tells five stories of plants, people, property, politics, peace, and protection. Each case study stands on its own, but they have interacted in important ways. The presence of socially complex dracaena in East and West Africa is, as described in Chapters 3 and 4, partly the result of the ancient Bantu expansion through which agricultural systems spread across the continent. The similar social lives of cordyline in Oceania are, as detailed in Chapters 5 and 6, the consequences of this plant having been a "canoe plant" for the ancient Austronesian seafarers who settled the Pacific. Finally, the description of cordyline in the Caribbean (Chapter 7) shows the effects of European empire-building and the emergence of Afro-Caribbean peasantries from the structural violence of plantation slavery. This book argues that in addition to their localized roles in forming landscapes and societies, these plants mark multiple boundaries and demonstrate deep historical connections across much of the planet's tropics. And because they are so deeply rooted in society and culture, these plants have also been the routes through which postcolonial agrarian societies have negotiated both social and cultural continuity and change. My narrow focus on two plant species unifies my discussion of cultural similarity and difference, social reproduction and revolution, and geographic linkage and separation, but comparing the case studies requires a more nuanced framework. This book demonstrates a multi-sited ethnographic political ecology of ethnobotanical institutions. This introduction unpacks this admittedly dense string of methodological terms in order to build the comparative scaffolding undergirding the chapters that follow.

Multi-sited ethnography

Anthropology's core method is participant observation. By sticking around long enough to learn how much of what people do matches what they say they do and why, the anthropologist eventually learns the right questions to ask and perhaps glimpses some answers. The secret ingredient in this recipe is that by staying put the anthropologist learns nuances and constantly asks themselves if observation A has some connection to experience B. To prevent projects from expanding infinitely in time and space, many anthropologists spend a year or two in a place they can get to know well on foot (Hannerz 2009). For the discipline's first century, this method produced in-depth studies of villages and localized groups, which then became comparative material. It also defined culture as the explanation for the texture of relationships in small spaces. This comfortable methodological assumption of bounded units became increasingly untenable in the 1990s as Cold War binaries and stabilities dissolved into the messy shifting dislocations of globalization (Gupta and Ferguson 1997a). Hybrid, deterritorialized, and processual social identities in shifting locations increasingly replaced the "bounded fields" that ethnographers had long tilled (Gupta and Ferguson 1997b). Multi-sited ethnography emerged as a methodological response to the fuzziness, fragmentation, and contradictions of capitalist globalization (Coleman and von Hellermann 2011; Marcus 1995). Instead of staying put, ethnographers investigated connections and relations by following people, things, and meanings.

Multi-sited ethnography allows new insights to emerge from globetrotting. By following migrant laborers (Holmes 2013), coffee beans (West 2012), or public health programs (Closser 2010), ethnographers can learn about connections beyond the prototypical village. Multiple sites encourage analytical innovation when the juxtaposition of concepts from one field site illuminates something about another far-off place (e.g., Myhre 2016a). Critics charge that multi-sited ethnography sacrifices depth for breadth, loses track of the native point of view amid Western modernity, and creates a false sense of sociocultural wholeness from fragmented observations (Boccagni 2019; Marcus 2016). The problem with accepting that "the field is no longer objectively 'out there'," and that anthropology's task is to study "regimes of living, global assemblages" (Marcus 2016, 193, 195) is that ethnographers must still choose swatches of the world's social fabric to stitch into meaningful accounts. Where and how do we make the cuts that create a "contingent framing" that re-weaves tangled threads into a comprehensible pattern (Candea 2007, 171)?

I made my cut by following two species of plants to places where they were socially significant and culturally complex. I had to deduce what stories I could tell about these boundary plants. I already had extensive experience in Tanzania, and the Cameroon Grassfields formed a dense cluster of significant dracaena in my initial research (Sheridan 2008). Papua New Guinea had to be on the list because of Rappaport's work. Tahiti made the cut it was where the British acquired cordyline for their Caribbean colonies. Finally, a regional survey showed that cordyline was deeply involved in landholding and religious practice in St. Vincent. Limitations of time and research funds meant that I had about two to three weeks at each field site. I relied on focus groups and interviews (approximately 30 in each location) to reveal the social lives of dracaena and cordyline. The resulting data set reflects several of the weaknesses of multi-sited ethnography, which I patched up with in-depth historical and ethnographic contextualization. I immersed myself in each region's literature to relate my field data to these accounts of the native point(s) of view. The resulting historical ethnographies retain vernacular terms as much as possible and emphasize direct quotation to maximize local flavor and meaning. All five case studies are postcolonial societies, so the analysis tacks back and forth between indigenous views of human-plant-property relations and colonial changes to landholding institutions. I avoid false holism by telling five boundary plant stories instead of two species' stories. These narratives are partial (depending on how much ethnohistorical material exists, and from what sorts of biased sources) and their interconnections tenuous. I was inspired by Peter Geschiere's comparative global study of witchcraft (2013). Instead of treating witchcraft as a single phenomenon with local manifestations, he contextualizes witchcraft's "trajectories" in different areas before drawing comparisons. Each of this book's case studies shows a plant's trajectory through time and space in order to describe how boundary plants have, like witchcraft, mediated "the impact of modern changes to long-term historical processes in the production and reproduction of society" (Geschiere 2013, xxviii). This book is not about two plants; it is a description of two plants' involvement in five distinct dynamics of continuity and change.[1]

My multi-sited method turns Gupta and Ferguson's advice for abandoning bounded cultural fields inside out. This book is very much concerned with boundaries in fields (and around houses). The crucial difference here is that these fields are not stable objects to be explained, but contingent windows into sociocultural complexity (Candea 2007, 179). Boundary plants' meanings of stability and order must be contextualized as fixed points for re-orienting land use and society in shifting historical contexts. In their book about globalizing economic integration and localizing cultural essentialism in the 1990s, Birgit Meyer and Peter Geschiere remark that

> in a world characterized by flows, a great deal of energy is devoted to controlling and freezing them: grasping the flux often actually entails a politics of 'fixing' – a politics which is, above all, operative in struggles about the construction of identities.
>
> (1999, 5)

Boundary plants are fixed points that organize the flows around them. This draws our attention to culture-specific processes of fixation. As the case studies demonstrate, each field location in this book is a fractured, incomplete, and fundamentally messy entanglement of people, plants, power, and politics. Each chapter untangles these threads, some fixed and some flowing, and re-weaves them into a narrative that allows comparison and juxtaposition. These chapters are partial maps of how and why fixed points in landscape, society, and culture came to be.

Political ecology

Political ecology is an interdisciplinary "community of practice" that invites anthropologists, geographers, historians, and ecologists into a collaborative intellectual space, not a bounded discipline (Robbins 2020, 17). The motivating assumption in political ecology is that there are no ecosystems, at least not on this planet, that are not also social systems. Relations of power keep this dialectic of nature and culture connected, so that every coffee berry harvested and every drop of gasoline burned are political and ecological objects and events. By re-framing ecological issues within political and economic contexts, better understandings of the causes of land degradation, climate change, and biodiversity collapse can emerge, and better solutions may be found. Political ecologists generally assume that power and conflict in social relations lead to transformations in ecology, landscape, and society, and I base my use of social-ecological systems as core analytical units on this approach. There is, however, a robust functionalist literature on social-ecological systems, which assumes that systems seek coherence and resilience (Colding and Barthel 2019). Chapter 8 revisits this contrast to explore how boundary plants could increase the coherence and resilience of tropical social-ecological systems.[2]

Political ecology relies on two analytical methods. Using a "hatchet," political ecologists chop up asymmetries of power in social-ecological systems to show why things are the way they are. By documenting creative, adaptable, and alternative ways of organizing environments and societies, political ecologists plant "seeds" from which healthier social-ecological systems might grow. Methodologically, political ecologists tend to focus on how social inequalities determine whose perceptions, uses, and claims to resources matter. Analysis can proceed from the classic factors of production in agricultural economics (water, labor, and capital), the elements of social organization (groups, networks, and categories), or cultural expressions of value and influence (power, wealth, and meaning) – or all three at once (Shipton 1994). Resources are not static objects embedded in human affairs and are better understood as "bundled" and co-constituting relations of control, access, power, and legitimacy (Ribot and Peluso 2003). Each chapter in this book wields a political ecological hatchet to reveal patterns of power and privilege, which then contribute to the overall argument that boundary plants have long been the seeds for creating and revitalizing social-ecological systems.

The Marxian anthropologist Eric Wolf coined the term political ecology in the title of a 1972 article about property rights in the Alps. He argues that

> The local rules of ownership and inheritance are thus not simply norms for the allocation of rights and obligations among a given population, but mechanisms which mediate between the pressures emanating from the larger society and the exigencies of the local ecosystem.
>
> (Wolf 1972, 202)

By locating land rights as the linkage between ecology and political economy, Wolf planted a seed for new scholarship and foreshadowed this book's examination of boundary plants. Wolf was a student of Julian Steward, who established cultural ecology as a way to organize ethnographic material by focusing on food acquisition and production. Steward's main idea was that a "cultural core" of food economics shapes a society's organization and cultural values. The differences among foragers, farmers, and industrial workers are the results of these economic specializations, not evolutionary steps. Wolf extended this approach by putting each system into a global historical context (see Chapter 2). Political ecology's next formative moment was Piers Blaikie and Harold Brookfield's *Land Degradation and Society* (1987), which reacted to simplistic Malthusian accounts of population growth causing environmental decline. They insisted that environmental change must be explained at multiple scales, from land managers' muddy boots to state policies and the structure of the global capitalist economy. This sort of structuralist political ecology was firmly materialist and showed how famine, soil erosion, and deforestation are ultimately caused by colonial histories and global capitalism. This framework produced robust explanations, but its economic determinism tended to cast local resource users as either the passive victims of state and corporate greed, or as active resistors and potential revolutionaries.

In the 1990s, scholars increasingly used ideas, symbols, and discourses to put flesh on these structuralist bones to produce poststructuralist and intersectional political ecology (Vaughn et al. 2021; Watts 2015). By examining environmental struggles and the frameworks of meaning that constitute them, this second generation focused on how the dialectic of ecology and society (also phrased as nature and culture) generates resource claims, social subjectivities, and legitimizing narratives (Paulson and Gezon 2005; Peet and Watts 2004). This analytical toolbox created new opportunities to investigate how environmental issues shape social categories and personhood. For example, feminist political ecologists showed that development agency efforts to formalize women's land rights often undermine already fragile terms of access and control, which then redefines power, gender relations, and the meanings of male and female (Elmhirst 2015). This generally constructivist approach to resources, landscape, and subjectivity exploded assumptions about static and stable units like trees, fields, and farmers.

It also demonstrated that intersectional social inequalities like gender, race, and class are the consequences of practices and struggles concerning resources and meanings, not simply their structural causes. The case studies in this book draw on poststructuralist political ecology to analyze the social history and cultural contexts of boundary plants.

Critics worried that this interdisciplinary community had too much narrow-minded discipline. Material dynamics like soil mechanics, tree growth, and water quality often seemed downplayed by political ecologists' focus on representation and discourse (Vayda and Walters 1999). Furthermore, the politics seemed to involve the same sorts of power and types of struggle. Were the politics in political ecology only about state oppression, capitalist accumulation, and hegemonic discourses controlling just about everything? What if power is plural and contextual instead of a coercive monolith (Biersack and Greenberg 2006)? In a recent review of anthropological theory, Sherry Ortner categorizes the ethnographic investigation of "the harsh and brutal dimensions of human experience, and the structural and historical conditions that produce them" as being driven by "dark theory" (2016, 49). If this sort of power is omnipresent, then it becomes the "intellectual black hole into which all kinds of cultural contents get sucked" and ultimately there isn't much else to talk about (Sahlins 2018, 20). To prevent power from flattening cultural difference, Ortner advises us to expand approaches to power and consider an "anthropology of the good." A political ecology of the good would focus on how environmental struggles and narratives relate to morality, well-being, imagination, empathy, care, hope, and justice, and not just their absence (Ortner 2016, 58). Although this book has a considerable amount of dark theory, it contributes to a political ecology of the good because the people in the following chapters consistently view their boundary plants as icons of social order, vitality, and moral personhood. My explanations of the project usually produced delighted smiles and glowing eyes, which I had not expected as a political ecologist marinated in dark theory. I had planned to write a book about boundary struggles, but instead these responses inexorably led me back to Malinowski's description of the greatest reward in ethnographic research as "realizing the substance of their happiness" (1961, 22). I am deeply grateful and humbled by this lesson in political ecology.

There is a conceptual challenge in this review of political ecology. The field is fundamentally about change and separating primary from secondary factors. A political ecology of the good is about continuities that integrate living beings into webs of life and hope. Which factors, then, have analytical, explanatory, and interpretive priority? Causation has become productively blurred in recent poststructuralist environmental ethnographies, which tend to describe entanglements and assemblages instead of structures and systems (Fontein 2015; Moore 2005). This terminology helps authors to focus on complexity, ambiguity, and indeterminacy, avoid essentialism and reductionism, and describe phenomena by examining how agency is distributed throughout a network (Guyer 2016, 23). I use these terms in this book but worry that a political ecology without clear causal principles is anemic. This book relies on poststructuralist and post-humanist insights, particularly those from the spatial, multispecies, and ontological "turns" in social theory (see Chapter 2) to get some analytical traction, but organizes them within a structuralist framework of economy, society, and culture. Each case study's social-ecological system is also a nature-culture entanglement, a socio-natural meshwork, and a multispecies assemblage, but I emphasize them as social-ecological systems in order to foreground how economic relations generate social organization and cultural meaning.

Ethnobotany

This book contributes to plants and people studies because it concerns the uses and meanings of plants in particular cultural contexts. I am a sociocultural anthropologist with an interest in plants, but I am not an ethnobotanist. My graduate school training did not include ethnobotanical methods, and I did not place voucher specimens of the plants I encountered in a national herbarium. Unlike most ethnobotanists, I did not investigate multiple plants and their significances. With a focus on just two species, this book offers narrow slices of a particular sort of ethnobotanical relationship instead of broad surveys of societies and flora. Its analytical themes, however, correspond with recent trends in plants and people studies.

Ethnobotany emerged in the 19th century as a method for documenting colonized peoples' knowledge about the economic, agricultural, and medicinal values of plants (Balick and Cox 1996; Nolan and Turner 2011). The discipline was broadly utilitarian, and focused on plants' uses as food, fiber, waxes, and so on. It was generally evolutionist, in that it investigated "aboriginal botany" and "primitive people" in sharp contrast to the knowledge system of Western scientific botany. Over the 20th century, ethnobotanists shifted from listing plant names and functional uses to address how plant knowledge and usage illuminate broader processes. Harold Conklin's study of farmers in the Philippines showed the sophistication of non-Western plant taxonomy, thus shaping the emerging field of ethnoscience (1957). Research on hallucinogenic plants generated insights into cross-cultural studies of religion and psychology (Schultes and Hofmann 1979). Finally, the discipline's roots in anthropology and ecology increasingly prompted ethnobotanists to address plants as parts of social-ecological systems using both natural science approaches, such as forest dynamics, and social science perspectives on meaning. Ethnobotanists expanded beyond the "tribal" academic niche by studying peasant farming communities and urban trade networks (Prance 1995). Ethnobotany has become increasingly critical, processual, and contextual instead of focusing on the functional aspects of plant knowledge (Voeks 2012).

Twenty-first-century ethnobotany is eclectic, with methods ranging from local taxonomies to aerial photography and ethnographic investigation of common property management via ritual and religion (Cunningham 2001). It has extended into the deep past with paleoethnobotany (Pearsall 2015) and studies of ancient human migrations (Pieroni and Vandebroek 2009). Most importantly for this book, this new phase of ethnobotany examines plants as sites of struggle and meaning-making in colonial and postcolonial contexts (McAlvay et al. 2021). Judith Carney, for example, has shown how African botanical knowledge about crops like rice and yams shaped New World landscapes (2001; Carney and Rosomoff 2009). Chris Duvall has demonstrated the similar racialized social history of a single plant genus, *Cannabis*, with particular attention to its role in colonial oppression and the social movements that resisted domination (2019). In these books, historical political economy meets ethnobotany to produce new insights into plant-people dynamics. Recent poststructuralist and posthumanist ethnobotanical work emphasizes how plant agency interacts with human intentionality. This plants and people scholarship argues that agency is distributed through human-plant networks, and that the botanical features of plants contribute to the form and meanings of these relationships because human and their societies have co-evolved with them (Besky and Padwe 2016; Daly et al. 2016). Chapter 2 continues this discussion of vegetative agency and its social science applications. Because the boundary plants in this book are both property rights institutions and representations of agency and personhood, these ethnobotanies of historical political economy and poststructuralist meaning-making will inform the ethnographic analyses in the case study chapters that follow.

Institutions

This book relies on the term "institution" as a way to describe boundary plants instead of custom, tradition, or trait (as in older scholarship). My reasons for this include ease of reading, suitability for translating what my ethnographic sources said about these plants, and the term's potential to catch the attention of policymakers. The English word comes from the Latin term *institutus*, the past participle of a verb for "setting up or putting in place." This word, in turn, comes from the proto-Indo-European root *sta-*, meaning "to stand, make, or be firm."[3] Boundary plants literally stand in place as landscape markers, and stand for social distinctions and symbolic categories, so calling them institutions is particularly fitting. The orderliness and stability of boundary plants within messy social and historical contexts is a major theme of this book, so it is necessary to define institutions very carefully with a review of social scientific uses of the term.

Classical social science views institutions as functional and structural arrangements that respond to human needs for biophysical, economic, social, political, and psychological order (Malinowski 1944; Parsons 1954).

Social institutions like laws, families, and government exist because we need them to, and together these building blocks of order form social and cultural systems. The task of anthropology, from this perspective, is to document and explain how particular institutions come to exist, why they fit together, and test whether these patterns illuminate social structure (Spiro 1965) and shape consciousness (Douglas 1986). Anthropology tends to take a wide view of institutions as persistent ways of doing things, from informal practices like handshaking to the strict formality of coronations. Institutional economists like Douglass North define institutions more narrowly as "the rules of the game" that constrain and direct social interaction (1990, 3). These approaches assert that institutions exist to get things done, and generally seek order and efficiency (Turner 2003). The people I interviewed for this book agree that boundary plant institutions exist to create social order and cultural continuity. This normative and functional view cannot, however, account for why these institutions take the particular forms they do, and why their roots entangle other institutions in different ways.

The critical response to this structuralism was, much like the reviews of fieldwork methods, political ecology, and ethnobotany showed, to refocus on process in the early postcolonial period. Colonial states had relied on a simple evolutionary assumption that traditional institutions would eventually become rational modern ones and spark economic growth. But modernization theory failed to predict the course of postcolonial change, and it turns out that the world's wealthy societies were never that modern in the first place (Latour 2012). A processual approach to institutions avoids tautological functionalism by viewing them as sites of social struggle and the outcomes of that action. In my own field of African studies, for example, Sara Berry redefined land tenure as a dynamic relationship among diverse social actors, who use different sorts of ambiguity and legitimacy to negotiate resource claims and the meanings of social categories (1989, 1993, 2017). Institutions are temporary solutions to old problems being experienced as permanent and inevitable. This seems obvious enough for technical institutions like the QWERTY keyboard. The lesson also applies to social arrangements like gender hierarchy, patrilineal kinship, and bridewealth exchanges. Today's East African societies are the institutional by-products of struggles over labor, resources, and meaning over 500 years ago (Schoenbrun 1998). This book presents boundary plants as similarly processual institutions which, despite standing fast and representing fixed order, have been conduits for change.

The most important institutional aspect of boundary plants is their role in property rights matters. Over approximately 40% of Earth's land area, two billion people hold land through indigenous institutions that are often ignored by their governments or regarded as problems to be solved rather than the foundations of livelihood and security (Alden Wily 2018a). In the 1990s, one of the central tenets of the so-called Washington Consensus was that private property rights and economic liberalism generate economic growth (Serra and Stiglitz 2008). Although interest in the potential of common property systems (Ostrom 1990) grew alongside this generally neoliberal trend, the policymaking focus was clearly on individualized and legally simple title deeds instead of collective and socially complex institutions. Hernando de Soto, for example, argues that poverty persists because poor landowners have only "dead capital" without secure title, which prevents them from converting land capital into financial capital (2000). Boundary plants are, from this point of view, economically dead – which, as we will see, is very much the opposite of how the people who use these institutions perceive them. The issue here is the role of the state. From Plato to John Locke, Adam Smith, and Karl Marx, the core question was how states should define property rights. Throughout most of this intellectual history, the hand holding property was an individual male's, and rights were a matter of how individuals relate to the state, not their relations to one another (Alden Wiley 2018b). The boundary plants in this book are, in contrast, held by gendered and socially categorized hands and reflect and assert relations within communities. Boundary plants are intersectional institutions with lively entanglements with other institutions.

This issue of institutional embeddedness is a central concern in economics. Do institutions of production, exchange, and consumption constitute a separate enough domain from kinship and religion that formal methods and equations can describe and predict behavior? Or is the economy so crosslinked with other social institutions and cultural meanings that it can only be understood contextually? The first "formalist" position is characteristic of mainstream neoclassical economics, which assumes that people are self-interested maximizers, while the second more cultural "substantivist" position locates economics as a collective phenomenon that operates in different ways in different societies (Gudeman 2016; Wilk and Cliggett 2019). This book takes a broadly substantivist position on boundary plants as institutions because, as the case study chapters reveal, these economic institutions are often more about inclusion and mutuality than exclusion and separation. But if boundary plant institutions are best viewed as processes, what sorts of processes do they contain and direct?

From the perspective of New Institutional Economics (NIE), institutions exist to reduce transaction costs, mitigate uncertainty, and incentivize efficiency (Baland et al. 2020; Galiani and Sened 2014). Once established, they make societies "path dependent" by limiting choice and allowing people to monitor others' behavior. This is how NIE defines social and cultural continuity. Boundary plants are what Douglass North would call informal and impersonal "third-party enforcement" mechanisms that let stateless societies without formal bureaucracies regulate individual behavior enough to achieve continuity (1990, 35). At the same time, actors make choices based on incomplete and flawed subjective models, and this eventually causes institutional change. For its proponents, NIE is a necessary correction to neoclassical economics and a way to synthesize it with a more substantivist approach

(Ensminger 2002; Ensminger and Henrich 2014). Its critics charge that NIE is a static and bourgeois answer to Marx's identification of institutional contradictions about labor and surplus as the primary causes of historical change (Platteau 2000, 10). In this book, I argue that NIE provides a useful macroscopic view of how and why property rights institutions matter, but that a microscopic lens is better for explaining how and why particular institutional forms shape social histories and landscapes. Understanding these institutions as processes among people instead of structural relations to the state requires social theories more grounded in actors' experiences (Leach et al. 1999, 238).

Political ecology has produced concepts that answer Douglass North's call for attention to informal cultural interactions with formal rules (1990, 140). Jesse Ribot and Nancy Peluso reconceptualized property and redefined institutions as social processes of access to material resources, persons, institutions, and symbols (2003, 153). By examining "bundles of power" in motion instead of structural bundles of rights, they outline methods for investigating and contextualizing the micropolitics of access to technology, capital, markets, labor, knowledge, authority, social identity, and social relations of friendship and trust. This approach brings discourses, power, and institutions together into a coherent framework. It also asks how particular access processes recursively produce specific identities and subjectivities. I follow this method and approach to institutions throughout this book. I complement this theory of access with Pierre Bourdieu's theory of practice (1990; Sheridan 2014). Bourdieu borrows the term capital from Marx and subdivides it into economic, social, cultural, and symbolic capital. Social actors both use and circumvent the "rules of the game" in an institutional domain (which he calls a social field) in an endless struggle to convert one sort of capital to another. A farm, for example, is a social field where farmers convert the economic capital of land into cash and the social capital of reputation. But this is never easy; social life is best understood as an enormous metaphorical market in which unequal people convert the products of their physical, social, and symbolic labor. With analytical tools like access as a bundle of powers and social life as capital conversions, we can understand boundary plants' particular institutional connections and entanglements with subjectivity. This helps to explain, for example, why some Vincentians access and accumulate social status through astral travel experiences, following the same plants that mark their yards, to the boundary of heaven (Chapter 7). The social-ecological systems described in this book consist of institutional struggles over access and control and constructions of subjectivity and personhood in specific historical and environmental contexts.

Outline of the book

This introduction defines the overall project of this book as a multi-sited ethnographic political ecology of ethnobotanical institutions. Each of these terms illuminates the topic of boundary plants when redefined as processes and relations instead of structures and things. Each term is a statement of method, and each contributes tools for the analytical toolkit that we will use to comprehend boundary plants.

Chapter 2 reviews the anthropological literature on boundaries, territories, and plant-people interactions. It introduces some freshly coined vocabulary to distinguish between relatively simple meanings of boundary plants and ones that are layered with significance and bristling with connections to other institutions. Summaries of the spatial, multispecies, and ontological "turns" in social theory lead to an argument that historical political economy has causal priority for our exploration of vegetative boundaries in social-ecological systems.

Chapter 3 begins the book's ethnographic journey with the story of the dracaena plant in northeastern Tanzania. I have been living in and studying this part of East Africa since 1988, and this case study of the Mt. Kilimanjaro social-ecological system represents the beginnings of my own encounters with boundary plants. The chapter examines the land tenure practices of peasant coffee farmers and their institutional connections to the social lives of ancestors in the landscape. On Kilimanjaro, the dracaena plant both symbolizes peace and embodies a sort of vitality and "vegetative gaze" that controls and disciplines Chagga people, so this chapter explores how boundary plants have a particular sort of moral personhood. Throughout this chapter, I show how the relatively "horizontal" power of kinship and gender relations find expression in the dracaena plant in contexts of colonial and postcolonial agrarian change.

Chapter 4 zooms across tropical Africa to the Cameroon Grassfields, where the same boundary plant does strikingly similar things, but in the context of a strictly ranked monarchy. I show how dracaena demonstrates land tenure in the Oku kingdom and describe how it expresses social organization among ranked kin groups. This is most apparent in the masquerades and anti-witchcraft activities of Oku's secret societies, which channel the life force that the people of Oku find so delightfully visible in the dracaena plant to create social and ecological order. The chapter closes with an argument that the plant directs this vital force of prosperity and well-being toward the symbolic and political center of the kingdom. It therefore represents the "vertical" dimensions of power in a ranked society.

Chapter 5 travels Papua New Guinea's Highlands Highway to the island's northeast coast at Madang. It presents evidence that New Guineans perceive aspects of themselves and define some of their social relations by reference to the botanical properties of plants. One of the major plants in this "vegeculture" is the cordyline plant, which is both a mechanism for access to land and a means for communication across linguistic boundaries. After a review of land tenure matters, I explore how Papua New Guineans use this plant to embody places and to emplace bodies. I discuss how and why islanders find the cordyline plant so aesthetically pleasing because of its vitality and its capacity to link bodies to landscapes with relations of order and protection. The chapter ends with an analysis of how cultural tourism is transforming boundary plant institutions. This ethnographic description has much in common with the Kilimanjaro case study because both are about socially "flat" societies in which social labor and moral personhood are organized mostly by gender and kinship.

Chapter 6 follows the ancient migration of Austronesian sailors across the Pacific to French Polynesia's Society Islands because cordyline was one of the plants they used to colonize new islands. The chapter explores how this boundary plant delimited sacred space, how cordyline mediated the colonial encounter, and how postcolonial French Polynesians are now using the plant to revitalize their culture. In contrast to the Papua New Guinea chapter, the "vegeculture" on display in this chapter was, until very recently, about social ranking, much like in Cameroon. Because indigenous economic foundations, social organization, and cultural practices were all thoroughly shattered by European colonization, this chapter is primarily concerned with boundary plants as mediators of change, often precisely because they represent lost continuity.

Chapter 7 describes another epic maritime voyage, Captain William Bligh's journey from Tahiti to St. Vincent in 1792, which happened to carry the cordyline plant. This plant is now significant as an Afro-Caribbean property rights institution and mechanism for accumulating cultural capital through religious practice. This probably does not express African meanings of dracaena or Oceanic uses of cordyline. Instead, I argue that boundary plants defined sites of struggle and creativity because cordyline showed the limits of oppression by the plantation system. Farmers used the plant to express their access to "provision grounds," and this became one of the cracks in the edifice of the racist plantation system where resistance could grow, and eventually produce an independent peasantry and a particular sort of moral personhood. Unlike the other case studies in which boundary plants demonstrate fixedness, the St. Vincent chapter shows how cordyline moves around a creolized cultural landscape. The chapter concludes with an account of how some Vincentians use cordyline to escape social marginalization through out-of-body experiences of travel to "Africaland." Unlike the other case studies that show mixtures of continuity and change, the St. Vincent story is nothing but change. Cordyline in St. Vincent is a story about creative creole efforts to build order and justice against the rigid and racist class hierarchy cemented in place by colonial capitalism.

Chapter 8 concludes the book by returning to the topical and methodological issues set forth in this introduction. It examines what boundary plants contribute to the multi-sited ethnography of land tenure, the political ecology of peace and vitality, the ethnobotanical study of social-ecological systems, and the comparative study of economic institutions. I review how boundary plants have shaped social forms and landscapes in precolonial and colonial pasts, in the postcolonial present, and extrapolate what the future of boundary plants may entail.

This introductory chapter is a knot of power, agency, place, comparison, social organization, politics, environmental knowledge, economic transformation, and land use. Each boundary plant story is a partly-untangled knot of ecology, society, culture, and history. Each is a bit like the knot of dracaena that I carried from Athumani's hardware store to Edgar's office at the secondary school in North Pare, Tanzania in 1998. Each encloses and expresses culture-specific definitions of peace and order, even in harrowing colonial and postcolonial conditions. Each ethnographic case study conveys glimpses of a culture-specific "substance of their happiness." I argue that these knots of life and vitality take the forms they do because boundary plants are institutions that connect economic, social, and cultural fields. These forms have a certain family resemblance across the case studies, but each presents a different face because, I argue, their particular histories differ in terms of their organization of social labor by kinship and gender, tribute and rank, and racialized capitalism. These stories of boundary plants matter not only because they express something significant to the people I met in 2014–2015. The struggles and narratives of people like Athumani and Edgar are important because boundary plants have been and likely will continue to be part of postcolonial efforts to achieve justice, fairness, and sustainability in social-ecological systems. The next chapter begins our journey by gathering up the tools necessary for the comparative study of boundary plant institutions.

Notes

1. George Marcus (1995, 101) notes that multi-sited ethnography often explores a series of monolingual contexts. All of my case studies were, at some point, parts of the British Empire. The languages I used other than English include Swahili (Tanzania; I am fluent), Eblam Ebkwo and Pidgin English (Cameroon), Tok Pisin (Papua New Guinea), French (Society Islands), and Vincentian Creole (St. Vincent). In most field sites, I had research assistants to clarify and translate because my language skills range from beginner's (French) to negligible (Eblam Ebkwo) and laughable (Tok Pisin).
2. See also Berkes et al. (2003), McGinnis and Ostrom (2014), and Ostrom (2009). Conflict-oriented critics of functionalism include Håkansson (2019), Hornborg (2009) and (2013), Sheridan (2012), and Widgren (2012).
3. https://www.etymonline.com/word/institution

2 Beating the Bounds for Boundary Plants

Laughing, the men threw one struggling boy into the icy stream at the parish boundary. A less fortunate child went face-down into a bed of stinging nettles. Most boys had their heads knocked against the village's ancient boundary stones while the men struck the rocks with thin willow branches. In some parts of the United Kingdom, this late medieval custom of "beating the bounds" has been revived recently as a more family-friendly event culminating with lunch at a local pub (Darian-Smith 1995). The painful ritual's purpose was to "remember" the community's territorial and moral limits into the boys' minds and bodies (Houseman 1998, 450). This hazing also defined a social-ecological system because these boundaries determined access to fish, pasture, and firewood (Whyte 2007). The custom declined as surveyors and maps carved up England's common lands in the 18th and 19th centuries to produce today's landscape of private fields and hedgerows. A low-stress version survives where I live in Vermont because state law (1 V.S.A. § 611) requires our attorney general to meet their counterpart from New Hampshire to "perambulate" the boundary between our states every seven years. In this spirit, this chapter beats the bounds of this book's intellectual territory.

Boundaries have always been anthropological staples. Understanding nationalism, globalization, or migration requires attention to spatial limits. Gender, race, and class studies focus on bounded categories and the work required to cross them. In environmental studies, ecosystem conservation, organic agriculture, and spiritual ecology only make sense when bounded by land laws, agricultural practices, and cosmological meanings. Anthropologists ask how some boundaries maintain relationships while others allow ecological, social, and cultural transformation. Yet despite the centrality of boundaries in social science, there has been little scholarship on plants that do these things. Boundary plants are ecological features imbued with social functions and cultural meanings, but often appear only in the ethnographic background. It is high time to bring living fences and vegetative borders in from the academic periphery.

This attention to botanical edges is important because land is fundamental to sustainable livelihoods (World Bank 2020), and a key policy issue is

DOI: 10.4324/9781003356462-2

whether the world's poor need stronger individual property rights (de Soto 2000) or strengthened indigenous systems (Home and Lim 2004; Lawry et al. 2017). By documenting what boundary plants mean and how they relate to landscape and society, this book gives smallholder farmers a tool for negotiating with state institutions. Boundary plants are analytically interesting because they exist at the intersections of landscape ecology, social structure, and cultural meaning-making. They are particularly good for relating social power to ecological dynamics (Widgren 2012). Boundary plants are, like trees, "good to think with" (Bloch 1998), because they require both materialist and symbolic approaches. I begin by discussing boundaries in classical anthropology, with particular attention to botanical examples. After introducing some terminology, I review examples of boundary plants in the social sciences and recent literature on space and place. Finally, I examine three theoretical "turns" and evaluate their usefulness for illuminating the social lives of boundary plants.

Structure, territory, and tenure

Defining, maintaining, and transforming land boundaries are classic anthropological topics. Anthropology tends to view boundaries as structures of rules and meanings, as elements within systems of ecological knowledge, and as sites of struggle and contestation. For 19th-century evolutionists, boundary formalization signified a transformation from communal ownership to private property (Marx 1915, 100; Morgan 1878, 217). To help organize local government and spur cash crop production in its colonial empire, British ethnographers documented diverse tenurial systems with a bewildering array of methods for demarcating land (Garson and Read 1892, 150, 224). Their construction of "customary law" often revealed more about European attitudes toward land than those of colonized people (Chanock 1991). Many ethnographers followed Bronislaw Malinowski's dictum that land tenure scholars should contextualize these relationships in terms of meaning and mythology as well as economics and law (2013, 766). Jomo Kenyatta, for example, showed how Kenya's Gikuyu people once marked boundaries with the pink *itoka* lily (*Crinum kirkii*) and the stomach contents of a sacrificial sheep (1965, 40). The anthropology of land tenure established that it is better to consider tenure as a socially distributed "bundle of rights" than a dichotomy of communal and individual objects (Bohannan 1963), and as "estates of administration" instead of personal advantages (Gluckman 1965). In practice, these meant that an African woman could harvest a tree's fruit, but only her husband could cut it for timber, and that neither could sell the land upon which it stood because this was a kin group's inalienable property. Overall, these classic land tenure ethnographies took a normative and jural approach, leading anthropologists to document functional rules and roles.

Land tenure studies became increasingly important as development agencies tried to lift former colonies to the production and consumption

levels of wealthy countries. Would markets or demography drive the evolution of farming from horticultural gardening to intensive agriculture,
and how should states plan for these changes? Economist Ester Boserup
persuaded planners that population pressure leads to both technological
innovation and the emergence of land markets (1965; Robertson 1984). In
rural Tanzania, for example, farmers creatively plant private eucalyptus
trees on the edges of what had been the fields of entire kin groups (Gillson
et al. 2003, 376). Concern about environmental degradation spurred a new
focus on common property regimes (McCay and Acheson 1987) in response
to Garrett Hardin's oversimplification of collective resource management
as the "tragedy of the commons" (1968). Scholarship in the 1980s showed
that societies often develop sustainable management institutions. The
most important design principle for successful common property systems
is "clearly defined boundaries" (Ostrom 1990, 90). These boundaries are
increasingly described as expressions of knowledge, interest, and social
organization by differently positioned social actors (Leach et al. 1999). In
general, the forms of boundaries, whether they consist of metal pegs, lines
of shrubs, or clay pots buried at the edges of fields, have mattered less than
their social and ecological functions.

The meta-theoretical issue at stake here is how continuity relates to
change. As social-ecological systems change, which elements and relations
are resilient enough to persist through demographic growth or colonization, and which are profoundly altered? Do these changes follow universal
rules, or is each case particularistic? Are these one-way transformations or
dialectics? Do systems continue or change because of structural conditions
or because of human choices? These were the founding questions of late
19th-century social science, when scholars were grappling with a strange
new world of industrial capitalism, colonizing nation-states, and secular
bureaucracies (Giddens 1993). These questions matter for our exploration of
boundary plants because these vegetative institutions often represent social
and cultural continuity yet are also critical sites for change. The case studies
in this book engage this paradox by showing landscapes and personhood
interacting through a triangle of political economy, social organization, and
ideas that transforms space into place. But each story of a social-ecological
system experiencing continuity and change through boundary plants is different because its economic foundations relate to society and symbols in
a unique way. The goal of comparison, therefore, is to let one case study
sharpen the questions for examining other examples of continuity and
change.

Colonial states treated land matters as structures instead of dynamic relationships, and in doing so they constructed systems of customary law which
often damaged landholding and resource management more than facili-
tating stewardship. In recent anthropology and agricultural history, land
tenure has been recast as a social process rather than a static arrangement
of rules and objects. The ongoing struggle to decide what land belongs to

which people led scholars to focus on negotiation and accumulation (Berry 1993; Peters 2009). New property rights institutions, such as freehold title deeds, often conflict with older ones like inalienable family land. As social actors strategize and struggle their way through this institutional tangle, they also generate and assert new identities and ways to legitimize claims (Peters 2013, 549). From this perspective, planting a boundary is one move in a grand game of making claims, defending access, and asserting control in a particular social context and historical moment. The lesson of this processual approach to land tenure is that boundaries are miniature arenas where people struggle for access to and control over resources and meanings. From this point of view, property is not a social relationship between people and things; it is "a vast field of cultural as well as social relations" and "symbolic as well as the material contexts within which things are recognized and personal as well as collective identities are made" (Hann 1998, 5). Boundary plants are prime examples of these sorts of social relations at the material and symbolic intersections of ecology and identity.

From structure to process

Anthropological units were once bounded social groups like tribes, villages, and lineages. The naturalness of these units came into question as scholars applied Marxist and constructionist approaches to these groupings (Guyer 1981; Mamdani 1996). The most influential work on boundary issues was Fredrik Barth's *Ethnic Groups and Boundaries* (1969), which argued that ethnicity results from accumulated transactions among individuals. The behavior that constitutes ethnic identity is therefore most apparent at group boundaries, where those differences most shape interaction. This explains why St. Patrick's Day parades are often more popular on the margins of the Irish diaspora than in Dublin (Cronin and Adair 2002). Barth repositioned identities and boundaries as consequences of behavior rather than preexisting conditions shaping behavior. This move shifted attention from the social units' internal structural characteristics to external relations among units, which then refocused anthropology on boundary-making, breaking, and crossing.

Recent work on boundaries has generally followed Barth, while also asking when and how boundary construction occurs, and what sorts of constructs result. Why do some become taken-for-granted foundations of social action and political conflict (like Balkan ethnic wars) while other constructions remain more or less quiescent (like Switzerland's lack of ethnic political parties despite its ethnolinguistic differences)? Why are some boundaries hard, closed, and sharply defined while others are soft, open, and fuzzy? In short, when and why do boundaries matter? One answer is that identities are social relationships between people, not ideas locked in our heads. This "relational" approach (Lewellen 2002, 108) stresses that socioeconomic inequalities can become so naturalized in the minutiae of daily life that constructed categories such as Palestinian or Tutsi seem natural (Comaroff 1996). The pace of identity construction accelerates when societies endure economic stress, and as the structural bases of group boundaries become unstable, the symbolic boundaries become more rigid (Cohen 1985). By defining boundaries as relationships, this approach emphasizes historical processes rather than stable entities in static structures. In this new work, the focus is less on identities and boundaries, and more on what kinds of boundary-making relationships are possible and the sorts of social action within them (Wimmer 2008). Instead of rigid boundaries and clear identities, this processual approach examines boundaries as flexible borderlands where ambivalence, contradiction, and hybridity complicate identities and the political economies that contain them (Anzaldúa 2012). In this book, boundary plants will be analyzed primarily as a relational phenomenon.

This generally processual turn in the social sciences can be summarized with a simple but challenging maxim: when you encounter a noun about people, treat it as a verb if possible. For example, American anthropologists used the term "culture" to describe group commonalities. "Because of culture" was therefore the answer to many anthropological questions. In the new processual mode, culture is less deterministic, and instead something that people do to and with one another, not a pre-existing cause. This book's social-ecological systems, boundary plants, and land managers are also processes of becoming, not static units. The list of dynamic institutions in this book includes familiar anthropological units like kin groups, royal clans, gendered bodies, and peasants, and less familiar ones like "yard society" and "vectorial person." To organize this material and allow comparison, I designed this project to examine boundary plants in three types of social systems: gender and kinship-based, ranked and tributary, and capitalist.

This typology comes from Eric Wolf's materialist approach to world history (1982). Wolf hammered together the concepts of culture and power to add cultural nuance to the core Marxian assumption that labor generates social and cultural forms, practices, and ideas. He approached culture as "a material social process" to make the point that culture is rooted in material matters like food supply (Roseberry 1989, 26). Culture is produced by people in particular economic contexts, not a pre-existing set of ideas and practices that determine interaction. From Wolf's insight that social and cultural action is a sort of work that people do every day, concepts like social and symbolic labor follow. But, he teaches us, these dynamics and their forms vary according to mode of production and place in the global political economy. In this book, the Tanzanian and Papua New Guinean case studies illustrate gender and kinship-based social-ecological systems, in which social labor is relatively "horizontal." The Cameroon Grassfields and French Polynesian examples show tributary social labor, through which surplus production supports elites in more "vertical" societies. The St. Vincent chapter demonstrates struggles over labor, land, and meaning in a capitalist post-slavery peasantry. Wolf's typology of modes of production brings the similarities among these boundary plant stories into focus, helps to recognize some important differences, and defines this book's overall analytical problem as the connections among property rights institutions, social organization, and culture-specific ideas about agency and personhood.

In his later work, Wolf redefined power to complement this reconceptualization of culture. He outlined four "modalities of power" (1999, 5) that can be described in scalar terms as increasingly abstracted social relations. These levels of analysis are useful because they avoid the circularity of finding that power is powerful. Instead, Wolf insists that we locate and interpret power, and powerful ideas in particular, in specific parts of society and in particular historical contexts. This approach illuminates the social actions of boundary plants because it pushes the analysis beyond the topic of vegetative agency. In the following chapters, we will encounter different forms and meanings of plant agency, and Wolf's scheme contextualizes these variations according to economic system. A boundary plant's institutional power to organize a landscape is qualitatively different in relatively egalitarian systems than in more hierarchical ones. A boundary plant has a different role as a symbol in a gender/kinship-based society compared to its structural power in a tributary monarchy. Wolf's culturally nuanced historical materialism is a good fit for understanding how and why boundary plants are often both property rights institutions and icons of cosmological order. This does not, however, mean that power is simply an ideological veneer on resource access and control. Power at each of Wolf's levels is the culturally defined process of meaningful agency and action, not just a matter of coercion (Fig. 2.1).

Modality	Description	Examples
Individual power	The potency or capability inherent in an individual; force of personality	Labor capacity, embodied skills, charisma
Interpersonal power	The ability of a person to impose their will on another	Weapons, the use of force
Institutional power	Social arrangements that control the contexts of interaction	Laws, rules, norms
Structural power	Ideas and ideologies that assign and legitimize resource allocation and the mobilization of social labor to particular people	Taboo, sacrifice, gender hierarchy, ethnicity, caste, sacredness, racism

Figure 2.1 Eric Wolf on power

These processual approaches offer tools for taking the study of boundary plants beyond simple markers of land tenure. The boundary plants in this book represent more than property rights; they are also emblems of group identity and expressions of the legitimacy of particular social orders in specific historical contexts. Landscape features like garden plants, hedgerows, and living fences contain social identities and cultural meanings, and are sites for the negotiations and struggles that comprise and construct them. For this book's tropical land managers, boundary plants are like hinges around which social and cultural continuity pivots as peoples and landscapes experience historical change. This multi-layered significance makes them "key symbols" that both summarize cultural values and allow the elaboration of those meanings in social action (Ortner 1973). Like a national flag or a rags-to-riches narrative, a key symbol is both good to think with and good to feel with. As the following chapters will demonstrate, boundary plants are emotionally and intellectually powerful because they mark landscapes and symbolic domains.

Symbolic boundary processes

Symbolic domains have been basic to ethnographic investigation since Durkheim argued that religion tends to create a category of sacredness that is "set apart and forbidden" (1915, 47). In her classic work on symbolic boundaries, Mary Douglas extends these ideas by asserting that the limits of the human body are analogous to the boundaries of the social group. This explains why concerns over blood, semen, excrement, and urine shaped the social organization of both ancient Israelites and contemporary Hindus (1966, 125). Anxieties about polluting substances that cross the body's boundaries reflect anxieties about the purity (and ultimately the existence) of the body politic and find expression in ritual prohibitions. Taboos are therefore not irrational ethnographic curiosities; they are attempts to maintain social order by keeping symbolic categories bounded. As we will see, boundary plants often involve both bodily protection and social order.

In Victor Turner's work on ritual process, we see these symbolic boundaries in action. Turner built an analytical language for symbols by focusing on behavior rather than thought, and perhaps his most famous example concerns an African boundary plant. Turner examines the multiple meanings of the *mudyi* tree (*Diplorhynchus condylocarpon*) for the Ndembu people of northwestern Zambia. Pubescent Ndembu girls spend a hot day wrapped in a blanket at the base of this tree. Because it has milky latex sap, the Ndembu associate the tree with breast milk and the ties between mothers and daughters. The Ndembu reckon descent matrilineally and reside patrilocally, so closely linked mothers and daughters often live far apart. In Ndembu thought, the *mudyi* tree stands for men's unity and "tribal custom" (1967, 22), yet in ritual practice, the same symbol separates mothers and daughters, the girl's mother from other women, and women from men. By showing the symbol in action, providing Ndembu explanations, and exploring his own interpretations, Turner established methods for symbolic analysis. The *mudyi* tree is a dominant symbol that condenses diverse significances into abstract ideological and pragmatic sensory meanings, which then become props in the "social dramas" of Ndembu life. This boundary plant does not confer property rights; instead, it mediates Ndembu cultural categories. The value of Turner's examination of the *mudyi* tree is that he contextualizes a boundary plant as part of a social process without reducing its significance to normative rules.

The problem with symbolic anthropology's theoretical toolkit for understanding boundaries in action is that the notion of boundary may not mean the same thing universally. As Barth notes, "'boundary' has consistently been *our* concept, made to serve our own analytical purposes" (2000, 34). In English, a boundary is simultaneously a marker of territory, a separator of social groups, and a divider of mental categories. This is just as much a cultural model as the Ndembu *mudyi* tree, but it has come to be a particularly hegemonic concept because of its place at the foundations of capitalism and the nation-state system. Barth cautions against assuming that everyone everywhere thinks of territory-group-category limits in the same way, and this opens up the empirical and ontological question of whether different cultural groups perceive boundaries differently. An English boundary limits and separates, but what if another people's boundary opens and unifies? Would that still be a boundary, or would such a translation into English completely distort its significance? The study of boundary plants can respond to Barth's challenge by testing, and perhaps transforming, the boundary concept.

These diverse approaches to boundaries in classical anthropology share some characteristics. All move from treating symbols as objective and static facts toward a concern for process and construction. This processual turn sets the agenda for the other turns that we will examine in this chapter. Land tenure had concerned sets of rules, but increasingly it is about contests among unequal actors (Juul and Lund 2002). Ethnicity and identity had been group characteristics, but are now interactions and relations of power (Comaroff and Comaroff 2009). Symbols had been units of meaning, but they became processes of negotiation with cosmological, social, and ecological aspects (V. Turner 1977). The core lesson for the study of boundary plants is that they are always embedded in economic, sociopolitical, and ideological processes and that their status as separators and/or uniters places them literally at the crux of these dynamics.

Monomarcation and polymarcation

Boundary plants usually appear only in the background in academic literature. In Kojo Amanor's account of land degradation in Ghana, we learn that farmers use dracaena to make live fences, but not the plant's other meanings (1994, 118). Wallace Zane's ethnography of an eastern Caribbean religion mentions that believers mark graves with dracaena plants (this is actually *Cordyline fruticosa*, see Chapter 7), but the plant's role in land matters is unexplored (1999, 48). In her examination of a community-based conservation program in Papua New Guinea, Paige West describes the violent consequences of someone cutting the cordyline plants around a conservation office – but the issue at hand is the fight, not land tenure (2006, 22). The ethnobotanical literature is even more limited in social contextualization. H.M. Burkhill reports that *Dracaena arborea* is used throughout West Africa to repel evil spirits (1985, 509). On the Oceanic island of Pohnpei, builders bury cordyline leaves under new houses to protect them from witchcraft (Balick 2009, 283). Of course, these authors were understandably more concerned with (respectively) farming systems, religious practice, conservation, economic botany, and ethnomedicine; however, I argue that boundary plants would, if layered into researchers' analyses, deepen their understanding of the economic, sociopolitical, and ideological processes that make up socio-ecological systems.

Some new terminology is needed to evaluate these aspects of boundary plants. Some of their roles seem rather simple, like the privet hedges (*Ligustrum ovalifolium*) that separate yards in American suburbs, while others carry complex cultural loads. I propose two neologisms to discuss this contrast. Monomarcation refers to the usage of a boundary plant in a single social domain, like confining livestock or preventing witchcraft. This designation reflects how scholars describe a plant, and is a starting point for further empirical investigation, not a classification. In the Ecuadorian Amazon, for example, many Runa people place hallucinogenic *Brugmansia suaveolens* plants around their houses to prevent spiritual attack (Swanson 2009, 41), but understanding the economic and political implications requires more research. Even monomarcation can be socially complex. When Tanzanian men plant eucalyptus on the edges of their farms, the reason for this economic monomarcation is not just because they want fuelwood, it is because they are planting a long-term crop that protects their sons' inheritance from land reform (Gillson et al. 2003, 376). Accordingly, although monomarcating plants such as *Euphorbia* spp., *Croton* spp., *Grevillea robusta*, *Casuarina* spp., sisal (*Agave sisalana*), Christ's thorn (*Ziziphus spina-christi*), *Newbouldia laevis*, and prickly pear (*Opuntia* spp.) line yards, farm plots, and roads from Sudan to Mexico (Thompson et al. 2010; Zuria and Gates 2006), their social, political, and ideological relations require further research. Polymarcation, on the other hand, describes boundary plants in multiple domains, such as when the same species appears on both property boundaries and graves (at the symbolic boundary of life and death). Polymarcation is a particularly fertile area for social science analysis because it illustrates the intersections of material, social, and symbolic dynamics.

One polymarcating boundary plant occupies a central position in environmental anthropology, but more as a prop in a social drama than an actor in its own right. Roy Rappaport's ethnography of the Tsembaga Maring people of Papua New Guinea foregrounds rituals using the *rumbim* plant (*Cordyline fruticosa*). Taking a village as his unit, Rappaport delineates the trophic flows of caloric energy (particularly taro and pork) throughout the system (1984, 121ff). As the pig population grows, the conflicts that result from pigs raiding another village's gardens escalate to war. The men uproot the rumbim at the village boundary, and after safeguarding their souls at their men's house behind another rumbim, the warriors go to battle. When both sides replant rumbim on their boundary, peace returns. If defeated enemies fail to mark the boundary, the Tsembaga invade and create a new boundary with rumbim. Rappaport interprets the rumbim as a ritual "transducer" that allows the socio-ecological system to adjust to shifting populations of pigs and people, redistribute people over the land, determine the frequency and intensity of violence, and create exchange relationships between the Tsembaga and their allies (1984, 3).

If the Tsembaga landscape were a written language, rumbim would be its punctuation. Its roots entangle agricultural economics, village politics, ritual practice, and ideologies about ancestral spirits and gender relations. Rappaport focuses on its first three aspects, but the rumbim is clearly symbolically complex. It stands for men's membership in territorial groups (Rappaport 1984, 171), represents the intersections of life and death (Strathern and Stewart 2001, 281), and signifies men's reproductive success (Lipuma 1988, 67). Is this polymarcating boundary plant primarily about land tenure, social group identity, or symbolic order? For the Tsembaga, rumbim is all of these at once because it stitches together the fabric of cosmology, social life, and landscape.

Another classic of boundary plant ethnography is Sally Falk Moore's work on law and society on Mount Kilimanjaro (1986). Moore examines social change by focusing on dispute settlement processes among Tanzanian farmers. One of her case studies concerns a boundary plant. The *Dracaena fragrans* plant is ubiquitous in the Chagga landscape (Chapter 3). Stalks of dark green sword-like leaves grow on graves and the boundaries of yards and farm plots, and it features in witchcraft detection and peacemaking rituals. In one 1968 land dispute, Richard uprooted the seedlings that his kinsman Elifatio had planted within a dracaena boundary (Moore 1986, 285ff). The boundary was on Elifatio's side of the path dividing their plots, so he considered replacing part of the hedge with fruit trees well within his rights as a landowner. Richard countered that no one could plant on a government-owned footpath, and that in any case, the ancestors' boundary plants should remain undisturbed. Elifatio failed to win reimbursement for his seedlings because he lacked the social network and legal finesse of his more educated kinsman. Moore's example demonstrates that contradictory norms coexist and that rules do not determine behavior. Instead, social life is a messy matter of working out interests, obligations, opportunities, and strategies using old and new cultural materials. In this case, the tenurial aspect of a polymarcating boundary plant has contradictory meanings because it allows individual strategizing but also represents timeless ancestral authority.

One benchmark of agrarian change is the increasing formalization of property rights, and boundary plants often express this shift. Glenn Stone (1994) argues that as farmers increase their labor investment in land, they also invest in boundaries to protect the improved soil from encroachment. "Perimetrics" like inert stone walls are traces of agrarian change, but living fences also delineate landscape history (Sheridan 2008). Perhaps the best example of monomarcating boundary plants expressing economic change is the British Enclosure Movement. Before 1750, much of Britain was a patchwork of fields, forests, and meadows separated by irregular hedges. Between 1750 and 1850, around 200,000 miles of hedges were planted, more than the previous 500 years' plantings (Rackham 1986, 190). The new single-species hedges (usually hawthorn, *Crataegus monogyna*) ran in straight professionally surveyed lines. The old ecologically diverse hedges, managed via a socially distributed bundle of rights, had been important as livestock barriers, territorial markers, sources of medicinal and food herbs, and refuges for wildlife. The new ecologically simple hedges lined large rectangular fields, facilitated individual rather than collective management, and demonstrated the new private property regime (Dowdeswell 1987; Pollard et al. 1977). This "primitive accumulation" of capital via the privatization of common-property systems required a legal foundation, and between 1750 and 1830 more than 4,000 Parliamentary Acts permitted the enclosure of about 6.8 million acres, or 21 percent of England's total area (Beckett 1990, 36; Bhattacharya and Seda-Irizarry 2017). Production of grain and meat increased dramatically in response to demand created by the French Revolution and the Napoleonic wars (Ross 1998), while the newly landless rural poor became the industrial labor force or immigrants to the New World. Along with the coal smoke of industrializing London, boundary plants were material manifestations of the construction of the modern capitalist system.

The medieval open-field system of meadows, woods, and farms with multiple uses did not make a smooth transition to an intensive agricultural system based on private property. Dispossessed and displaced smallholder farmers and squatters formed the Diggers and True Levellers movements in the 17th century. These agrarian populists uprooted the monomarcating hedges and filled in the ditches that defined the new mode of production (Kennedy 2008). With the increasing mechanization of agriculture in the 20th century, hedges again became the sites of rural class struggle. Bulldozers and government subsidies allowed farmers to remove hedges that blocked combine harvesters, so between 1946 and 1970 about 4,500 miles of hedge were destroyed annually in England and Wales (Barnes and Williamson 2006, 22), with another 10,000 miles lost by 1993 (Oreszczyn and Lane 2000, 102). A new discipline, field margin ecology, evaluates the role of hedges as ecological connectors in European rural landscapes (Barr and Petit 2001; Marshall and Moonen 2002), and legislators enacted laws to empower local governments to conserve hedges. Monomarcating hawthorn hedges have gradually become polymarcating. The hedged landscape that resulted from this struggle over land, livelihood, and modernity is now seen as quintessentially British (Barker 2012). As the value of hedges for controlling livestock and marking territory declined, their values for biodiversity conservation and sense of place increased, particularly among the non-farming public and policy-makers. These hedges now define what the British landscape should look like and serve as symbolic markers of the boundary between urban and rural.

Much of this scholarship treats boundary plants as structural landscape elements with biophysical, economic, social, political, and ideological functions, which then indicate larger processes. These plants, by definition, "do things" in horticultural, agricultural, and industrial production systems. This makes them particularly appropriate topics for political ecology, which originally applied structuralist political economy to land degradation and resource management issues (Blaikie and Brookfield 1987). As features that literally root social meanings in a landscape, boundary plants are also relevant for a post-structuralist political ecology that addresses both struggles over resources and the "symbolic contestations that constitute those struggles" (Moore 1993, 381; see also Paulson and Gezon 2005). Boundaries often involve non-territorial symbolic contests (Walker and Peters 2001), so the issue at hand is how territorial dynamics relate to other sorts of meaning-making. Addressing this while still obeying Malinowski's injunction to "grasp the native's point of view" (1961, 5) requires attention to culture-specific concepts of space, place, and power to avoid flattening the ideas and practices we seek to document and interpret. Recent anthropological work explores the spatial aspects of social and cultural phenomena, while a new-found interdisciplinary interest in the social lives of other species extends concepts like agency and cognition to plants. Together the spatial turn and several posthumanist turns offer new concepts for unraveling the ways that biophysical ecology, socioeconomic difference, and cultural meaning-making get tangled up in boundary plants.

The spatial turn

In the 1990s, a new approach to space and place emerged in the wake of post-Cold War geopolitical destabilization. The correspondence of peoples and places, long taken for granted, was reevaluated as a process. The recognition that ethnic groups did not match ethnic territories inspired new literature on the de-territorialization of ethnicity (Gupta and Ferguson 1997a), building on Barth's insight that cultural identities are dynamic relationships, not static units. Boundaries and borders were redefined as complex processes rather than simple edges through processual terms like "boundary-work" and "boundary crossing" (Lamont and Molnar 2002, 168). This approach de-naturalizes property and treats it as a "bundle of powers" that generates territory instead of a structure of rights that explains social outcomes (Ribot and Peluso 2003). Territorialization is how states re-order space, redefine access to and control over resources, and disenfranchise local land managers, usually in pursuit of economic development and capital accumulation (Vandergeest and Peluso 1995). Local land-use practices, such as how Thai farmers and NGO activists ordain trees as Buddhist monks to resist state-sanctioned forest extraction, can then be contrasted as examples of counter-territorialization (Isager and Ivarsson 2002; Peluso 2003). This approach to the social and political production of space cross-fertilized with the "spatial turn," which relies on spatial metaphors like "positionality" to interpret social phenomena (Silber 1995; Warf and Arias 2009). Boundary plants are sites where state territorialization collides with local counter-territorialization, often with destructive or creative results.

Two themes dominate this diverse literature. The first concerns how space becomes place. Landscape was often a static backdrop in ethnography, behind the social action. Eric Hirsch redefined landscape as a social and cultural process by distinguishing between the "foreground" of actual social life and the potential cultural "background." The actual is the way things are, the potential is the way that things could and should be. Landscape is a process by which people work to "realize in the foreground what can only be a potentiality and for the most part in the background" (Hirsch 1995, 22). Agricultural fertility rituals and ecosystem conservation are examples of management practices that turn the potential into the actual. Boundary plants are prime locations for foregrounding the cultural background because they involve property relations, group identities, and domains of meaning. With these concepts, we can consider how Rappaport's Tsembaga work to turn cultural potential into social actuality, and space into place, using boundary plants.

But what forms does place-making take? The literature suggests that people follow two overlapping avenues – narrative performances and strategic struggles. The narrative approach takes a social constructionist view of how storytelling, music, poetry, and ritual practice comprise a "local theory of dwelling" that explains how a people inhabit a landscape, and how it inhabits them to shape personhood (Feld and Basso 1996; Low 2016, 68). As these narrative performances make landscapes meaningful, abstract and objective spaces become pragmatic and subjective places. Thus when Australian Aborigines speak about totemic ancestors' journeys in "the Dreaming," they re-create a cultural landscape, rehearse social identities, and assert land rights (Myers 1991). Strategic struggles among social actors with differential access to resources and power create contested spaces. Examples include protest marches and inheritance disputes, but the concept can be extended to slave plantations, factory shop floors, and even American gated communities (Low 1999). In her account of a 1966 contested space in

Swaziland, Hilda Kuper develops Turner's ideas to argue that these areas are condensed symbols within political dramas that summarize the relationship of colonial states to colonized people (2003). From this perspective, Rappaport's account of cordyline as part of a homeostatic socio-ecological system could be re-interpreted as condensed symbol in the ongoing drama of a decentralized political system full of reciprocity and alliance-building, and the Tsembaga stories and rituals around the plant as landscape-forming narratives. Every time that the Tsembaga uproot, plant, or just talk about the rumbim, these strategies and narratives re-create place as background potentials intersect with foreground actualities.

The second theme asks who constructs place, and to what ends (Low 2016, 34). Whose agency counts, and to what degree are spatial relations by-products of structural inequality? Two French philosophers set the agenda for examining the social production of space, power, and control. Michel Foucault explored how modern architecture, particularly in forms like prisons, hospitals, and asylums, produces "docile bodies" that internalize discipline imposed (often literally) from above (1975, 198). With his metaphor of the Panopticon, a hyper-efficient prison with cells organized around a central watchtower, Foucault argues that modernity is about self-regulation because actors know that they are being watched. Knowledge coexists with power, as everyone going through airport security knows. This awareness of power ("governmentality") therefore shapes subjectivity and personhood. The most prominent applications of Foucault's ideas about space have been in critical urban geography and sociology (Baudrillard 1994; Harvey 2009; Soja 1996), largely on the hidden mechanisms of economic and cultural control in modern nation-states. These Foucauldian concepts are applicable to the study of boundary plants in agrarian societies because these plants embody control mechanisms and create disciplined landscapes through a sort of "vegetative gaze." As the next chapters illustrate, the application of Foucault to non-European contexts requires some analytical stretching and redefinition.

The second strand of French thinking about the social production of space comes from Henri Lefebvre. He argues, following Marx, that just as money is abstracted labor, spaces and places are abstracted social relations representing the hegemonic interests of the dominant class (Merrifield 2013, 106). For Lefebvre, three interacting elements produce space: representations of space by elite technocrats (like monuments and housing developments), spaces of representation in which everyday life actually occurs (like bedrooms and yards), and spatial practices that provide conventions for recognizing and organizing space (like following routes and obeying boundaries). This approach to dominant ideologies, ordinary life, and social practice is similar to Pierre Bourdieu's theory of practice, but Lefebvre adds that each mode of economic production has its own mode of spatial production, and that "the shift from one mode to another must entail the production of a new space" (Lefebvre 1991, 46). This is an excellent description of the

Enclosure Movement, through which industrial capitalism remade the British landscape to suit its own needs. Rectilinear hawthorn hedges represented technocratic and legal spaces, which became the utterly ordinary agricultural spaces of representation for the production of wool, meat, and oats. The spatial practice of maintaining those hedges became increasingly difficult, however, with the advent of monster farm machines. The current English nostalgia for the hedged landscape demonstrates Lefebvre's argument that socially produced space masks the contradictions of its production (Low and Lawrence-Zúñiga 2003, 30) – so that many English people prize hedges for their aesthetic and conservation values, not their history of class struggle. Analytically, we can conclude that old strategic struggles produced new narratives of meaning within a new mode of economic and spatial production.

These authors have influenced scholars of class-based industrial societies with centralized bureaucratic and corporate institutions. Urban geographer Edward Soja, for example, draws on both Foucault and Lefebvre to analyze social justice in the Los Angeles landscape (2013). Making their ideas applicable to rural smallholder farmers requires attention to different axes of social differentiation, particularly relationships of gender, kinship, and status ranking. Among Rappaport's Tsembaga, for example, women never touch the rumbim plant (1984, 149). In a strikingly similar way, Moore's Chagga dracaena users are consistently men. These boundary plants represent men's spaces within societies with a gendered division of labor, and both warfare and farming produce gendered places. Given that every social group on the planet is now part of the globalized capitalist world-system, however, we cannot analyze boundary plants' roles in the social production of space only in isolated localities and the ethnographic present. Boundary plants were also aspects of precolonial migrations and colonial empire-building. Gillian Hart argues that socially produced spaces should be seen "not simply as the effects of global flows but rather as constitutive of them" (2002, 294). In Africa, Oceania, and the Caribbean, the fact that boundary plants constituted past global flows continues to shape how people turn space into place. But what material properties make these particular plants so important for place-making? Answering this question requires following several recent poststructuralist and posthumanist turns.

The plant and multispecies turns

To what degree are non-humans social actors, strategic agents, and intelligent thinking and doing beings? How do social science concepts like agency and personhood apply to plants? These thorny issues have united a loose coalition of popular science writers, philosophers, and anthropologists to articulate the vegetative point of view and give voice to these intelligent and capable organisms (Pollan 2001; Ryan 2012; van der Veen 2014). Whenever a plant produces a seed, this shows that it wants to live, and this scholarship explores what that wanting might mean. By treating plants as passive objects, these scholars argue that we repeat the same injustices behind colonialism and other sorts of oppression. By rejecting anthropocentrism, the authors of the plant turn seek a new sustainable future in which all of Earth's species can co-exist (Haraway 2016; Myers 2015).

The plant turn is part of a larger project in "multispecies ethnography," which moves beyond anthropology's traditional concern for people to examine other living selves (Kirksey and Helmreich 2010). The search is on for a theory of hope for a world reeling from social inequality and ecological decline. This quest has inspired a new generation of scholars to examine "life's emergence within a shifting assemblage of agentive beings" (Ogden et al. 2013, 6). Generally, this scholarship seeks empirical examples of people, animals, and plants as "companion species" and alternatives to domination and extinction (Haraway 2003; Koenig 2016). For example, Anna Tsing finds life and value emerging from the capitalist ruin of Oregon's damaged forests through the mushroom industry, and argues that new multispecies landscapes are necessary for life on our planet (2015). Sarah Osterhoudt's account of vanilla farmers in Madagascar presents agroforestry work as a way that memory, emotion, and morality become landscape features (2017). When a farmer carefully pollinates vanilla flowers by hand, they also cultivate their sense of self and moral personhood. Teresa Miller examines a community of people and plants in the Brazilian savannah (2019). In this "sensory ethnobotany," the indigenous Canela people experience love for and kinship with plants, and delight in calling beans their "crop children." By extending anthropology beyond the human, multispecies scholars are re-envisioning life on Earth and how to sustain it.

The plant turn uses post-processual theory to study plants as persons and document how meaning emerges from relations with non-humans. The analytical units are assemblages, not structures or systems, and their connections are tangled webs, not causes or consequences. These language choices allow authors to say new things with old words. Eduardo Kohn, for example, describes how an Amazonian rainforest "thinks" as a "multispecies assemblage" (2013, 83) that communicates through growth and decay. The Runa people's Amazonian home is not a socio-ecological system for Kohn, it is a semiotic-ecological network, with trophic flows of meaning in addition to nutrients and energy. He insists that trees and animals express meaning through non-linguistic means (such as the sounds of falling branches and birdsongs) and that human language and symbols are just parts of this larger semiotic community. Caitlin Berrigan (2014) asks what dandelions (*Taraxacum* sp.) mean, in terms of "microbiopolitics," when she waters them with her own blood and makes medicinal tea from their leaves. Her work pushes Foucault's approach to power into new domains and asks where the human person ends and the plant person begins. John Hartigan explores what crop breeding facilities and botanical gardens look like from a plant's point of view (2017). He argues that human representations of non-humans, such as the species concept itself, are drenched with anthropocentrism and distortion. By exploring an entangled assemblage of care, growth, and interaction, and depicting human purposes as problems rather than assumptions, Hartigan redefines society like Kohn redefines communication. Multispecies ethnographers regard being human as an "interspecies collaborative project" that lets them tell new stories (Galvin 2018; Ogden et al. 2013, 14) and undermine humanity's arrogance and complacency as we devour the planet. Without so much objectification and exploitation, these authors argue that life on Earth might have a chance.

In many of this book's case studies, people attribute personhood to boundary plants. In anthropology, personhood is usually divided into a single human's sense of self and the "moral person," a shared cultural definition of a person in terms of social relations and cultural representations (Jackson and Karp 1990, 15; McIntosh 2018). Personhood is not synonymous with human existence. In many societies, people slowly achieve personhood as they pass through their society's life cycle rituals and accumulate economic, social, and cultural capital (Comaroff and Comaroff 2001). Finally, the relations and representations that constitute personhood can lead to different ways of being. For the Amazonian Wari', personhood emerges from exchanges of bodily fluids (Conklin 2010), while Papua New Guineans become relational persons through gift-giving (Strathern 1988). These interpretive models about variable personhood (not types of persons, Strathern 2018) prompt observations about boundary plants that move beyond the issue of vegetative selfhood. Boundary plants are often tangled up with, and even exemplify, moral personhood and cultural models of agency. Much like states, boundary plants shape subjectivity and the embodiment of spatial relations through territorialization (Peluso and Lund 2011). They mark the edges of landscapes and of life cycle transitions. In some case studies, boundary plants are wise elders, in others, guardians or guides. These social roles correspond to cultural assumptions of proper and orderly personhood, which I argue (following Wolf) are rooted differently according to the ways that these societies mobilize social labor.

The assemblages of the plant turn are unlike environmental anthropology's social-ecological systems because they are open-ended instead of deterministic and decentered instead of anthropocentric (Orr et al. 2015). For example, when Hartigan interviews an herbalist about her remedies (2017, 275), he avoids the verb "uses" because that would place the plants into an all-too-human structure of functional value; instead, these plants have "multispecies enlistments." Sophie Chao provides a multispecies account of how indigenous Papuans think of the African oil palms on Indonesian plantations as "violently unloving" because they have sharp thorns (2018, 624). She shows how the Marind people perceive the palms as antagonistic persons and applaud the native species that prey upon them (2021). These authors treat non-humans as having agency and subjectivity *before* becoming involved in human structures of value and meaning.

This work is useful for the comparative study of boundary plants because it starts with vegetative agency, and only then asks how people elaborate this capacity into institutions and symbols. But does a focus on plants' agency and subjectivity twist these terms into shapes a social scientist can no longer recognize?

In the multispecies and plant turns, agency is a relationship among entities, never an internal capacity (Dwiartama and Rosin 2014, 3). Defining and locating agency, therefore, requires attention to loose assemblages, networks, and tangles. Theoretical concepts from Bruno Latour and Tim Ingold provide the grammar for this scholarship. In Latour's Actor Network Theory (ANT), non-humans are actors and not "the hapless bearers of symbolic projection" (2005, 10). ANT is not a theory of causation and consequence; instead, it asks how society and meaning emerge from the ways that people, plants, objects, and ideas "jostle against each another" (Hitchings 2003, 100). In ANT, there are no objects, only relationships, and entities like plants and people exist through their relationships with other entities (Gershon 2010). Latour uses the term "actant" to describe anything with the agency to affect other nodes of its network, which helps him explain how Louis Pasteur isolated the agency of anthrax microbes using Petri dishes (1988). Part of what makes particular actants influential in ANT scholarship is their material physicality; a dandelion is different from a sequoia. Furthermore, some actants are "privileged." Pasteur interacted with French government officials in a way that microbes, sheep, and farmers could not. In ANT terms, boundary plants are privileged actants whose botanical materialities bring particular sorts of agency to bear on social institutions.

Tim Ingold's environmental phenomenology relies on fibrous metaphors to describe how life consists of "lines" of movement through time and space. His focus is not on agency, but rather on the ways that actions become forms. Instead of theorizing the agency of actors surrounded by environments, Ingold presents a theory of life as movement and relations through entanglement in a "meshwork of interwoven lines" (2016, 106). Some of these tangles and weaves involve "knotting" that secures objects, organisms, and knowledge in relatively stable forms (2015, 14). Ingold's redefinition of social-ecological systems focuses on the experience of living instead of an analytical sequence of causation and influence. Because boundary plants often form actual lines as living fences and property rights markers, Ingold's vocabulary informs how and why botanical properties ramify beyond gardens. Both dracaena and cordyline grow from cuttings; one simply inserts a stalk into the ground, and it takes root. It is easy to make lines using plants that propagate vegetatively. Lines of dracaena and cordyline enmesh, entangle, and knot together farming, houses, social groups, moral personhood, and life itself to produce distinctive cultural meshworks. Tanzanian farmers tie a knot in a dracaena leaf from their property line and use it to make an apology witnessed by their ancestors. Ingold provides the analytical language to show how this knot of property relates to ties of kinship and ideas about vitality. But what happens when a society's knots slip and its meshwork becomes full of holes? Many of this book's chapters concern frayed meshworks with disentangled and unraveled lines, and how people perceive these transformations.

The ontological turn

If we accept that plants are agents, it follows that they experience being alive. Although their slow growth, rootedness, and lack of central nervous systems make this task difficult, is it possible to understand, describe and empathize with plants' perspectives? What is the most productive way to interpret culture-specific ideas about vegetative agency and personhood? The "ontological turn" in anthropology developed largely from the work of Bruno Latour, Philippe Descola, and Eduardo Viveiros de Castro to build new conceptual tools for social scientific studies of reality itself (Kohn 2015). This turn tries to completely escape the gravity well of scientific reductionism and discover completely new worlds. Its critics counter that the ontological turn is just a flight of fancy, leading to more controversy than insight (Heywood 2018).

The ontological turn insists that anthropologists should not impose Western scientific assumptions of "what is real" onto non-Western peoples in order to judge and explain their mistaken points of view. Standard concepts such as structure, system, and culture are all mechanisms to re-colonize, commodify, and objectify indigenous perspectives. It is therefore politically progressive and ethically responsible to take the native point of view completely seriously and avoid categorizing their ideas as false. For example, when a Vincentian person's soul follows a boundary plant to the edge of heaven (Chapter 7), an ontological turner should accept the account as a legitimate experience instead of an intellectual error, and avoid explaining what the story "really means." Instead of a single shared objective reality viewed from many culture-specific positions, ontological turning means investigating the reality of multiple co-existing worlds (Viveiros de Castro 2014). Because modernity got the planet into its current precarious state, the non-moderns hold the keys to staving off the apocalypse (Danowski and Viveiros de Castro 2017). From this position of intellectual humility and epistemic openness, social scientists can collaborate more respectfully with their subjects to build a more diverse and humane "pluriverse" composed of many worlds (Daly et al. 2016; Escobar 2018; Sullivan 2017). It also means using indigenous concepts to analyze data, instead of explaining them from a positivist standpoint or interpreting them with a hermeneutic approach. Along with the plant and multispecies turns, the ontological turn puts anthropology on a trajectory toward a theory of hope.

The major strength of the ontological turn is that it allows new questions to generate new insights. It takes the anthropological strategy of cultural relativism to its logical conclusion of letting the native point of view grasp the ethnographer, reversing the usual anthropological goal of grasping the native point of view. Although many authors use Philippe Descola's ontological categories (animism, naturalism, totemism, and analogism, Descola 2013) to demonstrate contrasts between indigenous perspectives and outsider exploitation, the emerging consensus is that studying ontology is more method than topic. In Martin Holbraad and Morten Pedersen's assessment of the ontological turn and its applications, they define it as a "technology of ethnographic description" (2017, ix). Their point is that asking ontological questions is more constructive than documenting differing ontologies. This turn is therefore about writing differently, not showing differences. Holbraad and Pedersen provide examples from their own ethnographic work on divination and shamanism. Cuban diviners use sacred powders to make the gods visible, and Mongolian shamans wear gowns covered with knots, strings, and flaps (2017, 220–238). These authors use the objects' material properties (powder floating in the air, knots binding and connecting) to draw conclusions from the newly sharpened concepts of motility (for the powder) and mutability (for the gowns) to illuminate the nature of spirits. By letting these objects "emit concepts," Holbraad and Pedersen find more precise analytical tools based on a native point of view (2017, 236).

The vital flow of life is a concept repeatedly emitted by the boundary plants in this book. The idea that life is better understood as an endless process of becoming than as cell biology has resurged in recent ontological scholarship (Crépeau and Laugrand 2017; Duarte 2021). This neo-vitalism connects 19th-century European notions about a vital spark in living things to ideas about the relatedness of all things often found in non-Western cosmologies (Bird-David 1999). By extending vitality to objects and territories, scholars ask us to reconsider life and agency (Bennett 2010; Brighenti and Kärrholm 2020). For example, investigating how indigenous peoples in Russia and Canada use the metaphor of "shared breath" to experience the vitality they have in common with animals, plants, and stones offers far more analytical depth than abstractions like belief and ritual (Siragusa et al. 2020). Examining how objects in American storage facilities get so charged with an emotional force that they become active agents leads to new ideas about symbols and objects (Newell 2018). In the following case studies, there are many narratives about the existence, flow, and control of vitality and immanent life force, based on the fact that dracaena and cordyline are remarkably durable plants. They take root from vegetative propagation, resist drought, and (as people told me repeatedly throughout my project) "have more life than other plants." Following the ontological turn means letting these peoples' vitalist perspectives propel the analysis, and I use this method repeatedly. But are all of these turns leading anthropology in a circle? How much do we really learn from plant agency, multispecies assemblages, and ontological inquiry?

Re-turning to political economy

Critics have charged these turns with leading anthropology astray. Their critiques form three arguments about the logic, analytical value, and limitations of posthumanist inquiry.

First, posthumanist approaches often posit a binary of modern vs. indigenous without addressing complexity, hybridity, and creolization. Ontological turners present non-modern and indigenous perspectives and cosmologies not as viewpoints about a shared world, but fully independent worlds (Bessire and Bond 2014, 442). This sidesteps issues that matter deeply to non-moderns, such as life-saving medicines, labor migration, and secure access to resources. By focusing attention on how agency is distributed in assemblages of plants, animals, and objects, posthumanists blur and obscure matters of causation and responsibility (Hornborg 2021, 755). At issue here is the distinction between purposive and purposeful behavior. A seedling is purposive when its roots search for water and nutrients, but the farmer who planted it is purposeful because she has multiple ambitions (Suzman 2021, 41). Furthermore, there is little room for non-human subjects to experience ambiguity or doubt, because their agency leads to clearly non-Western or non-human states of being, not vague indeterminacy (Graeber 2015, 11). Finally, if plants, animals, and inanimate objects are all agents and actants, this is universal. How, then, does a constant generate so much difference? For critics, these logical flaws undermine the posthuman project from within.

Posthumanist scholars aim to "inspire novel maneuvers that enable the mutual flourishing of humans and our more-than-human environments" (Keleman Saxena et al. 2018, 58). The critics' second critique is that posthumanist analysis flattens sociocultural diversity to a single non-modern alterity (Bessire and Bond 2014, 443). By focusing on discrete cultural worlds instead of one shared natural world, ontological scholarship prevents comparison, which is ultimately a conservative stance. If one culture cannot challenge another's assumptions, there is little for activists to do (Graeber 2015, 7). For example, accepting the prohibitions against women touching boundary plants in Papua New Guinea (Chapter 5) as ontologically valid would hinder feminist analyses of gender, land tenure, and power. When an ontologically distinct way of being distributes resources by race, caste, and class, should this be investigated as an elite ideology, or remain unquestioned? Posthumanism may be interesting waters to wade in, but for critics, it lacks analytical depth.

The third critique is that posthumanism imposes limitations on itself. It ignores and therefore mystifies the global capitalist system. "Although presenting itself as subversive," Alf Hornborg writes, "posthumanism in this sense turns out to be an extension of the instrumentalist logic of capitalism" (2021, 758). Non-human entities like carbon dioxide and oil cross ontological and territorial boundaries and are generally unimportant in posthumanist scholarship (Bessire and Bond 2014, 446). Boundary plants in Oceania and the Caribbean have strikingly similar social and cultural biographies as property rights markers and mystical protectors, but a posthumanist would avoid tracing these similarities through colonial land law. As Scott Simon asks about the multispecies assemblage of dogs, pigs, and indigenous Taiwanese, "how do these relations articulate with postcolonial dynamics of gender, class, and ethnicity?" (2015, 695). Without attention to the contexts of empire, states, and capitalism, the sanguine posthumanist search for the beating heart of a vital living planet seems rather anemic.

The posthuman turns often investigate relationships between the physical materiality of things and metaphysical abstractions, not the intervening economic, social, political, and ideological relationships, institutions, and meanings. Focusing on the agency and personhood of non-human actors is a "politics of recognition" without a "politics of redistribution" (Fraser 1995). Recognizing agency is important, but the transformation of structural inequality requires the redistribution of costs and benefits.[1] For example, Katharina Schneider insightfully outlines a multispecies ethnographic approach for understanding pigs, fish, and birds in Papua New Guinea. She extends Marilyn Strathern's (1988) exploration of gender relations to species, and concludes by rethinking agency as co-species, cross-species, and same-species relations (2013, 32). This is a useful intervention that recognizes relational sub-categories. Applying this perspective to this book's case studies would generate an analysis of the co-species relation of a plant growing on the edge of a garden, the cross-species relation of a farmer cutting a stalk to plant a new boundary, and the same-species relation of that farmer taking a leaf for a fully human purpose like dancing, apologizing, or sacrificing. But which of these relations has analytical priority? Which relations matter, to what ends, and for whom? Which help to explain poverty or sustainable land management? Recognizing interspecies agency would probably not help that farmer take a land-grabbing corporation to court.

Posthumanism inspires method, and this book treats its insights as tools in the analytical toolbox, alongside political economy, social and cultural capital, and constructionist concepts like place-making (Besky and Padwe 2016). Multispecies and ontological approaches allow new questions about why metaphors of belonging, territory, and place take particular forms (Schoenbrun and Johnson 2018). As Eduardo Kohn concludes, these are concepts that augment but do not replace constructivist and materialist anthropologies (2015, 322). Following Roy Ellen's advice for ethnobiologists, I use the term ontology as sparingly as possible (2016, 12). Instead of labeling beliefs and practices about boundary plants as intellectual mistakes, I contextualize them despite my limitations as an imperfect witness, translator, and writer (Lloyd 2019). I approach this work as a critical realist who, like Eric Wolf, believes that there is one real world that needs to be explained, that anthropology must study power, and that structural inequality is the motor of social and cultural continuity and change

(Marcus 2003; Wolf 2001). This world can be known empirically and studied historically, and this search for understanding needs attention to narrative constructions and power-drenched regimes of truth and hegemonic meaning (Sullivan 2017). I reject the binary of viewing plants as either false representations or true agents in their own right. As described throughout this text, certain plants' agency for vegetative propagation is a source for producing social institutions and constructing cultural meanings. My argument is that material contexts and the organization of labor have analytical priority for showing how and why boundary plants become entangled in particular social-ecological systems. This book argues for the importance of the historically specific interaction of structure and agency to produce particular institutions, practices, and meanings.

Conclusion

This chapter has "beaten the bounds" of social science for this study of boundary plants. The review of boundaries and plant-people interactions has demonstrated that they involve diverse analytical concerns as territorial landscape markers, social group definitions, and symbolic edges of cosmology and meaning. I introduced new terminology to describe the thickness of these relations, and the following chapters show how complex polymarcating boundary plants are becoming more monomarcating. The survey of anthropology's processual, spatial, plant, multispecies, and ontological turns summarized and evaluated concepts and analytical methods that illuminate the messy social histories of dracaena and cordyline in Africa, Oceania, and the Caribbean. Three levels of analysis unravel boundary plants' tangled intersections in the case studies: landscape, society, and meaning. Throughout this book, I use concepts from anthropology's turns to consider how and why these systems, meshworks, and assemblages differ, even when they all concern similar plants in similar environments.

Boundary plants are a rich topic for environmental anthropology because they stand firmly at the intersections of material and symbolic phenomena and are fundamental aspects of migration, settlement, colonization, exploitation, resistance, and place-making. These plants show how botanical characteristics, especially vegetative propagation from cuttings, acquire economic, socio-political, and cultural power and significance. If we take property to be living bundles of rights and relationships rather than dead objects, boundary plants are often the rope that ties these bundles together. They are both parts of the political economy of the global world-system and key aspects of local histories. Finally, boundary plants can enhance local land stewardship with diverse indigenous concepts of place, property, and morality instead of the dull uniformity of capitalism's definition of land as a commodity. If states gave boundary plants legal recognition, these polymarcating plants could contribute to land security and cultural distinctiveness for smallholder farmers. Boundary plants are not only good to think with because their rootedness connotes permanence and vitality, but also because the ways they socially and symbolically demarcate land make them particularly lively material for cultural elaboration. This review of social theory argues that the ascription of agency to boundary plants reflects the ways that people conceptualize power and legitimize the distribution of social labor in different social-ecological systems. Ethnographic attention to these plants' complex social lives redefines the boundary concept by demonstrating how historical, spatial, and social contexts shape how boundaries divide and unify.

Note

1. For an application of these ideas about the tension between posthumanism and structural approaches to the study of rainmaking in Africa, see Sheridan (forthcoming).

3 Tanzania
Knots of Peace on Kilimanjaro

My research assistant and I were walking toward Marangu when we came across a tailor working on a pair of pants. It was a hot afternoon, so he had set up his treadle-powered sewing machine outside next to the ridgeline footpath shaded by bananas and avocado trees. Spotting me, he quirked an eyebrow but remained focused on pinning an inseam. Taking the eyebrow as my cue, I greeted him with "*Shimbonyi mbee!*" (How are you, old man?). With a smile at the novelty of an American speaking Chagga and with glance at the touch of grey in my beard, he replied, "*Nachisha, mbee!*" (I'm fine, old man). After I explained my research project, he launched into what I came to recognize as the standard explanation about dracaena on Kilimanjaro:

> You can't do anything without *masale*. It's the origin of the Chagga people. Its uses and meaning have remained exactly the same since the time of our forefathers, although there are some things we don't do as much, like sacrifices. It is powerful. It brings blessings and fertility to the earth and makes the crops grow well. It is respected. Without it, you don't have anything. What it does most is protect the crops from other people. It doesn't do anything until they violate the boundary. The three major uses of masale – power, closing things, and peace – are all one thing. The origin of the Chagga was banana plants, masale, and farms. Then coffee came in. A woman cannot put boundaries on the land, she must find a man. She doesn't hold masale, the hand that usually holds masale is male. There are a lot of regulations about masale. Now you have it!

As he finished this explanation, the boy that he had sent to buy a small plastic packet of gin returned with the alcohol and his change. I added a 1000-shilling note to the boy's handful of coins, and said in a mixture of Chagga and Swahili, "*Haika sana, mbee! Hii ndiyo kitorchi kidogo kuponesha kiu chako!*" (Thank you very much, old man! This is a little gift to cool your thirst!). He grinned and returned to his work.

This chapter contextualizes this narrative about boundaries by locating it in political and economic history, social organization, and cosmology. Three plants define the cultural landscape of Mt. Kilimanjaro, Tanzania:

DOI: 10.4324/9781003356462-3

bananas, coffee, and masale (*Dracaena fragrans*). The Chagga call themselves the *wandu wa ndenyi*, "people of the banana gardens" (Philippson
and Montlahuc 2006, 476), and it is difficult for a Chagga person to feel at
home without bananas and masale. Even in Dar es Salaam, where the heat
and sandy soil make it difficult for dracaena plants to thrive, Chagga keep
them in their homes as potted plants. The ideal Chagga homestead, called
a *kihamba* and often glossed in English as a "homegarden," is a house nestled into an intercropped plot of bananas, coffee, maize, and beans. This
describes a classic smallholder farm similar to those found throughout the
tropics (Kumar and Nair 2004). One element that makes kihamba a distinctly Chagga social-ecological system is the masale hedge that surrounds
it. These dark green dracaena plants line the shadowed roads and footpaths
that weave the homegardens together into neighborhoods.[1]

This chapter explores the changing roles of dracaena in the homegardens,
in Chagga social relations, and in Chagga thought. This plant was once
a "total social phenomenon" (Mauss 1967) that connected and integrated
Chagga society. It was a polymarcating boundary plant because it was the
basic economic fact of land rights, the key facilitator for social relations,
and the living embodiment of Chagga ideas about peace, prosperity, and
ethics. Today dracaena still defines the kihamba as a place for food production, social reproduction, and cultural construction, but it is less polymarcating because its meanings are becoming primarily economic and less
social and cosmological. This chapter analyzes this disentanglement and
explores dracaena's persistence as a knot in the Chagga social and ecological meshwork despite sweeping ecological, economic, social, political, and
ideological changes on Kilimanjaro.

The explanations and interpretations that I heard about dracaena's many
uses and meanings on Kilimanjaro during my fieldwork emphasized that
this boundary plant expresses customary law and unchanging cultural significance.[2] When I asked annoying anthropological questions about social
change and transformation, my respondents patiently asserted that "we use
it just the same as our ancestors did" and that "other things may change, but
masale stays the same." On Kilimanjaro, this boundary plant is rooted at
the eye of a paradox. It is an icon of social order and cultural continuity, but
it has mediated historical disruption and change. This chapter analyzes and
contextualizes the history of dracaena on Kilimanjaro in order to show how
Chagga discourse about their boundary plants emphasizes the continuity
of an important symbol while its content, meaning, and functions actually
shifted. The customs and rules that entangle these plants with landscape
and society are better investigated as active responses to change instead of
cultural practices that were simply untouched by it.

Power seems to be eclipsing culture as the discipline's core concept. Power
is a good way to answer social science riddles like whether culture, ideas,
and practices shape society, organization, and patterns, or is it the other
way around (Barrett 2002). It allows an analysis to focus on socially unequal

actors collectively producing cultural concepts and normative practices, and systems of symbols informing how and why social inequality exists (Dirks et al. 1994). The problem is that if power explains everything, then it explains nothing (Sahlins 2018). We must specify what power means in a particular context and how it affects (and reflects) continuity and change. This is what makes the study of boundary plants so compelling. Their roots tap into both physical and mental terrains, and they mark the contours of economic relations, social organization, and meaningful experience. On Kilimanjaro, boundary plants embody culture-specific notions of agency and power, which makes them critical resources for continuity and order amidst the messy disorder of economic, social, and cultural change. In the Kilimanjaro homegarden system, dracaena is a living cosmology, an idiom of power in narrative and ritual performances, and a strategic resource for struggles about Chagga landscapes, social relations, and meaning.

Kilimanjaro as a social-ecological system

In terms of the regional ecology, the southern and southeastern slopes of Mt. Kilimanjaro form a sweet spot. The mountain peak looms over an arid region at 5895 meters (19,314 feet) above sea level, making it a "sky island" (Bender 2019, 16) that wrings moisture out of the dominant easterly winds as the warm air cools with increasing altitude. The shoulders of this enormous snow-capped dormant stratovolcano are, compared to the hot and dusty plains around it, a green oasis of lichen-draped cloud forests, whispering waterfalls, and the shaded Chagga homegardens. The mountain's southeastern face has six distinct ecological layers (Molina-Venegas et al. 2020). Kilimanjaro National Park occupies the alpine desert, subalpine heath, and cloud forest zones above 1800 meters (Fig. 3.1). Below this

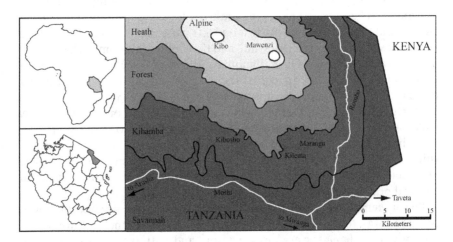

Figure 3.1 Southeastern Kilimanjaro

lies the densely populated kihamba belt of homes, coffee, and bananas between 1800 and 1100 meters (Hemp 2006a, 2006b, 2006c). According to the Chagga oral histories I collected, the banana canopy was once so thick that little sunlight reached the ground. This is why Chagga call stepping off a sunlit road into the kihamba area "going inside." From 1100 meters down to the main highway around 900 meters lies a savanna zone of inter-cropped maize, beans, and millet, known in both Swahili and Chagga as *shamba*. Below this altitude rain-dependent farming becomes precarious, so the lowlands are primarily for livestock and irrigated crops. These lowlands get about 400–900 mm of rain annually, the shamba zone receives about 1000–1200 mm, and the kihamba zone 1200–2000 mm (Hemp and Hemp 2009; Soini 2005a). These three zones form a single social-ecological system because many Chagga live permanently in the kihamba zone but work in the middle and lower areas. Residential villages do not exist on Kilimanjaro; instead, homesteads are distributed throughout the approximately 100 kilometer-long and 12-kilometer wide kihamba belt, while shops, churches, schools, and other institutions line the mountain roads. With an average population density over 600 people per square kilometer (Maro 2009), the kihamba zone is the heart of this system because it maximizes food security and also allows access to the other ecological zones.

The kihamba agroforestry system is a model of sustainable permaculture by an autonomous smallholder peasantry oriented toward food security (Fernandes et al. 1985). The archaeological, historical, and linguistic evidence for the kihamba system's antiquity suggest an age between 300 and 1000 years (Stump and Tagseth 2009). Dozens of banana varieties (*Musa* spp., with specific uses for cooking, brewing, and livestock fodder) shade the *arabica* coffee trees that have driven the mountain's economy since the 1920s. Together with timber trees and intercropped potatoes, sweet potatoes, yams, taro, beans, maize, pumpkins, and eggplant, these gardens produce food year-round by mimicking a multistory rainforest, intercepting sunlight at different levels, and minimizing rain splash and erosion. This biodiverse layered vegetation structure led the Food and Agriculture Organization to designate the Chagga homegardens a Globally Important Agricultural Heritage System (Kitalyi et al. 2013). With reliable rainfall and indigenous irrigation canals, this system maintains Kilimanjaro's volcanic soils. Livestock live in zero-grazing pens, which concentrates their manure and allows farmers to invest those nutrients in their farms. The soil also benefits from the ability of bananas to mulch themselves with dropped leaves, and wet fibrous banana stems provide excellent erosion control when placed across the slope. To the untrained eye, this system looks chaotic, but to Chagga farmers this intricate dance of energy, resources, and labor is the essence of permanence and order. Dracaena defines the perimeters of each dancefloor as the quintessential emblem of a legitimate homestead. In the Chagga language, the axiom describing how dracaena is the hub around which their society revolves is *sale kikoro kilanza kichagga* – "dracaena is the foundation of the Chagga."[3]

This somewhat static description of the kihamba system emphasizes the continuity of a uniformly functional structure. Like most African farming systems, understanding the Chagga system requires attention to diversity, change, and power dynamics. Many kihamba crops came from the New World, which shows that Chagga farmers have long been experimenting with and adopting new plants (Soini 2005b). Even the prototypical Chagga crop, bananas, arrived from southeast Asia thousands of years ago and revolutionized East African farmers' ability to settle the humid highlands (Neumann and Hildebrand 2009). The point here is the kihamba zone has always been a specialized system in flux, not an evolutionary step toward intensive agriculture and functional stability (Widgren and Sutton 2004). This was a dynamic farming system oriented toward exchange and accumulation, not subsistence. In the 19th century, the homegardens supported an economic system organized around cattle, long-distance trade in iron, salt, and ivory, and eventually slave caravan supplies and guns (Håkansson 2009; Kimambo 1996). All of this exchange mainly benefitted patrilineages of related men. The system was largely supported by women's work farming and carrying fodder grass from the lowlands to feed the livestock kept safe in the kihamba. This unequal division of labor by gender persisted into the colonial and postcolonial periods (Allan 1965, 161; Lema 1995a). Resource access and control on Kilimanjaro typically lies in male hands, and this means that the stability of the kihamba system, from its soil characteristics to agricultural outputs, has long been underwritten by gendered relations of power.

Coffee and livestock are male concerns; women focus more on bananas, seasonal crops, and milk. The kihamba zone is where the gendered division of cash crops and food crops overlaps spatially when men tend their coffee trees and women spread manure beneath the banana plants that shade them. Age determines who gets what sort of farm because although the patrilineal kinship system allocates land to sons, many middle sons get no land. Instead, the oldest son inherits most of his father's fields, while the youngest son gets the homestead and cares for his elderly parents. Middle sons get squeezed out and migrate to Tanzania's towns and cities (Setel 1999). Women are dependent on men, in their roles as mothers, sisters, wives, or daughters, for access to land despite their disproportionate share of agricultural labor. These long-standing patterns of labor, residence, and property have constructed today's kihamba zone, and the landscape is a map of Chagga social relations. The green ink with which this map gets drawn is, as described below, dracaena.

Kilimanjaro's approximately 400 landholding patrilineages are processes, not structural units. A processual approach emphasizes that kin/gender interactions constantly construct Chagga society. Chagga patrilineages are exogamous (meaning that men find wives outside of their own groups) and segmented (so that households eventually branch into separate patrilineages as brothers become fathers and elders). These tightly bonded men's networks are cross-cut by other relationships like age-sets (groups of men from different patrilineages but of the same generation), territorial membership (once *mtaa* districts and chiefdoms, now villages and divisions), and irrigation canal user groups (S. Moore 1977, 2016, 118). These forms of social capital connect Chagga men to one another, and the principle of male resource control means that at marriage women get detached from one patrilineage and attached to another. The resulting social-ecological system gives social organization a topographic form (Myhre 2019a). Brothers' houses often line a ridge, which makes blood relations generally vertical while non-kin networks are more horizontal relationships across the volcano's slope. Dracaena plants typically mark both the external boundaries where one patrilineage abuts their *wamamrasa* ("people of the boundary") neighbors and the internal divisions within a patrilineage (particularly if the brothers do not get along). Viewing this meshwork as a process rather than a structure allows an analytical focus on power, domination, and authority as relationships that continuously generate society and culture (Håkansson 2003). Kin-based and gendered power are rooted in dracaena because this plant determines how Chagga ritual practices and cosmology relate to the division of labor in the homegardens.

Despite their long-term stability, the Chagga homegardens face serious challenges today. Rapid population growth over the 20th century put increasing pressure on the kihamba system. There were about 100,000 people living on Kilimanjaro in 1913 (Winter 2009); nearly a century later the population had increased tenfold to 1.1 million (Government of Tanzania 2013). In the 1920s, a typical highlands kihamba was 2.2 hectares and benefitted from irrigation and regular manure application, but now land fragmentation means that most households farm about 1.5 hectares divided into several plots. Of this the residential plot is only 0.7 hectares. This fragmentation results from subdividing the land with every generation and land commercialization because of coffee. About half of Kilimanjaro's farmers rent plots, and many travel up to five kilometers to reach their lowlands farms (Devenne et al. 2002; Soini 2005a, 2005b). Kilimanjaro's agricultural productivity is threatened by long-term rainfall decline, increasing regional climate variability (Nicholson 2016; Williams and Funk 2011), and the loss of approximately 30% of the mountain's forest since 1929 (Hemp and Hemp 2009, 254). With less forest, the altitude at which clouds condense has risen, which has shrunk the size of the rainclouds on which everything depends. Gradual desiccation and warming, along with the reduced water storage capacity in the montane zone due to forest fires, will soon put the homegardens under increasing stress (Craparo et al. 2015; Hemp 2005, 2009). Finally, persistently low coffee prices have made the youngest generation of Chagga increasingly oriented toward urban occupations and upward social mobility. The result is that the kihamba zone is now largely occupied by children, adult women, and elderly men, which makes agricultural labor scarce. In 2015, many coffee trees were clearly (as many farmers told me) "on life support." They had been pruned back severely, so that the trees remain alive but do not shade the food crops growing under them or require much management. Poverty compels some farmers to sell the trees that once formed the upper canopy of their kihamba for timber, and to re-plant with profitable crops like tomatoes and cabbages that require more light, which means that the homegardens of the poor are no longer shady multi-layered systems.

These challenges do not mean that the kihamba system is doomed. Observers have worried that the specialized permacultural agroforestry system was at its limit since the 1970s, yet Chagga creative adaptation and flexibility have shown the system's elasticity instead of brittleness (Fernandes et al. 1985, 34). Kilimanjaro is increasingly a class-stratified mountain of salaried commuters living in the highlands and casual laborers in the lowlands instead of the sky island of a farming peasantry. As one man in Marangu summarized the situation,

> No one is very serious about coffee. The economy is now based on our children going to a city to follow a job, and this means that we don't have the labor to do real farming. We just hire day laborers to make it look like a real kihamba.

The shifts in what this social-ecological system is, does, and means have been shaped by boundary plants and how they link economic to social capital. The rest of this chapter tells the story of these transformations and suggests how dracaena has formed a uniquely Chagga vision of ecological, social, and moral order despite these changes. The history and current state of the Kilimanjaro social-ecological system offer lessons for the construction of sustainable and meaningful agrarian livelihoods far beyond East Africa.

Living land tenure

Farmers on Kilimanjaro told me that the first thing a person says when visiting a farm is "show me your masale." This boundary plant is the "fixed point" upon which Chagga construct other social facts (Meyer and Geschiere 1999). It is a social institution that assigns resources to people and clothes these relationships with culturally meaningful legitimacy. The botanical characteristics at the core of this significance are the plant's capacity for vegetative propagation and its durability. In a Chagga version of the Actor Network Theory approach used by plant turn scholars, dracaena would be a privileged actant because economic, social, and cosmological matters intersect in its roots, stalk, and leaves, and because Chagga emphasize the plant's agency and subjectivity. Its multiple meanings and connections to kinship and religion mean that dracaena's role in the Kilimanjaro social-ecological system cannot be isolated to its function as a territorial marker. These intersecting lines of significance make dracaena the Chagga "ritual attractor"

Figure 3.2 Kihamba path lined with masale

(Fox 2006) parallel to the symbolic elaboration of cordyline we will encounter in Chapters 5, 6, and 7. Property in Africa is a living "bundle of power" and an ongoing negotiation over resource access and meaning, not people acting upon inert objects (Bohannan 1963; Moore 2016). On Kilimanjaro masale is the green cord that binds this bundle of rights, claims, and meanings together (Fig. 3.2).

Many Chagga insist that Kilimanjaro's first settlers brought masale from the northwest hundreds of years ago (Winter 2009, 273). Some elders said that each Chagga settlement was founded around a big *mvumo* strangler fig (*Ficus thonningii*) with a dracaena plant marking the adjacent *kiunguu* sacrificial area. "There was no way to make a new village without using masale!" one declared. Although the plant grows wild in the mountain's deep river gorges (Hemp 2006a, 2006b), even there it signifies purposeful human action for today's Chagga. As they say, "it shows the hand of someone." The species is widely distributed across tropical Africa, so it is likely that dracaena was already growing on Kilimanjaro's slopes long before sedentary agriculture. Most Chagga oral histories and genealogies are only about 300 to 400 years deep (Stahl 1964), and the scanty archaeological and linguistic evidence suggests 1000–2000 years of occupation (Odner 1971; Stump and Tagseth 2009, 109). Dracaena may have been part of the botanical toolkit that Bantu-speaking farmers used to expand eastwards from Central-West Africa (Sheridan 2008), but for most Chagga the plant's role in forming society is unquestionable.

The first scrap of documentary evidence about dracaena on Kilimanjaro comes from Johannes Rebmann's description of the kihamba zone as "isolated inclosures ... always covered with banana-trees. Each yard is occupied by a single family, in several huts, protected by hedgerows of growing bushes" (Krapf 1860, 244). When Rebmann met Chief Masaki of Kilema in 1848, both held "grass" in their hands (Krapf 1860, 238), likely dracaena as signs of peace. In 1861, Karl Klaus von der Decken observed that on Kilimanjaro "every property is enclosed... by high fences" (1978, vol. I, 270), and by the 1890s several authors had described the homegardens' banana-dracaena complex (Abbott 1892, 392; Johnston 1886, 151; Le Roy 1893, 244). Today's dracaena hedges effectively block livestock and prying eyes, but in the late 19th century the plants were placed so tightly together that a gun barrel could not be inserted between the thick stems (Volkens 1897, 289; Widenmann 1899, 63), and thin horizontal masale rails were woven between these living posts. These fences provided excellent security as a green labyrinth of homesteads and footpaths. The kihamba zone was, in the late 19th century, a carefully managed cultural landscape, and its spatial form was a direct reflection of Chagga social organization and values.

Building on this decentralized economic base, Kilimanjaro's chiefs increasingly dominated the regional economy of cattle, caravan provisioning, and long-distance trade in luxury goods at the time of European contact (Håkansson 2009; Moore 2009, 51). Land tenure was very different in the kihamba highlands than in the mid-altitude shamba area. In the kihamba, land was the permanent property of kin groups and organized with dracaena, but shambas were allocated by chiefs for annual cropping and marked by stones, trees, or streams (Gutmann 1932, 273; Moore 2009, 61). The kihamba therefore represent the part of Kilimanjaro not directly controlled by precolonial chiefs. This was a system in flux. Instead of a static landscape structured by customary law, the hedges and fences that so impressed European explorers were social innovations calcified into custom (Figgis 1958, 7) during a period of demographic growth and economic change.

European practices of space, society, and power increasingly dominated this landscape over the 20th century. The ecological, social, and moral order that masale manifests on Kilimanjaro is, for many Chagga, an inversion of colonial and postcolonial changes. The polymarcating significance of dracaena is not the static relic of an old social-ecological system. Change has flowed through this plant because it contains Chagga ideas and practices. The colonial reorientation of the economy away from livestock, iron, and ivory and towards coffee made land security the top concern of an emergent peasantry, and the rootedness of masale in Chagga social practice made it a solid institutional foundation for the new system. Even during Tanzania's socialist experiments in the 1970s, Chagga patrilineages continued to organize land with dracaena despite the state's centralization of authority. Today's kihamba belt is, compared to the land tenure muddles and ambiguities elsewhere in the Tanzanian highlands (Sheridan 2008), a beacon of indigenous order based in part on the ways that the Chagga transformed the institutional roles of their boundary plant.

Coffee lay at the center of this heady brew of continuity and change. It was an excellent addition to the 19th-century kihamba system (O'kting'ati and Mongi 1986). Shaded *arabica* coffee slotted neatly into the layered artificial forest ecology of the kihamba, right below the bananas and above the ground crops. The resulting banana-dracaena-coffee complex made the Chagga relatively wealthy in East African terms, and the export value of their cash crop insulated Kilimanjaro from state intervention. Both colonial administrations (Germany 1891–1919, Britain 1919–1961) let the Chagga organize the kihamba belt but did not build land policy on their indigenous management institutions because officials were more concerned with the facts of land ownership than its meaning and social context.

The evidence was available but ignored. The major ethnographies of the period by Dundas (1924) and Gutmann (1926) both show the centrality of dracaena for delineating Chagga land rights and social obligations. Where the tenurial aspects of this boundary plant do appear in the colonial record, administrators were ambivalent about its role in the evolution of the private property systems assumed in English common law. The first comprehensive report on land matters defines a kihamba by its masale fence and suggests that formalizing and registering coffee farms under freehold tenure would be easy because they were already well-defined units, both visually and socially (Griffiths 1930, 92). On the other hand, in the Kilimanjaro District Book (a compilation of social history, agricultural data, and demography for British administrative officers' reference), boundary plantings demonstrated disorder rather than the order that Chagga farmers intended:

> the Chagga generally plants up his boundary with the local masale (Dracaena) hedge and any planting of masale, often the only evidence of a boundary, beyond the extent of a kihamba will generally cause a court action to arise very quickly. That the Chagga does not delineate his boundaries with more care is surprising since he always clings to the smallest vestige of right to a kihamba which he can adduce … Nor is the position clarified by the frequency with which the Chagga subdivides his kihamba with masale for what is now no very apparent reason at all…
>
> (Tanzania National Archives, ca. 1937, 83–84)

The 1947 Report of the Arusha-Moshi Lands Commission ignored boundary plants, and instead focused on the Chagga expansion into the dry lowlands to relieve population pressure in the kihamba zone (Wilson 1947). This had become acute as the few remaining *mbuga* grazing areas had been carved up into dracaena-lined homegardens (Tanzania National Archives, ca. 1937, 86). A 1957 land tenure analyst identified Kilimanjaro's masale hedges as an exception to the general rule that East Africans did not use permanent boundary marks, and remarked that "the Chagga are so jealous of these boundaries that it is customary to fine trespassers Shs. 5/= per footprint"

(Oldaker 1957, 142). Property rights were not considered a problem in the kihamba zone in the 1958 Chagga Land Tenure Report because the ubiquity of coffee and masale had already made the area secure enough that no registration process was necessary (Figgis 1958, 162). For these administrators, it was the economic and social functions of land tenure that mattered, not the forms it took and the entanglements that made these boundaries culturally salient. Throughout the colonial period, dracaena was a staple of Chagga narrative performances and strategic struggles over land in the local courts, but was taken for granted as an unimportant cultural detail rather than a central plank for building Chagga landholding institutions.

This was only a vacuum in terms of colonial policy. Chagga oral histories and documentary evidence agree that missionaries, administrators, and settlers quickly adopted dracaena as a boundary plant institution throughout the upland areas of the Kilimanjaro region. Settlers typically relied on surveyors' maps to demarcate their land, but then planted dracaena fences in order to mark their land rights in African terms (Munson 2013, 164). The main market hall in Moshi had a perimeter of dracaena (IMP n.d.), and missionary societies throughout the region organized their rectilinear compounds with flowerbeds, trees, grass, and dracaena (Fassmann 1902, 345; IFL 1907; Müller 1897, 269). In the German colonial town of Arusha about 75 km west of Kilimanjaro, the 1912 town plan banned dracaena hedges from the central business area to prevent it from appearing overly African (Munson 2013, 169).

Colonists adopted Chagga mechanisms for order, but an unruly land struggle was emerging just downhill of the kihamba zone. It was here that dracaena appeared most vividly as the medium of change. Precolonial chiefs were not involved in kihamba land issues, but they allocated short-term access to shamba land below the 1100-meter contour. German settlers took much of this prime farmland for their own coffee plantations, and after World War I, the new British administration re-alienated this land instead of reserving it for Chagga expansion (Lema 1995b; Wilson 1947, 20). The Chagga were, in effect, trapped between the settlers and the upper limit of banana/coffee cultivation at 1800 meters. The chiefs still controlled access to the remaining shamba lands, and dracaena shaped the ensuing struggle for land rights among the middle sons being pushed out of the kihamba zone by the Chagga inheritance system. These men planted masale to convert rights of access to shamba land into rights of control. A 1948 legal case established that dracaena indicated whether a piece of land was shamba or kihamba and affirmed the chiefs' lack of authority over kihamba land. A farmer alleged that the chief had no right to re-allocate the mid-altitude plot that he had used for many years because it was a kihamba, not a shamba. The District Commissioner determined that because the plot lacked a house, bananas, coffee, and masale, it had no kihamba characteristics and therefore dismissed the case (James and Fimbo 1973, 79). The oral histories I collected about the settlement of the shamba zone in the late colonial period described these struggles. Elders in lower Kibosho showed me the old dracaenas in their yards and explained that

> We used masale to begin settlement here in the lowlands. You can bring some masale from the banana highlands, and you must put the masale on the corners of the farm, and a masale at the center of the compound, because if you don't, you'll never get good results there!

In many areas, half of these shamba fields were rental plots in the late colonial period. The landowners planted masale to formalize rental agreements and retained the right to plant bananas because a tenant's bananas would make that plot a kihamba (Figgis 1958, 76). Rental arrangements became increasingly contractual rather than minor favors with token payments of agricultural produce, and the meanings of dracaena shifted from kin group matters toward individual control over lowland plots. Dracaena was not just a highlands custom, but a strategy for indigenous re-territorialization that clothed frontier land with an older institution.

After independence in 1961, Julius Nyerere's pursuit of "African socialism" abolished freehold tenure, established state ownership of all land, and displaced Tanzanians into modernist Ujamaa villages (Scott 1998; von Freyhold 1979). This often led to ecological, economic, and social crises rather than sustainable development. In terms of land law, Tanzania has had a "tyranny of rules" rather than the rule of law, which was only a marginal improvement on the despotic "rightless law" of colonialism (Shivji 2000, 39). But there was relatively little disruption on Kilimanjaro because the state relied on coffee for foreign exchange (Coulson 1982). In my interviews with retired Chagga magistrates, they emphasized that Ujamaa did not affect the kihamba area because "coffee and masale were all one thing." Cash crop production and secure land tenure were so thoroughly linked that the state could not risk using coercion to reorganize production as it did in the parts of Tanzania without coffee. In the 1960s, the transformation of mid-level and lowland shamba to highlands-style kihamba continued, and many open fields became covered with trees (Holand 1996). The shamba belt has effectively moved downhill closer to the main Arusha-Moshi-Himo road and is now mostly rain-fed maize. At the same time, boundary plants continued to shape land matters in the expanding kihamba zone.

Sally Falk Moore's tale of the "cursed banana garden" provides a compelling postcolonial case study of this process (2016, 109). Two brothers, an older well-to-do teacher with children and a younger and childless subsistence farmer, subdivided a kihamba when their father went to live in the lowlands. Problems began when the farming brother's wife suspected that her sister-in-law had made her infertile. The teacher told his wife to pull up some of the masale between their plots, plant vegetables, and place a new boundary. The younger brother was enraged and called for the father to settle the issue. After the old man ripped up the dracaena and restored the old boundary, his first daughter-in-law was undeterred. She sued the father for violating a customary law preventing a father-in-law from entering his daughter-in-law's kihamba without his son's presence. She lost the case, but her father-in-law was deeply insulted by the entire affair. The teacher's wife again uprooted and replanted the masale; her father-in-law did the same. She sued again and had him jailed (and shamed) for three days – but soon after became ill and died just after the father cursed her and the land. The younger son moved to live near his father in the lowlands, and neither would speak to the teacher, who lived as a socially isolated widower fearing that he would die if he ate bananas from his cursed land. This story demonstrates how Chagga boundary plants both represent orderly landholding and the means of conflict. It also shows how women are not the passive victims of male machinations on Kilimanjaro, but are able to creatively strategize customary law to negotiate resource access.

The overall impression of equilibrium and the continuity of custom in the homegardens applies only at broadest temporal and spatial scales. In practice, Kilimanjaro's dracaena adjusts the landscape to suit changing social needs. Masale is a fixed point in institutional and ideological terms, but is really a living transformational process. Farmers regularly trim their hedges and patch holes with the cuttings, relying on dracaena's knack for vegetative propagation. This becomes particularly important when a man dies, because his sons and neighbors must re-evaluate their boundaries to avoid future disputes. As one elder explained,

> Having masale does not mean that the boundaries never get touched. Instead with every death, it is time to take another look at the boundaries, and sometimes all of the people concerned will decide to straighten them, especially if the boundary winds all around like a snake... Each death is a new adjustment.

This is a group effort because ideally the only hand that moves a boundary is the same one that planted it, and any deviation from this norm requires witnesses. One interviewee described how his neighbor died in 2014, leaving five sons. He went to the brothers and said, "I know our boundary, this is not something in which anyone can cheat little bit by little bit." He made the brothers stand on the boundary, and together the six men planted a new line of masale at one-foot intervals. "I made them into each others' witnesses!" he exulted. Violations of these rules are staples of community gossip, and everyone I interviewed described the consequences as catastrophe or death for the someone who uproots a dracaena plant improperly. "Unless it's done by the same hand that put it there, removing masale always brings madness, illness, or other problems," a focus group told me. These hands are usually those of elderly men. Like the unhappy teacher in Moore's story, today many Chagga insist that crops grown on kihamba land acquired without a male elder's permission will produce poorly, and if such a reprehensible farmer did manage to harvest, the food would not be nourishing or its sale profitable.

For the Chagga, masale does not regulate land matters simply because it symbolizes legal and social institutions, like metal fences or concrete block walls. It works because it has its own active vegetative agency, as both guardian and witness. Chagga farmers told me that "masale has eyes" watching land use. If a man moves a boundary just a few centimeters into his brother's land and gets someone to bear false witness, both risk their lives. One retired craftsman told me how he developed his own attitude toward masale after his neighbor died:

> This man had two wives. After he died, the second wife moved the masale in order to take some of the land of the first wife. She planted three stalks and then fell down unconscious right there in the farm. She lasted for three days before she died. That's why I respect masale for its power – but I don't touch it and I don't do any sacrifices.

Another narrative performance illustrating the power of masale shows how Chagga concepts of human agency are entangled within its fibrous stems:

> Masale is the witness of truth. Masale is like a soldier or a guard, carefully watching how you are taking care of the land. A man had two sons. One was at home and the other worked far away in Dar es Salaam. One day the father announced, "I see that I am too old to work, so I have decided to divide the farm." The old man called two other elders from the neighborhood to come bear witness. The old man told them that he was too old, and all three of them sent their first sons, and whichever other sons were around, to go cut some stalks of masale. The father stood there in the farm, and in front of the witnesses, set the boundary between the two brothers. Each one to his own side. But remember, through all of this, the second brother, the one working in the city, was not there. Five years later the father was dying and unable to get out of bed. So the first son decided to move the boundary because he thought the old man was not watching. After the old man died, the witnesses had to testify about the boundary. The first witnessing elder said that the boundary was here, right where the old man had set it. But the second witness was greedy, and only looked after his stomach. He testified falsely because the first brother had given him a bribe. Soon after the false witness died suddenly, and it was clear that the masale had done its work. There was no way for the liar to clean up his sin, and so the first brother died too. Masale has eyes, meaning that it watches both land users and witnesses. Masale needs witnesses to start this work of watching.

This power is a latent force; as another man in Mamba said, "masale sleeps there peacefully until someone violates it and wakes it up. No other plant in the world works like this!" Similarly, in Kilema I heard that masale "stays silent there on the boundary until someone violates it, then it wakes up and does its work."

Dracaena is a moral person who intervenes in Chagga land matters to provide what Douglass North would call "third party enforcement" (1990, 35). Such a neoclassical economic explanation for this boundary plant would miss the substance of land tenure matters on Kilimanjaro. Masale represents the agency of particular people (nearly always men) to inscribe social meanings on the Kilimanjaro landscape, but it also, from a Chagga point of view, has its own agency. It does not simply legitimize property claims as social facts, it constantly reaffirms and defends an emphatically Chagga way of organizing rights to land by gender and kinship. It watches everyday social activities and punishes misbehavior. Michel Foucault wrote extensively about how Europeans acquired self-discipline because they were being watched by powerful institutions embodying the power of the state, science, and medicine. He calls this "governmentality," the mentality one has of being governed (1997, 341). The kihamba zone is an area with a vegetative governmentality materializing the power of patrilineages to tie together the bundle of power that is land tenure. Chagga landholding is a collective phenomenon, and Chagga patrilineage membership typically stretches back three or four generations, so it is perhaps not too surprising that this sense of being disciplined by boundary plants has direct connections to the participation of the dead in society.

Ancestors in the landscape

Within a kihamba, nestled somewhere among the banana plants, there is a *mbuonyi* skull shrine marked by masale. The ones that I visited in 2015 were less than a meter in diameter, with six to twelve stalks of dracaena shading male ancestors' skulls. The form of an mbuonyi varies; in Mwika the skulls lie in clay pots turned on their sides, while in Marangu four flat stones form a similar protective structure (Dundas 1924, 192; Hasu 1999, 52, Stahl 1964, 211). At the mbuonyi that I visited in Kibosho, masale formed a literal family tree because each stalk located a particular ancestor's skull.

Until the 1960s, rituals for incorporating a man into the company of his ancestors relied on dracaena. According to both oral history and early 20th-century ethnography, the surviving sons would use masale stalks from their boundary hedge to dig their father's grave inside his senior wife's house, while other kinsmen replaced the corpse's jewelry with dracaena leaves. They sealed the grave with mud and clay and set a fire to drive off unpleasant odors. About a year later, the family exhumed the grave, using the same dracaena digging sticks, and moved the soil and most of the bones into the shade of their boundary hedge. Tying together the skull and the right humerus with masale leaves, patrilineage members carried these bones to the mbuonyi, installed them alongside the other ancestors' skulls, and planted a dracaena cutting (ideally from the stalk that marked the skull of the dead man's father) next to the new member of the ancestral community (Dundas 1924, 180; Gutmann 1926, 50; Marealle 2002, 65). A senior wife's skull lay in the shade of her husband's dracaena, and junior wives formed a slightly separate group within the mbuonyi. The result was that patrilineage skulls formed a cluster, and the skulls of women who had married into the kin group formed another. Uninitiated youths and unmarried women were buried within the kihamba's boundary hedge but did not have their skulls transferred to the mbuonyi, while the corpses of childless men and women were thrown into the bush beyond the settlement. The mbuonyi was for socially legitimate adults, the perimeter masale for the socially liminal people, and the unsocialized space beyond for those who had failed to accumulate social capital. Because the homestead is usually inherited by the youngest brother, he becomes his parents' caretaker both in their old age and after their installation in the mbuonyi. The other brothers each start their own mbuonyi at their homes, while still going to their junior brother's kihamba for lineage rituals. The Chagga sense of place in the homegardens and their experience of segmented patrilineages were, in many ways, the results of these mortuary practices that carved social capital into the landscape with masale.

The third major landscape feature made with dracaena on Kilimanjaro, after perimeter hedges and the mbuonyi, is the *uani*, a densely screened private area for men to slaughter livestock and wash down roasted meat with banana-millet beer (*mbege*). The elders I interviewed emphasized that this feasting had to be hidden from women because "they would become rude" if they saw the blood. And because livestock were rarely killed simply for their meat, but instead to placate the ancestors for a variety of taboo violations, these places defined one end of an exchange relationship. Early 20th century Chagga sacrifices were diverse ritual performances by different kin groups with a shared cultural repertoire, not a formula, but masale was always part of this symbolic language. Dundas describes how a ritual leader used two plants to sprinkle milk on the head of the sacrificial animal, while saying, "you, great-grandfather, I pray to you with milk and Dracaena... that you will receive this my offering" (1924, 139). After the elder roasted the meat, he chopped it into tiny morsels, arranged some into a kinship diagram of patri-lateral and matrilateral kin on a bed of dracaena stems, and offered them to his great-grandparents. Petro Marealle describes how elders wrapped particular cuts of meat for male and female ancestors in banana-leaf packets and carefully placed them under dracaena plants (1965, 59). If the meat was missing the next day, this signified that the sacrifice had been accepted. Many sacrificial offerings were (and are) explicitly gendered, and not just directed to gendered groups of kin. Men pour banana beer for the dead (three times to a male ancestor, four to a female) while women offer fermented milk with a wooden spoon. The bull sacrificed to a male ancestor gets three masale leaves tied around its neck; the cow for a woman has four (Hasu 1999, 502).

Current Chagga sacrificial practices continue many of these technical, gendered, numerological, and botanical practices (Clack 2009; Hasu 2009a, 2009b; Myhre 2016b). But not just any masale leaf will do for a proper sacrifice. One ritual expert in Marangu showed me that

> You take the third leaf from the middle of the "head" of the masale [the terminal inflorescence at the end of a stalk]. You look at the shape of the broken edge of the leaf. If it is a clean straight break, the sacrifice can proceed... You take that leaf and hold it up to your mouth and explain the problem for which your sacrifice is intended.

The substance of a sacrifice – it could be a white sheep with a black face or a calabash of banana beer – is usually determined by a diviner. Each diviner has their own method for checking what the ancestors demand, but their techniques always rely on masale leaves (Myhre 2006). A typical consultation with a diviner in Marangu follows this script:

> Usually you go to their place early in the morning with a fresh masale leaf. When you arrive, they wordlessly take your leaf and inspect it carefully and then they tell you exactly what your problem is. The leaf carries the message. You have to be careful to throw that leaf away back at home in your own land, or throw it away on the roadside so that it doesn't get into someone else's farm. That would be completely inappropriate!

Masale is the conduit for ancestors to communicate with the living and adjust relationships. If a Chagga person gets buried far away from Kilimanjaro, the family can go to that gravesite and collect a bit of soil inside a packet of masale leaves, and deposit it back at the kihamba with a freshly planted stalk of dracaena. Chagga sacrifices involve the movement of words and objects (usually meat, milk, and banana-millet beer) from the social space of the living to that of the dead, and dracaena is the botanical technology for emplacing these exchanges. It both encloses the uani and extends relationships along the pathways defined by gender and kinship. Masale is a communicative and spatial event, and this is the reason that everywhere on Kilimanjaro today, mourners decorate the four corners of vehicles carrying coffins with dracaena. They are communicating to both the roadside public and the ancestors that a social transformation is under way. Overall, the bodies of the dead become places through ritual actions involving masale.

These funereal and ritual uses of boundary plants to convert space into place shifted as Chagga became increasingly Christian (Fig. 3.3). In the early colonial period, missionaries adopted the Chagga custom of marking graves with dracaena. Less than a decade after its establishment, the Lutheran mission at Mamba had a Christian cemetery surrounded by dracaena (Althaus 1903). In 1896, two Lutheran missionaries were murdered near Mt. Meru,

Figure 3.3 Grave with masale on the corners

and in 1905, the missionary Arno Krause located their remains and interred them under a pile of stones marked by a cross and a dracaena hedge (1905, 446). A few generations later, churches rejected what these early missionaries had embraced. None of the Chagga cemeteries that I visited in 2015 had hedges or grave markers of dracaena, and elderly informants recounted that churches across Kilimanjaro rejected this use of masale in the 1960s. Now that many church cemeteries are full, more burials are again in the kihamba. Most Chagga build tiled concrete graves, with little fragments of mirror or glass to catch the sun and a masale stalk at each corner of the rectangle. "You can trim these," one focus group agreed, "but never remove them." Many old mbuonyi now co-exist with these newer graves, but are either neglected and trash-littered, or simply left untouched. One woman showed me the taro growing on her grandfather's mbuonyi, where a hedge of masale had surrounded a big fig tree. The tree fell down, but new trees are growing there now. "I won't touch those trees, I won't eat that taro, and I won't sacrifice there – but I will leave it alone because it is our mbuonyi," she said.

In the 1960s, particularly under the influence of Lutheran Bishop Stefano Moshi, Christians abandoned traditions that involved direct interaction with ancestral spirits. This sharply curtailed public usage of dracaena, particularly sacrifices (Bailey 1968, 166; Hasu 1999). Many Chagga now get meat from supermarkets and alcohol from *mbege* beer bars rather than uani sacrifices. Dracaena rituals proceed quietly because some Christians, especially Evangelicals, oppose them so strongly that sacrifices can lead to precisely the sort of discord that they are supposed to prevent. At the same time, kin group rituals became politically incorrect private events during the socialist government's policy of Ujamaa, which envisioned economics as a communal extension of family sharing. Sacrifices became smaller household affairs instead of public events.

The simplest sort of sacrifice that I heard about repeatedly on Kilimanjaro followed this format:

> Take a fresh green leaf of masale, from the top tuft of leaves so that it has a white base, spit on the leaf telling ancestors to accept the gift, then put meat and some blood on leaf along with a calabash of banana beer. Go back in two days to check that the spirits have eaten and drunk the food and drink. Then the good things will come.

More serious sacrifices require livestock, with the value of the animal corresponding to the size of the problem. Most of the uani that I saw in 2015 did not have fences of masale any longer, but instead were three-stone fireplaces between old mbuonyi and new graves. This is where the butchering takes place.

> To settle a dispute, you have to strangle a goat and put masale in its mouth, and then take some of the meat from each part of the goat – heart, liver, brain, etc. – and put it on a banana leaf and tie the whole thing up with masale, then take the package to the uani.

The degree of simplification of sacrificial procedures depends on whether an area was Christianized by Lutherans or Catholics. In predominantly Catholic areas like Kilema and Rombo, Chagga use three plants – masale, *kinghera* (*Commelina africana*), and *wambi* or *ngambi* (an unidentified ground herb with tiny leaves along thin brown stalks, like oregano) – in sacrifices. Catholic elders explained that these three are "brothers who cooperate" in a sacrifice. Similar to the transformations in land tenure, dracaena's roles in practices like marking gravesites and communicating with the dead have changed while remaining icons of continuity.

Christians on Kilimanjaro now face serious moral dilemmas in land management. They decorate churches with masale and carry it in religious processions, but these performances are emphatically not about ancestors. The masale inside a kihamba can be safely ignored (unlike those on the perimeter), but what is to be done when a plot is inherited, bought, or rented? Farmers showed me old Christian graves on plots they had inherited from other families, like when a childless couple's plot went to the husband's sister, who had married into another patrilineage. They described how they struggled to manage the land ethically. They could not simply farm within these masale-lined graves, could not afford the rituals to move them, and were anxious every time they cut back the dracaena to keep it from overshadowing their crops. If they did move such a Christian grave, they would have to relocate the bones, stones, crosses, and the masale shrubs.

This is even more difficult with a mbuonyi because it requires ritual specialists and a truck.

> After a lot of sacrifices to begin, you take a cow hide and wrap up all of the bones and pots, and then put one masale plant from that mbuonyi on top. Then you move it all to the new village and plant that masale at the new place, and sacrifice a goat.

Alternatively and much more economically, a Christian can sell or rent out his inherited plot in order to avoid interacting with a mbuonyi at all. Whenever someone buys or rents land on Kilimanjaro, therefore, several elders explained,

> if you see some masale on land you want, you must ask if it's a grave even if there is no other evidence of a grave. Even if it is not, when you do rent or buy the farm, you do not remove that masale, you just leave it alone.

These moral boundaries also complicate plot boundaries for Christian Chagga; can one adjust plot perimeters or create new ones without sacrificing to the ancestors? Many Christians confided to me that they felt trapped by the contradiction of masale as a basic land management institution and their faith's ban on ancestor veneration, and therefore did nothing. "We want to sacrifice but cannot, so we cannot manage their farms properly by resetting the boundaries when things change!" one man lamented. At the crux of this contradiction lies the social fact that from mbuonyi to concrete graves, burial within a kihamba is the strongest stick in the Chagga bundle of powers for land ownership, and the strongest of the "bundle of responsibilities" for Chagga relations with the dead. These economic structures and spiritual meanings remain firmly entangled in masale on Kilimanjaro because both relate to social order and power.

Masale, gender, and kinship are so thoroughly interwoven on Kilimanjaro that they form a single meshwork. A proper marriage, for example, begins with the *mburu wa sale*, the "dracaena-goat" with a long leaf tied around its neck, which a prospective groom presents to his future father-in-law (Gutmann 1926, 91; Hasu 1999, 264).[4] The father-in-law may then anoint his new daughter-in-law with a bit of butter on a leaf of dracaena from his boundary markers (Myhre 2017, 115). These leaves are the conduits for socially legitimate transformations among the living just as they mediate relations with the dead. Every use of masale materializes Chagga organization through segmented patrilineages, the gender hierarchy that this demands, and the membership of the dead in the social groupings of the living. This repeated emphasis on the power of patrilineages to shape Chagga society is why so many male elders told me that "women do not hold masale, the hand that holds masale is male." Graves, divination, and sacrifice are places and events that manifest this gender/kinship system, and are sites of contestation, negotiation, and adjustment. Dracaena is not only something that Chagga use for their own social transformations, but has also been the medium through which they have experienced exogenous social change. The dynamism of this web of relationships among people, plants, skulls, and livestock shows that viewing Chagga kinship as a set of rules and roles is too mechanical. Instead Chagga social organization is an open-ended process of using masale as a ritual attractor that creates, negotiates, and legitimizes relationships of power, all of which relate directly to resource access and the meaning of the Kilimanjaro landscape.

Knots of peace, order, and meaning

Ask anyone on Kilimanjaro where to see dracaena, and they will point you toward a boundary or a grave. Ask what it means, and the universal answer is "peace." This meaning overlaps thoroughly with, but does not reduce to, dracaena's significance for land tenure and social organization (Myhre 2006). For many Chagga, peaceful relations with one's neighbors and the blessings of benevolent kin and ancestors are what these domains are ideally all about. As an abstraction, peace is best seen in its manifestations, but for the Chagga it is visible, transferable, and demonstrable in "the old Chagga plant of peace and pardon" (Stahl 1964, 27). Masale is an object that "emits concepts" on Kilimanjaro (Holbraad and Pedersen 2017, 236). The Chagga term for peace is *uforo*, and the masale plant materializes, contains, and directs it. When I asked one elder to describe its power, he exclaimed,

> The top meaning of masale is peace! All of its other meanings and uses follow from this! Masale is able to chase away any sort of discord. Its power is peace, it is like a flag of peace for all to see, that binds [grabbing my wrist tightly for emphasis] whoever meets it to behave this way and not that way!

Another argued that masale is a "tool for fixing peace, the complete opposite of a weapon that cuts." Chagga repeatedly answered my questions about masale's meaning with terms like binding, fixing, attaching, and closing to express its permanence and agency. When I asked why dracaena endures so well, they patiently explained that "it never dies!" Its ability to take root from cuttings makes masale stalks particularly useful for binding peace and order into the landscape. The way that its glossy dark green leaves resist drying out make masale especially good for creating, mandating, and enclosing peace and order for several days. This elaboration and transposition of the plant's material properties accounts, to a degree, for the symbolic extension of masale from everyday economic and social uses to esoteric ethnographic details, like Gutmann's note that when a Chagga person encounters spilt milk on the road, they spit onto a dracaena leaf and throw it on the dangerous spot (2017, 159).

Anthropologists refer to peacemaking institutions as "dispute settlement mechanisms." In the absence of a centralized judicial system, like in 19th-century Kilimanjaro, such mechanisms are crucial because they operated as centripetal forces to draw people together when centrifugal forces threatened to pull them apart. Masale settles disputes, prevents them, and makes cooperation binding. To borrow something valuable, like a cow or a truck, a Chagga person presents the request with a dracaena leaf in a bottle of fresh milk. When a Chagga chief wanted to prevent a developing conflict, his sign of peace was a sheep wearing a masale necklace (Dundas 1924, 78; Gutmann 1926, 475; Stahl 1964, 122). When the resolution involved a new territorial boundary, "the new border line was solemnly marked with dracaenas" (Gutmann 1926, 389). In the 19th century, a man walking between opposed Chagga armies with a stalk of masale could make them lower their spears. It could also show submission to political domination, like when Chief Marealle of Marangu conquered his Mamba neighbors and required them to uproot their masale and replant them in Marangu (ca. 1892, Stahl 1964, 325). My interviewees recalled his grandson, Thomas Marealle, scratching a message onto a leaf from his palace's boundary hedge to summon people to his court. Conflict within a kin group could be resolved when an elder made quarreling people bite a dracaena leaf to "clear things up quickly" (Gutmann 1932, 185). Finally, as the ultimate dispute settlement mechanism, one could not start trouble with dracaena. You can hit a goat with a masale stalk, I was told several times, but never a person (Dundas 1924, 182). Dispute prevention may be why the entrance to the bus station in the city of Moshi has a single forlorn stalk of dracaena growing in an old tire – to prevent trouble in this important transport hub.

Most importantly, this dispute settlement mechanism is mandatory. Even if someone isn't ready to settle their dispute, masale forces reconciliation. The most common practice mentioned by my research participants was tying leaves in knots (Fig. 3.4). This extract from an interview is typical of the stories I heard about knots on Kilimanjaro:

> If a young man robs my house, his father will come in the morning to my place. He will come slowly, giving all the proper greetings until he is very close to me, and then he puts the hidden knot of a masale leaf into my shirt. Only then does the father explain why he has come, because any use of masale requires an explanation. Now, the father of the thief will beg me to keep the whole affair secret, and that I should not go to the police and make a formal case out of it. This gives me the right to demand a goat with a masale leaf around its neck, paid for by the thief. The thief's father brings the goat, and with other elders as witnesses, and then explains exactly what happened in front of everyone.

Figure 3.4 Chagga ritual expert with a knotted leaf of masale

> The witnesses take the goat, slaughter, and eat it. It was the masale that
> made it possible to resolve the case without the use of the police, and to
> get peace again. So the way that masale makes peace is an active force,
> and it keeps things local. This happens even today as much as possible.
> This is all the work of masale.

Many interviewees corroborated the detail about slipping the knot inside
someone's shirt and showed me the knots stored in the rafters of their houses
as evidence of settled disputes. But how to settle very serious disputes, like
those involving physical violence? In such cases, the apologizer sends a tod-
dler holding the knot, with instructions to give it to the aggrieved person
when they pick up the obviously innocent child. "Inside of the knot there
are blessings fixed there," explained an elderly lady, and this ties together
the damaged social relationship. Several of the people who demonstrated
this apology ritual for me emphasized that when you smell the clean vegetable aroma of the masale knot on your chest, "you know that the bad odor
of the wrong is being driven away." This compulsion could also be used
strategically as a way to start legal action. Gutmann describes a vengeful
man eavesdropping to hear his drunk neighbor malign the chief. Hearing
the ill-chosen words, he put a knot of masale above the neighbor's doorway
and called another neighbor to co-witness both the drunkard's words and

the knot (Gutmann 1926, 555; Steiner 1954). Inevitably, the malefactor paid a fine of livestock to the chief.

I learned to entangle people in these knots during my fieldwork. The Chagga term for a token gift in return for a small favor is *kitorchi*, and I found that by wrapping a few banknotes within a dracaena leaf and tying it into a knot, I always got rewarded with a knowing smile. When I asked what my kitorchi would mean if it were verbalized, my interviewees explained that my message had been "please accept this modest gift." My gift was worth a few US dollars, an amount that is appreciated but not a windfall on Kilimanjaro. By wrapping the money in a masale leaf, I was forcing acceptance. Dracaena also bound the ancestors to accept an improper gift in colonial-era Kilimanjaro. If a poor man could not afford an important sacrifice, he could knot dracaena leaves onto a bull's neck, horns, and legs and then transfer the leaves to a sheep (Dundas 1924, 149). For the Chagga, masale glues relationships in place. Untended resources like women's bundles of fodder or men's beehives can be marked with knots of masale to proclaim "do not touch!" The two barrels of banana beer at a Chagga wedding each have a knot of dracaena on them to restrict usage; one for the men and the other for women. Overall, these masale knots are more than a dispute settlement institution. They are also a norm imposition mechanism that ties together and naturalizes rights, responsibilities, and relations of gender and kin-based power.[5]

The Chagga concept of patrilineage power extends beyond kinship, and these social elaborations revolve around the spatial metaphor that masale opens and closes paths. The most common image presented by my interviewees was that of a farmer (male or female), being frustrated by petty theft or foot traffic in their kihamba, planting a stalk of masale (with a few leaves tied into knots) where a path enters their land. In colonial Kilimanjaro, a dracaena at the door of a house limited the visitors to an ill person or an expectant mother (Dundas 1924, 180, 197; Marealle 2002, 93). When a mother nursed a newborn, she planted a stalk outside (with three knotted leaves for a boy, four for a girl) to restrict her husband's visits for the six months of a postpartum sex taboo. Several people insisted that a stalk of dracaena redirects army ants away from houses or sweeps them out better than any other broom. Road crews on Kilimanjaro redirect traffic away from rockslides with masale stalks. The first animal in a traditional Chagga marriage exchange, the *mburu wa sale*, gets its necklace of leaves from the kihamba boundary. When giving this goat, the presenter says "*nifunga mburu*" (I am closing the goat). Chagga ritual experts said that this goat "opens a path between the two families" and "closes the path to that girl." None of the other livestock in marriage transactions need such decoration because "you only have to close a path once for it to be closed." Two masale, their leaves tied together into an arch, mark the entrance to the groom's compound in a traditional wedding to show two families bound together and the closure of the way to that bride. Finally, Otto Raum's description of

Chagga children's play includes a game in which children free captives from "door-keepers" with masale leaves (1940, 270).

As I learned about how dracaena opens and closes relationships on Kilimanjaro, I struggled to find a metaphor to express these ideas and their connections in so many contexts. I asked if doors, lightswitches, or plumbing valves were appropriate images. Interviewees rejected my efforts until I tried locks and keys. "Now you have it!" some said, "it's THAT sort of opening and closing!" But what is it that masale locks so well? When I asked a focus group of elderly male and women to identify dracaena's power, they responded with the term *mbora*, meaning "blessings and good results." Knut Myhre's (2017) ethnography of Rombo, the eastern face of Kilimanjaro, provides some vocabulary to contextualize opening, closure, and power more deeply in Chagga thought. Myhre focuses on how they conceptualize life (*moo*) as the effect of how the vitality of life force (*horu*, the Eastern Kilimanjaro equivalent of *mbora*) is transferred and transformed through the Chagga experience of "dwelling" in kihamba homesteads (*ikaa*) (2017, 5). For Myhre, Chagga economic and social forms result from the interplay (and inversion, and reflection) of these ideas and experiences. Staple topics in East African ethnography, such as kinship, ritual, divination, cursing, and rainmaking all stand revealed as expressions of the unifying logic of keeping the life force circulating in order to "return life" from one part of Chagga society to another. Myhre follows Marilyn Strathern's approach to "partible personhood" in Papua New Guinea (see Chapter 5) to develop the concept of a "vectorial person" to explore what Chagga do with life force. Such an experience of personhood concerns how Chagga direct the flows of life-force-containing substances (like blood, semen, milk, and food), and this helps Myhre to interpret esoteric ethnographic data like rituals for the fictive closure of men's anuses in and around Kilimanjaro (Moore 1976; Myhre 2019b; Sheridan 2002). Vectorial personhood is useful for examining boundary plants on Kilimanjaro because it helps unravel the ways that dracaena entangles particular relationships and practices.

When they say that "it has eyes," Chagga express the idea that masale is a sort of person; Myhre's analysis helps to identify what sort of person this boundary plant is. Masale is vectorial because it directs and legitimizes the flows of resources and meanings through relationships and practices by opening and closing apertures. This attention to the regulation of life force explains, for example, why the Chagga strangle livestock (and put masale leaves in the animals' mouths) instead of slitting their throats – because these actions contain the life force so that it may be redirected. Traditionally, a newborn child's first food was a mixture of bananas, milk, and butter with a bit of dracaena leaf "to open the mouth" (Moore 1977, 56). This inverts dracaena's embodiment of place in burial practices by emplacing an orderly landscape into bodies. When I asked one elderly carpenter why the vertical posts, the attic sub-ceiling, and the crossbars of a bed in a traditional Chagga house were all made of dracaena stems, this is why he responded, "because that way peace and safety were closed in the house!" If an unauthorized person uproots the masale on a farm, my interviewees agreed that (like Moore's cursed banana plot) the farm would be unproductive because its flow of blessings and fertility would be interrupted. I asked one man in Rombo for an example of how masale "opens and closes life." He responded that

> Masale protects life. A new mother gets a bracelet made for her by her mother-in-law from masale and banana fiber, and she wears it for three months. The mother-in-law puts a masale branch with knots in it – three knots for a boy, four for a girl – across the doorway of the newborn's house. The guests jump over it to see the baby, and this cleans any *iha* that could harm the baby or spoil the milk. This lasts for three days.

Iha is a sort of "evil eye," not witchcraft. It occurs when someone unintentionally startles when seeing evidence of someone else's good fortune, which then can damage or ruin it. Masale prevents iha (Marealle 2002, 5). The *kichong* adoption ritual is another example of closure. Bringing someone into a new patrilineage, one ritual expert explained, requires opening the kin group and closing the child inside. The baby's sponsor must take the three innermost leaves from the terminal cluster of a stalk of masale from the family boundary, making sure that the base of each leaf has a straight tear. After sacrificing a goat or sheep, the sponsor puts a strip of its skin around his right hand, picks up the child, places the knot of leaves on the child's head, tops the load with a haunch of meat, and brings the infant into its new home. "The masale is the key in the lock for bringing someone inside the patrilineage," he summarized. These examples of ritual practice all demonstrate how masale determines the directions that life flows while watching as a vectorial person. This plant-as-person is, like the experience of personhood across Africa, composed of how it relates to and participates in other beings (Riesman 1986).

From a political ecology perspective, the opening and closing of relationships flowing with life force is a kind of symbolic capital restricting movement and designating people with socially legitimate rights. This is how masale ties together the Chagga bundle of rights that includes matters of land tenure, ancestor veneration, kinship/gender relations, and ritual action. The knotted leaves of boundary plants are binding because they form the metaphorical string that keeps Chagga social order bundled.

Conclusion

Why have these particular relations, practices, and meanings persisted while so many other aspects of life on Kilimanjaro have changed radically? Why does the cultural form of dracaena show such continuity? When I asked one man how a typical Chagga person feels when looking at a masale plant, he answered, "he thinks of the past and how this should continue into the future." Dracaena is so entangled in land use, social and political organization, and ritual practice that the fabric of Chagga life would not have its pattern without it. Along with bananas and coffee, it defines what it means to be Chagga. Its uses and meanings are shared across Kilimanjaro, but this does not necessarily make it a fossilized and functional social fact. The dynamism and transformation evident in the expansion of the kihamba zone into the shamba lowlands, the changes in relations with the dead and struggles to resolve the implications for land management, and old ritual practices in new contexts all show that boundary plants are often a medium of change even when they mean continuity. Much of what contemporary Chagga do with masale appears similar to older practices on Kilimanjaro, but it is not simply a tradition. The plant is something that people use to adjust to historical change, not a static cultural commitment.

Chagga narratives about stability and normalcy focus on the social implications of dracaena's botanical properties. To wish someone a long and healthy life, Chagga say *wara finya che sale,* "may you live as long as a dracaena." In Gutmann's account of how an older brother gives permission for his younger sibling to marry, the prayer to their deceased grandfather focuses on vitality and growth. "I beseech you to let the younger one rise like the fermentation of beer, let him prosper like a dracaena, let him expand like a swarm of bees" (1926, 62). During a colonial-era blessing for an age-set circumcision,[6] the ritual leader sacrificed a sheep, planted a dracaena, and prayed that the boys would "shine like the fat of the elephants, so that they endure like the dracaena at this site, which does not dry out" (Gutmann 1926, 294). When I asked "why *this* plant?", people responded that "masale holds onto life more," and that "has a power to grow and stay alive forever." It is a remarkably durable plant, and its growth habit of vegetative propagation from cuttings makes it a prime candidate for symbolizing human agency and creating order in both landscape and society. Agency, social order, and cultural meaning on Kilimanjaro have all been radically transformed by decades of economic, political, and ideological change, yet dracaena persists as an icon of continuity.

Currently, the main concerns of many Chagga on Kilimanjaro are upward social mobility, cell phones, and cash crops. They have abandoned their 19th-century cattle economy, age-set organization, patron-client chiefdom institutions, and many life-cycle rituals (Håkansson 2009; Moore 1986, 298ff). Economics, politics, and cosmology had been overlapping domains of intensely local Chagga-ness, but are now increasingly separated into global commodity markets, national bureaucracies, and Christian churches. Masale means less than what it once did, but it has endured while other cultural commitments have faded away completely. For example, one of the practices reported in Rebmann's first encounter with the Chagga, the custom of creating alliances with non-kin by fastening skin rings from livestock to a partner's fingers (Krapf 1860, 238; see also Dundas 1924, 144;

Gutmann 1926, 43; Moore 1986, 71), did not survive. Why has dracaena remained a symbol and technology for creating and legitimizing social relationships, but skin rings have not? Comparative work on African resource management shows that traditional customs are most likely to last when they allow local elites to access and control flows of economic and social capital (Haller 2010). The symbolic capital of masale met new interests and got incorporated into new institutions as Kilimanjaro shifted from a mountain of fiercely independent patrilineages and chiefdoms competing for land, labor, and cattle to a coffee-growing district in a nation-state. Masale continued to create and legitimize relationships because it was rooted so deeply in the Chagga social-ecological system via land tenure, social organization, and symbolic practice; the skin-rings lapsed because other social relations eclipsed livestock exchange. This boundary plant has remained a paragon of Chagga identity and meaning during a period of rapid population growth and economic change by inscribing Chagga concepts of power into the landscape. It has been, in brief, a vegetative means for enacting Chagga place-making narratives and embedding their strategic struggles in an orderly Kilimanjaro landscape.

Dracaena was a polymarcating boundary plant that tied together a culturally specific set of rights and responsibilities, but these meanings have eroded as new institutional waves have washed over Kilimanjaro. "It used to be full of meaning, but now it's just about economics and property," complained the elders in one focus group. "Masale doesn't have the heaviness it once did," "it used to be wide, but now it's narrow and mostly about land boundaries," and "everything is being ruined without masale!" said others. "You can't tell kids these days anything about masale," Chagga elders lament, "but they will know its importance when they get their own land!" Tanzania's experiments with "African socialism" and neoliberalism (Coulson 1982; Schroeder 2012), its struggle to enact land reform (Shivji 1994, 2000), and its reorientation from local religious systems to global faiths like Christianity and Islam (Ndaluka and Wijsen 2014) have all introduced new ways of doing things the Chagga had once done with dracaena. As one man put it when discussing how judges take bribes, "now money defeats masale." Many elders insist that the decline of dracaena's complexity is an index for decreasing peace and prosperity on Kilimanjaro. The newer institutions certainly do build relationships, they say, many of them good ones, but the problem is that "there is no lasting, permanent peace like we used to make before, by using masale." In terms of New Institutional Economics, the transaction costs are higher with the new institutions. The problem is that their boundary plant has become disentangled from many domains, and the Chagga cultural ideals of peace, order, and the flow of life force have been compartmentalized, forgotten, and repressed. Masale is still a polymarcating boundary plant on Kilimanjaro, but less so. In the 19th century, many of Kilimanjaro's dracaena hedges were tightly interwoven; now many stalks stand alone in fields. The disentanglement has been physical, social, and symbolic.

This case study of Kilimanjaro's boundary plant demonstrates how culture-specific concepts of space, place, agency, and power become manifest in landscape, and how landscape shapes social history. It is not analytically useful to conclude that masale is an agent on Kilimanjaro; the recognition that Chagga conceptualize and experience agency and change through their engagements with this plant offers more insight. Anthropologists could call dracaena a "multivocal key symbol" (Ortner 1973; Turner 1969, 52) or a "total social phenomenon" (Mauss 1967), depending whether we were focusing on symbols or society. Understanding what masale is from a Chagga point of view requires going beyond the interpretive vision that it "means something" and the functional perspective that it "does something." On Kilimanjaro, dracaena embodies Chagga notions of purposefulness and mediates historical change in a culture that once focused more on corporate identities than individuals. Dracaena watches with a Foucauldian "vegetative gaze" to legitimize social action, represents the hand of man (in the fully gendered sense of the term) on the landscape, and compels social and moral order. Dracaena plants are non-human yet fully agentive and purposeful social actors from a Chagga point of view. But they are not just any member of society – they are benevolent male elders wisely witnessing and re-directing their descendants' behavior into particular vectors. Dracaena contains a relatively "horizontal" cultural model of agency that is different from this book's more hierarchical constellations of meaning and practice. For Chagga, human agency was until recently about being a member of a corporate group by virtue of being male or female, young or old, and kin by blood or by marriage (Moore 1986, 310ff). When we visit western Cameroon in the next chapter, we will see how dracaena relates to a more "vertical" sort of agency within a hierarchical palace structure, which demonstrates how symbols and power relate to African social forms in different ways (Monroe 2013). On Kilimanjaro, then, boundary plants now simultaneously concern population pressure on the land, the changing nature of Tanzanian society, and the shifting meanings of personhood, community, and power.

Dracaena connotes the limits of power and its pathways on Kilimanjaro. Power is a useful way to frame how and why particular people drive history in a particular direction, while others are just as surely driven (Wolf 1999, 67). Yet these coercive movements are often subtle because power works best when hidden in physical, social, and ideological constructs that make people discipline themselves (Foucault 2012, 86). Studying power as a universal flattens, diminishes, and misinterprets the cultural differences that we seek to understand (Sahlins 2018). Power needs to be broken down into types and contextualized as social relations that construct ecology, society, and ideology. Of Wolf's four types of power (see Chapter 2), "structural power" expresses how the Chagga perceive masale (and, in the 19th century, cattle). Wolf defines structural power as "the power manifest in relationships that not only operates within settings and domains but also organizes and orchestrates the settings themselves, and that specifies the direction and distribution of energy flows" (1999, 5). When Chagga use masale to organize land matters, legitimize social relations, and create binding peace, they are directing flows of energy and resources by replacing generic (and fallible) human agency with specific (and incorruptible) plant agency. Every time a Chagga ties a knot in a green dracaena leaf, they recreate and communicate a Chagga sort of structure and order. It is a materialized cultural model of exchanges and transformations of economic, social, and symbolic capital organized by gender, age, and kin group. The benefits of this place-making order are not equitably distributed, and this meshwork of land use, social organization, and ideology tends to reinforce the influence of adult men. This is the social category, that, as the local elite, drives and disciplines people by managing land tenure, organizing marriages, and interacting with the ancestors. As Henri Lefebvre argued (1991), space is everywhere, but place is an emergent property of social inequality; in this case the kihamba zone manifests the structural power of patrilineages.

The other case studies in this book show how boundary plants operate as moral persons, but with different forms and meanings of structural power. The trajectory of the Kilimanjaro case study is ultimately about order in a kin-based and socially "horizontal" society. But the Chagga are now not a kin-based society. Their increasingly class-based society is now fully integrated into the world system via global connections like coffee sales, higher education, and Facebook. The relationship between Chagga individuals and collectivities is now very different from its 19th-century version. The ways that Kilimanjaro became integrated into larger systems through coffee, colonialism, and labor migration have introduced new practices and meanings into Chagga society, and this is why the significance of this boundary plant has transformed. As new legal, political, and religious institutions multiplied the means for Chagga to pursue and accumulate economic, social, and symbolic capital beyond the kihamba, they no longer need dracaena to bind contracts, confirm relationships, and ensure moral order. But within the social field of the kihamba, they do need it to demonstrate land rights, and the other meanings of this multivocal plant can then be invoked in both cultural narratives and strategic struggles in other social domains. The bundle of rights and responsibilities that define Chagga society and culture is still tied together with masale, but the knot is looser than it had been a century ago. Tightening that social-ecological knot would, according to my Chagga friends, strengthen their quest for sustainable agricultural livelihoods, help them build more effective institutions, and negotiate a better relationship with the Tanzanian state. 3000 kilometers away in the Cameroonian Grassfields, dracaena also ties together power, agency, place, social organization, and land use. But these are different knots, with different entanglements and different holds upon people and landscape. Unraveling those knots requires the analysis of boundary plants in a tributary African kingdom.

Notes

1. There are different terms in the three major Chagga dialects for the banana-coffee-dracaena complex; it is *masangaa maruu* in the western Mashami dialect, but *mndenyi* in the central Old Moshi and eastern Rombo dialects (Montlahuc and Philippson 2002, 62). I use the term *kihamba* because it is in common usage on Kilimanjaro. I use the term *masale* for dracaena, particularly when I express the Chagga point of view. A single dracaena stalk or leaf is a *sale* or *isale*, but I use terms like "masale leaf" and "masale plant" to avoid confusion with the English economic term "sale."

2. My 2015 fieldwork consisted of 36 interviews (in a mixture of Swahili and Chagga) over three weeks in the south-central Kilimanjaro communities of Kibosho, Kilema, Marangu, Mamba, and Rombo. I also draw on several decades of research experience in Kilimanjaro Region.

3. Dracaena is the focus of this chapter, but it is not the only boundary plant on Kilimanjaro. Monomarcating exotics like *Eucalyptus* spp., *Grevillea robusta*, *Agave sisalana*, *Jatropha curcas*, *Lantana camera*, and *Brugmansia arborea* also mark boundaries and form hedges on Chagga lands. Other dracaena species grow on the mountain, such as *D. afromontana* and *D. steudneri*, but only *Dracaena fragrans* is culturally complex and polymarcating. *Cordyline fruticosa* (see Chapters 5, 6, and 7) was introduced to Kilimanjaro around 2005 as an ornamental plant. It is called *masale ya kizungu* in Swahili, "the foreign dracaena."

4. This garlanded goat is still a sign of betrothal on Kilimanjaro today, but many Chagga now substitute an envelope of money tied up with a dracaena leaf (while still calling it a goat).

5. One elderly lady said that when Chagga women seek forgiveness from other women, they do not use knots. Instead "she would take a wooden dipper spoon (not metal!) with milk in the right hand along with a dracaena leaf (so that the leaf is between the hand and the handle), and poured the milk onto the other woman's three-stone cooking fire. Men did their business with *mbege* beer, we women did ours with milk."

6. Circumcision was the major public ritual establishing an age-set, often involving up to 100 boys. Their circumcised penises were wrapped in dracaena leaves and banana fiber (Widenmann 1899, 46). This method for inscribing Chagga social organization onto male bodies declined as the colonial states introduced new mechanisms for social control, such as taxation and identity papers (Moore 1977, 58).

4 Cameroon

Bounded Vitality and Rank in the Oku Monarchy

I was at an Oku traditional healer's house to interview him about boundary plants in this mountainous West African kingdom.[1] Although Oku is a remote rural area in the heart of Cameroon's Grassfields region, with an agricultural economy based on maize, potatoes, kola nuts, and honey, it is a major destination for traditional medicine (Koloss 2012, 47). Oku has more than 100 traditional healers, with overlapping and sometimes contradictory ethnomedical and ethnobotanical knowledge and practices (Bartelt 2006; Kelly 2012). The healer ushered me into the "medicine room" on the side of his mud-brick house. Bags and bundles of herbs hung from the walls and rafters, and a clay pot with a red and white bullseye occupied a shadowy corner. Just inside the entrance stood a calabash with a stalk of dracaena. I began by asking him about his healing career, what he had learned from his father, and what he had discovered for himself. He took his expertise very seriously and narrated this history dispassionately. Then I asked how the piece of dracaena (*Dracaena fragrans*, known in Eblam Ebkwo, the Oku language, as *nkeng*) at the doorway was important for his medical practice. His face erupted into a big smile and his eyes shone as he seized my hand and spoke emphatically and excitedly in Pidgin English, stating:

> Nkeng make all man be FRESH because it is so very green! This wash the bad spirit, this wash your eye by the nkeng, make your eye be shine! This for very fresh air, very good idea in this house. It keep it FRESH!

Until that moment, I had focused on dracaena's institutional and symbolic aspects, but clearly the plant was triggering strong emotion. After this, I asked each interviewee how it *feels* to encounter nkeng in Oku. Multiple informants agreed that it evokes powerful emotions because "for an Oku man, when you see nkeng, you feel something good, because it is green and fresh." They explained that nkeng has more life than other plants, and this vitality delights their eyes. This chapter interprets these powerful emotions as emotionally resonant "technologies of power" embedded in a particular historically constructed social-ecological system. When Oku people encounter boundary plants in their homes, farms, and communities,

DOI: 10.4324/9781003356462-4

they experience a culturally specific practice of power in a botanical idiom. Analytically, the political ecology of freshness helps to explain the Oku cultural landscape.

I chose Oku as the second African case study for this book because its boundary plant is as socially and symbolically "thick" as masale on Kilimanjaro. Dracaena is a hedge, boundary marker, and peace symbol throughout West and Central Africa (Bos 1984; Dalziel 1937), and in Cameroon's Grassfields it is the "tree of peace" (Depommier 1983). Many scholars do not contextualize dracaena beyond fencing or "ritual use," but in Oku, dracaena is complex, multivalent, and conspicuous. In local terms, it metaphorically drips with power and agency. It's what people talk about; it's what you see in nearly every nook and cranny of social life; and it's one of the threads that ties disparate aspects of life in Oku into a shared cultural fabric. As detailed below, Oku's plants, animals, and people form a social-ecological system pulsing with vitality and significance, and dracaena is one of the keystone species in this meshwork of life, meaning, and emotion. Dracaena is a technology of power that inscribes meaning, memory, identity, and belonging onto both objects and bodies in Oku, and connects political economy, social organization, and cosmology.[2] Like the Chagga, the people of Oku are farmers on side of a volcano who use boundary plants to organize land access, social relations, and the meanings of peace and life. An all-too-simple diffusionist argument would assert that an ancient Bantu wave of social change from Cameroon deposited this cultural trait on Kilimanjaro. An ontological and posthumanist approach would insist that each case study represents a distinct reality composed of people and plants. This chapter relies on comparison to demonstrate that although the usage and meaning of dracaena in Oku do resemble those in Tanzania, in Cameroon these elements are arranged differently. This case study is about social ranking and hierarchy in a tributary chiefdom, whereas the Chagga chapter was about a flatter social system of kin groups and ancestors.

Studying boundary plants in the Grassfields is challenging because many Cameroonians consider dracaena a timeless tradition beyond history and social change. This chapter replies by showing old institutions and symbols being deployed in new ways for new purposes. As was the case on Kilimanjaro, narratives and practices emphasizing continuity are responses to change. This processual view explains why Cameroonian women are addressing the civil war in the Grassfields' anglophone area (described further below) with this boundary plant. When Joseph Ngute, Cameroon's Prime Minister, visited the city of Bamenda on May 9, 2019, women marched, sang, and danced to protest violence and death on all sides. As they shouted "you have destroyed us!" and "we need our rights!" they waved stalks of dracaena; when talking to the media, the women displayed branches on camera next to the speaker (Associated Press Archive 2019; Budji 2020). They were using a Grassfields institution of intensely localized social order to demand better governance on the national political stage. This chapter visits Oku, a mountain kingdom northeast of Bamenda, to demonstrate why dracaena is a socially effective, culturally legitimate, and emotionally rich response to change, and not simply a static custom.

The chapter begins by reviewing the ecological context and social history of Oku and the Grassfields region. The second section focuses on how dracaena has shaped land tenure and social change. The next part describes how it relates to domestic and kingdom-wide social organization. The fourth and fifth sections examine how Oku's ranking system draws on vitality and life force. The chapter concludes with the themes of agency and emotion in the context of ecological and political crises in Oku.

Oku as a social-ecological system

From Bamenda, the capital of anglophone Cameroon, a shared taxi takes about two hours to reach Oku on a recently paved road. Located on the northern shoulder of Mt. Oku (at 3011 meters, mainland Central Africa's second-highest peak), Oku is the highest of Cameroon's kingdoms. From the dramatic Ibal valley, where thin waterfalls leap over sheer cliffs, the route climbs 1200 meters over 12 kilometers. Near the crest of the ridge, the road enters the Kilum-Ijim Mountain Forest Reserve and passes a dracaena-decorated signpost for Lake Mawes, a volcanic crater lake where royal installation rituals occur. The road winds down to the densely settled agricultural zone at Jikijem, and traverses a series of ridges and valleys before arriving in Elak-Oku town. The market and taxi stand in the middle of town are surrounded by the Oku palace, shops, bars, photocopy stalls, welding workshops, women selling palm oil at the roadside, schools, and some governmental offices, including a hospital and police station. From here, motorcycle taxis spread out over the highlands with passengers and unwieldy loads of timber planks, plastic jugs of kerosene, or bags of potatoes.

The Grassfields is an ecologically and culturally distinctive area on the rugged highland plateaux of Cameroon's Northwest, Southwest, and West Regions (Fig. 4.1). It got its name from the tall grass (*Hyparrhenia* spp.) covering much of the area at the time of European colonization. The Grassfields is according to archaeologists and historical linguists, just north of the homeland from which Bantu languages, technologies, and farming transformed much of tropical and subtropical Africa over the past 2500 years (Nkwi and Warnier 1982; Warnier 2012, Saulieu et al. 2021). It is famous for its kings, palaces, and masquerades. Grassfields societies are strongly hierarchical, in contrast to the more egalitarian peoples of Cameroon's southern forest belt. Life in the approximately 150 kingdoms revolves around sacred kings, palace institutions, and a dazzling variety of secret societies with specializations in divination, healing, and witchcraft eradication (Warnier 2007). A Grassfields king is known as a *fon*, and his kingdom as a *fondom*. Throughout the region, a repertoire of social deference rehearses this hierarchy with minute markers of rank. For example, the male elites of each

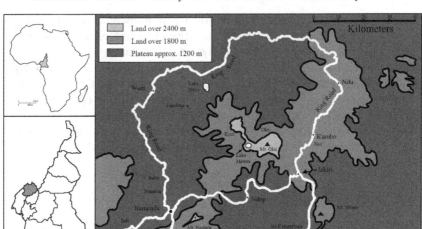

Figure 4.1 The northern Grassfields of Cameroon

fondom, known as "notables," wear special knit caps to demonstrate their status. These caps are brightly colored and distinctively shaped enough that one can spot a notable from a distance, even when they are zipping by on a motorcycle taxi. A Grassfields fon's special status is apparent in how people greet him. One does not shake his hand; instead you must bow your head, clap your hands twice, and say "mbey!" in response to whatever he says. This is the voice of a West African Dwarf Shorthorn cow, so in effect you are telling the fon that he is your shepherd, or perhaps your bull, in a single onomatopoeic term for a complex status/gender hierarchy.

Oku's social-ecological system exists because the mountain gets about 3000 mm of rain annually, enough to support biological complexity and a densely settled agricultural society. Oku subdivision (a political unit coterminous with the fondom) has an extremely high population density for rural Africa, in part because it is located in a formerly malaria-free zone over 1700 meters. Below this belt of farms and homes, shrub savanna is quickly becoming maize fields, and until recently the montane forest above it started at 2000 meters and transitioned to an alpine grassland at 2800 meters (Nyanchi et al. 2020). Mt. Oku is part of the Cameroon Line, a series of volcanoes over a mantle plume stretching from Sao Tomé and Príncipe, through Mt. Cameroon, and culminating in Cameroon's western highlands (Sunjo 2015). Crater lakes like Mawes punctuate this line, including the infamous Lake Nyos which emitted a cloud of carbon dioxide and suffocated 1746 people in 1986 (Shanklin 2007). The highlands around Bamenda form a rectangular Afromontane ecoregion, about 180 by 625 kilometers, characterized by a high degree of species endemism (Bergl et al. 2007). The Kilum-Ijim forest, which encompasses Mt. Oku, Lake Mawes, and the Ijim ridge separating Oku from Kom (a neighboring fondom), is now the only large tropical montane forest in the region. It is the best and perhaps only example of a *Podocarpus*/bamboo forest in Central and West Africa, the western limit of bird species like the African hill babbler (*Sylvia abyssinica*), the best chance for the survival of Bannerman's turaco (*Tauraco bannermani*), and likely the only home of the Lake Oku clawed toad (*Xenopus longipes*) (Cheek et al. 2000; Collar and Stuart 1988; MacLeod 1986). This forest had chimpanzees and duikers until the 1980s, but now most mammals over 200 grams are extinct due to hunting for bushmeat (Maisels et al. 2001). Of Oku's 372 square kilometers, the farming area expanded from 178 km² in 1980 to 264 km² in 2018 as wetlands, grasslands, and the lower slopes of the montane forest were converted to residences and fields (Frederick and Nguh 2020, 7). Oku's status as a biodiversity hotspot threatened by deforestation because of severe population pressure propelled its forest to the forefront of regional conservation efforts.

In Oku's rain-fed agricultural system, farmers grow maize, beans, potatoes, taro (cocoyam), cowpeas, and plantain in the area's moderately fertile volcanic soils. Unlike on Kilimanjaro, cattle are rare in this system. Most farmers do not benefit from the agro-pastoral synergy that results from incorporating livestock and manure into a farming system. Fresh milk is not prominent in the Oku diet, and red meat is scarce in the market. Chickens are common and are often slaughtered for ritual occasions. Fulani pastoralists (known locally as the Mbororo, Dze-Ngwa 2011) migrated to the area in the 1920s from northern Nigeria and graze their herds on Mt. Oku during the rainy season as part of a patron-client relationship with the Oku fon. When I asked farmers why they keep chickens instead of cows, they said that "cattle are only for the Mbororo Muslims and rich men." This separation of the livestock economy from the main current of life in Oku is evident in marriage practices; unlike the pattern across Africa's tropical savannahs, Oku marriage exchanges consist of firewood, palm oil, and salt, but not cattle.

Precolonial Oku was a regional center for the production of kola nuts, iron, and finely carved masks, and it imported palm oil and salt (Chilver and Kaberry 1967, 26; Fowler 1995; Nkwi and Warnier 1982, 41; Warnier 1975, 290). The colonial government introduced coffee in the 1950s. Like on Kilimanjaro, coffee became the quintessentially male cash crop and fostered an economic boom that lasted until coffee prices collapsed in the early 1980s. This shift to cash crops exacerbated the inequalities in an already gendered economic system; in Oku as in the neighboring fondom of Nso' (and indeed across much of patrilineal Africa), "men own the fields, women own the crops" (Goheen 1996; see also Bah 1998). In Oku, men control land and cash crops, while women gain access to land through relations to men and control food crops (Azong and Kelso 2021). Today most coffee trees are severely pruned while farmers wait for coffee prices to rise enough to make them worthwhile. Oku's honey is famous throughout Cameroon for the taste and creamy white color that it gets from the bees' diet of montane forest flowering trees like *Schefflera abyssinica* and *Nuxia congesta*. Oku produces about 11,000 liters of white honey annually, earning its apiculturists approximately USD $60,000 (Bainkong 2014). Oku White Honey was one of the first products that received a Protected Geographical Indication from the African Intellectual Property Organization (Ingram 2014, 168; WIPO n.d.). Oku honey is valued for its unique citrus-like taste and reported medicinal properties. Oku farmers also began exporting potatoes after the World Bank constructed a gravel road from Oku to Ibal in 1983. Many farmers now grow three crops a year and send the tubers as far away as Gabon. The agricultural lifestyle that undergirds the social organization and cultural imagination of Oku has, in brief, long been integrated into a regional system, and this system is becoming increasingly global.

An intricately ranked social system sits atop Oku's ecological and economic foundations. The fondom is an inward-looking community compared to Kilimanjaro, and Oku residents tend to focus on the palace and its village-level institutions. The core contrast is between the royal Mbele clan, who make up about half of the total population, and the commoner Ebjung and Mbulum clans (Koloss 2000, 77). These groupings matter for determining minute gradations of social rank, not collective action, and do not form residential clusters. The adult male heads of the oldest, wealthiest, and most powerful Mbele families form the *Kebei Kesamba* council, which selects the next fon from dozens of eligible princes (Koloss 1992, 35; 2000, 82). The Ebjung and Mbulum clans have special ritual duties, such as cleansing the land after a suicide and sacrificing to maintain Oku's soil fertility (Ndishangong 1984, 48). Each clan, as well as five smaller families too small to be considered clans, is an exogamous corporate patrilineage led by a man whose status as *fai* must be approved by the fon (Koloss 2008, 74). The *Kwifon*, a parliament based on kinship, gender, and rank, is a congress of these men. Princes and the fon's immediate relatives are excluded from the Kwifon, which creates a delicate balance of power among Oku men. Dignitaries within the ruling Mbele clan include the *enontock* queenmothers (s. *nontock*) and the "fathers of the palace," the *ebaantock* (s. *bantock*). There are one *nontock* and one *bantock* for each of Oku's 17 fons. All advise the fon, but are not members of Kwifon (Koloss 2000, 86ff). This limits competition among the Mbele and distributes some of the fon's honor and authority. The princes collectively form the Ngele secret society, an institutional innovation from the larger neighboring fondom of Nso' that the Oku Kwifon only recognized in the 1980s (Fjellman and Goheen 1984; Gufler and Bah 2006).

Grassfields fons are sacred kings, members of their Kwifon, and *de jure* chief executives, but Kwifon are the Grassfields' *de facto* governing institutions (Kaberry 1950; Warnier 2007, 36). The everyday business of a Grassfields Kwifon is making decisions, settling disputes, and punishing rulebreakers. Structurally, it also regulates competition for rank and status among its constituent kin groups (Goheen 1996, 147; Vansina 1992a; Warnier 1975, 450). The Kwifon is ultimately responsible for promoting and defending the health and vitality of Oku's land and people, as expressed in the axiom "Kwifon rules the child, the food, and the meat" (Koloss 2000, 112). It performs these basic tasks of ensuring human reproduction and agricultural production by ritually expelling witchcraft and promoting fertility. Gifts of alcohol and food flow upward through the system to ask them to perform these duties. Alcohol is the language of tribute in the Grassfields. When I first arrived at the Oku palace, I brought a bottle of whiskey. A *nchinda* royal messenger met me in the courtyard, accepted my "palace water" on behalf of the fon, and ushered me into his reception room. I soon learned to carry around bottles of palace water (cheap French Bordeaux from a shop near the Elak-Oku taxi stand) so that I could offer a little gift at the conclusion of each interview.

The people of Oku say that their Kwifon has "three strong arms," each of which is a secret society. These are the Ngang medicine society, the Manjong military society, and the Fembien women's society (the sole institution for Oku women). There are also more than 100 smaller secret societies in Oku, each with 50 to 100 members. All adult men belong to at least one society, so that the Oku social fabric is woven from strands of agnatic kinship knotted together with secret affinal networks of male cooperation and reciprocity. Each society (except the Fembien) owns a meeting house and a collection of masks, musical instruments, and medicines that it regularly deploys at funerals, blessing ceremonies, and installation rituals. Each society has an elaborate internal hierarchy of secret knowledge and ritual obligations, and a society's day-to-day business consists of members climbing this social ladder with tributes of food and drink and public performances of group membership. Dracaena is a fixture in these narrative performances and institutionalized contests.

Dracaena, known as the "life-plant" or "*l'arbre de la paix*" throughout the Grassfields, is just part of the area's vocabulary of power. Grassfields power is strongly vertical, and much of the region's ethnography examines hierarchies in domains like land use, gender relations, kinship, witchcraft, and healthcare (Feldman-Savelsberg 1999; Fisiy and Geschiere 1996; Goheen 1996; Maynard 2004; Tardits 1980). The monarchies of the Grassfields are neither Disney-like benevolent autocracies, nor examples of pre-modern "despotic states in Negro Africa" (Murdock 1959, 37). A fon's life is scripted down to minute details (such as the dictum that he may never step on someone else's shadow, Bah 2004, 439), and no one in Cameroon is quite as socially disciplined as a fon (Warnier 2007). The proverb "the fon rules the people, but the people hold the fon" expresses the ideal that these kings do not rule command-and-control structures. A fon is not divine, but rather a sacralized person who stands at the eye of a paradox between ongoing group cooperation and conflict (Vansina 1992a, 23). In Oku, the fon is called a "good devil" to express this omnipresent contradiction. A fon both summarizes core social practices like ranking and elaborates a cultural script for actions that resolve these contradictions, particularly dispute settlement and generalized fertility rituals. Not every Grassfields fondom uses the same elements in these scripts, but together they form a distinctive regional repertoire and social texture (Warnier 2012). The list of regional key symbols indexed to hierarchy includes:

1 Spider divination using burrowing tarantulas (*Heteroscodra crassipes*) to predict the future (Gebauer 1964; Zeitlyn 1993). In Oku, this practice is particularly associated with the ruling clan, and only the Mbele use the spider motif on stools, carved posts, and cobblestone courtyards.
2 Fertility rituals that connect fraternal or identical twins to the generalized fertility of women, landscape, and kingdoms (Diduk 1993; Maynard 2004; Mbunda-Samba et al. 1993). Becoming a twin mother is one of the few ways for Oku commoner women to achieve high status (Bah 2000).
3 Animal transformation powers that enable fons to change into fearsome creatures (Chilver and Kaberry 1967, 47). The Oku fon can become a leopard.
4 Elite material culture like double bell gongs, woodcarving styles, masks, and house construction techniques that demonstrate a kingdom's internal ranking (Notué and Triaca 2005). In Oku, only Mbele houses have a diagonal crisscrossing grid of *Raphia* palm stems (a reference to palm wine consumption).
5 Secret societies with military, medicinal, and witchcraft eradication responsibilities (Eyongetah and Brain 1974, 36). Oku is famous in Cameroon for the masquerades that accompany nearly all major life cycle events (particularly funerals).
6 The use of dracaena to mark boundaries and establish peaceful and orderly landscapes, social relations, and authority. Grassfields fons mark documents with images of dracaena on rubber stamps, and royalty put stalks on their cars to make soldiers wave them through highway checkpoints.

Because of these shared symbols and practices, the intensely local fondoms of the Grassfields form a remarkably unified region, both as a state of mind and a social landscape.

I visited eleven Grassfields fondoms in 2011 to collect baseline information about dracaena. Fig. 4.2 shows how the plant's names and usage vary across the region, but also cohere together as a shared Grassfields model of purposeful agency. As one man summarized, "in the Grassfields dracaena is always a sign that human activity must have taken place." If the shared symbolic repertoire of the Grassfields were a mosaic, this boundary plant would be the glossy green tiles that repeatedly form images of orderly gender, kinship, and status relations.

The challenge is that the cultural side of this social-ecological system is explicitly dedicated to harmony, but is experiencing increasingly acute

Fondom	Bafut	Bali	Essu	Kom	Mankon	Nso	Oku	Bandjoun	Bangangte	Foto
Region	North west	North west	North west	North west	North west	North west	North west	West	West	West
Name for dracaena	*nkeng*	*nkeng'keng*	*ekong'e*	*nkeng*	*nkeng*	*kikeng*	*nkeng*	*funkang'*	*fa'keng*	*n'kung*
Boundary marker, land tenure	Significant	Significant	Significant	Significant	Significant	Significant	Significant	Significant	Significant	Significant
Protection from evil and witchcraft	Significant	Significant	Significant	Significant		Significant	Significant		Significant	Significant
Shrine for sacrifice, blessing			Significant		Significant	Significant	Significant	Significant	Significant	Significant
Twins and fertility	Significant	Significant	Significant	Significant	Significant	Significant	Significant	Significant	Significant	
Cleansing and healing rituals	Significant	Significant	Significant		Significant	Significant	Highly significant	Highly significant	Significant	
Peacemaking		Significant	Significant	Significant	Significant	Significant	Significant	Significant	Significant	
Foundation of a village, center of fondom			Significant	Significant	Significant	Significant	Significant		Significant	
Message from the fon showing authority			Significant	Significant	Significant	Significant	Significant			
Graves of powerful men						Significant	Significant	Significant	Significant	

Key: Significant; Highly significant

Figure 4.2 The regional pattern of dracaena in the Cameroon Grassfields

stress. High population growth rates, ecological decline, and civil war are undermining this agrarian system. Oku's meshwork of resources, social relations, and meanings is now tattered. The people of Oku are finding that their reliable and relatively closed social-ecological system is becoming less systemic and more open to new and unfamiliar dynamics.

Much like Kilimanjaro, Oku's population increased more than tenfold over the 20th century, and this changed the structure and meaning of land use. But unlike Tanzania, in Oku this involved postcolonial institutional stretching, not a colonial transformation. In other parts of the Grassfields, the major population growth period was in the 1970s (Fisiy 1992, 79), but in Oku this occurred in the 1980s and 1990s after the construction of a gravel road improved access to health services. Fig. 4.3 summarizes the available population data for Oku. Extrapolating these figures forward leads to a conservative estimate of 190,000 people in Oku in 2021. After subtracting the portion of Oku subdivision that is an uninhabited forest reserve, this figure leads to an average population density of about 720 people per square kilometer – an astonishingly high figure for rural Africa.

In 1986, the International Council for Bird Preservation (ICBP, now BirdLife International) identified Mt. Oku as a critically endangered forest due to rapid population growth, illegal grazing, and resource overexploitation (Collar and Stuart 1988; MacLeod 1986). Land-hungry farmers were abandoning coffee and clearing the forest to expand production of potatoes

Date	Population estimate	Source
1906	6000	Glauning 1906
1925	2625	Drumond-Hay 1925, cited in Gufler 2009-2010
1953	9173	Official census data, Chilver and Kaberry 1967, 130
1969	26,898	ORSTROM 1976, cited in MacLeod 1986, 12
1976	32,000	Official census data, cited in Koloss 2000, 17
1986	30,242	MacLeod 1986, 12
1995	75,000	Koloss 2000, 17
2002	65,000	Argenti 2002, 499
2005	87,720	Official census data, NIS 2005
2012	144,800	Elak-Oku Council 2012

Figure 4.3 Oku demography

and beans (Gardner et al. 2001). Most importantly, Oku residents began harvesting bark from *Prunus africana*[3] trees, which a French pharmaceutical company made into an herbal prostate gland treatment (Page 2003, 362). A sustainably harvested *Prunus* tree provided about USD $10–20 of bark, but a felled and completely stripped tree yielded $200. In the 1984/85 season alone, 8000 *Prunus* trees in Oku's forest were debarked while the Kwifon seemed powerless to intervene (Fisiy 1994, 17). Even Fon Ngum III was himself a *Prunus* trader (Ingram 2014, 120). Clearly, something needed to be done before a critical forest habitat and many endemic species disappeared completely. ICBP set up a Community-Based Natural Resource Management Project (CBNRM) to institutionalize forest management in 1986 (Thomas et al. 2000).

Prior to 1961, the Kwifon regulated forest access by declaring particular forest areas sacred, restricting firewood collection to fallen deadwood, and sending its fearsome Mabu *juju* (masquerade performer and powerful spirit) to capture tree cutters (Argenti 1998, 759). The postcolonial state's management was largely ineffective and corrupt. ICBP's Kilum-Ijim Forest Project formalized user groups as the owners of particular areas, set strict rules for beekeeping and bushmeat trapping, banned livestock from the forest, evicted the Mbororo pastoralists from the Mt. Oku summit, and established clear forest boundaries (Foncha and Ewule 2020; Ngum III 2001; Nurse et al. 1994). From 1986 to 2004, the project spent about USD $2.35 million to stabilize the remaining forest (Ingram 2014, 128). Soon after the project ended, *Prunus* extraction surpassed the 1980s harvesting rate (K. Stewart 2009), leading the central government to ban Oku's *Prunus* trade completely in 2011. Since then a local environmental NGO has worked to sustain project gains and provide non-forest-based livelihoods for Oku's young people (CAMGEW 2016). The most lasting legacy of the ICBP project, other than the clear forest boundary (painted tree trunks and a few dented and peeling metal signs), is that the Cameroon Forestry Law of 1994 established CBNRM as the preferred management system for the entire country's forests, all based on the example of Oku. The illegal harvesting of bark has continued despite these struggles over access. As of 2018, 99.99% of the *Prunus* trees in Oku with a diameter greater than 10 cm had been debarked (Frederick and Nguh 2020, 14), and the Mbororo pastoralists had returned to the mountain's summit.

Although the Kilum-Ijim forest is so disturbed that it is only secondary forest (Solefack 2017), there are some signs of recovery. A local NGO is planting thousands of *Prunus* seedlings in the forest (each marked with nkeng, of course), farmers are increasingly adopting agroforestry techniques, and stall-fed goats, rabbits, and guinea pigs are increasingly popular (Foncha et al. 2019). The threat of ecological collapse remains very real in Oku because most of the forest's tree species need animals to disperse their seeds, and most mammals and many birds have become locally extinct (Maisels et al. 2001). Climate change is already shortening the Oku rainy season, and the long-term forecast is increasing unpredictability and declining rainfall by 2100 (Azong 2021; Azong et al. 2018; Ngute et al. 2021). Both the forest and the Oku agricultural system will become increasingly stressed. The fon and Kwifon have re-established rules for firewood collection and hardwood extraction for woodcarving, and try to regulate the livestock herders, *Prunus* collectors, and bushmeat hunters whose activities damage the forest. Finally, the ongoing Anglophone Crisis, a civil war between English-speaking separatists and the largely francophone Cameroonian military, is preventing effective institutional action in Oku (Amin 2021). Fon Sentieh II was kidnapped by these rebels in August 2018 and released several days later. In August 2021, the rebel leader "Field Marshall No Pity" attacked and burned the Elak-Oku Council building, the mayor's house, the market building, and a police station (and then posted video of the attacks on YouTube). With no end of the Anglophone Crisis in sight, violence and the displacement of over half a million people are likely to forestall the institutional and land-use innovations necessary to cope with stressors like population growth, extractive capitalism, and climate change in Oku and across the Grassfields area (Atabong 2018).

This story about ecological transformation, postcolonial demography, agrarian capitalism, and political violence shows the declining efficacy of local institutions as a relatively closed community became increasingly open to the regional and global political economy. The critical conjuncture was that transportation infrastructure spurred cash crop production at the same time that the postcolonial state was undermining local social organization (Feldman-Savelsberg 1999; Goheen 1996). The forest had been managed by palace institutions, not the vague "community" invoked by CBNRM. As state bureaucracies and international corporations overshadowed the royal tributary system, the forest became an open-access resource that individuals could exploit for short-term gain without social sanction. Oku's goods and services have escaped the Grassfields region and become national and global commodities. This pattern is not confined to cash crops and woodcarving. Divination and healing had been community-oriented and local matters, but Grassfields traditional healers now serve individual clients from all over the country (Maynard 2004).

While these new relationships allow more individual gain and social mobility than Oku's social-ecological system ever did, they also generate distrust, disorder, and ambiguity. The former mayor of Oku's secular government summarized that "as the importance of money goes up, the peace goes down." Oku farmers repeatedly told me that "the peace is disturbed by cash crops" and "there are more and more land problems that the Kwifon cannot control." The conjuncture of population pressure, postcolonial politics, and capitalist commodification means that most people in Oku have very clear ideas about what their home *should be* like, but little experience of actual social-ecological order. The "systemic" part of the Oku social-ecological system is what has eluded Oku farmers, fons, and development agencies for decades as a relatively closed rural peasantry became increasingly open to outside forces. In the Grassfields' long-term search for harmony, dracaena serves as a culturally defined compass that unerringly points toward order. This chapter reviews how and why dracaena guides the people of Oku toward their particular vision of sustainability and prosperity. This boundary plant operates at three social scales: the farming household economy, the fondom political system, and the web of symbols that legitimize and make those relationships meaningful.

Boundary plants and land tenure in Oku

Much like the Chagga on Kilimanjaro, Oku farming households use dracaena to form a living landscape of relational land tenure – an organizational process, not a physical and economic structure. Land management and agricultural production in Oku revolve around nkeng because people use it as living fences and on the corners of their land. Tight nkeng hedges line the footpaths and dirt roads that tie Oku together. Most compounds have at least one dracaena stalk on each corner, often next to a surveyor's concrete marker at a wealthy man's property. Other plants, trees, rocks, and streams can also serve as recognized markers, but only nkeng actually means a boundary and demonstrates what belongs to whom. Oku tenants and landlords exchange calabashes of palm wine for stalks of nkeng to formalize the contract (Ingram 2014, 117). It is the basic institution of the Oku political economy. Much like the "bundle of powers" that organizes social relations on Kilimanjaro, in Oku nkeng ties a bundle of claims and wealth (in both land and people) together. However, it is a different bundle because the polity that nkeng binds in Oku is a tributary monarchy, unlike the less centralized kin-based society on Kilimanjaro. For the Chagga, neighbors witness plantings of dracaena; in Oku this is the work of the *fai*.

Despite efforts to maintain the Oku fondom (like the Ngang society's village boundary rituals and the new fon's Facebook page[4]), wealth accumulation in Cameroon is shifting from gender, age, and status hierarchies to class (Peters 2004; 2013). Older institutions of wealth, identity, and value, like the Oku monarchy, persist because they determine access to resources and organize kinship and marriage. The result of these changes is that in Oku, as in much of Africa, there are multiple institutional arenas for claiming and legitimizing property, and this makes land tenure a never-ending negotiation process instead of a stable economic structure (Berry 1993). The male farmers of Oku can now choose which land matters to take to the Kwifon for settlement, and which to refer to government authorities (female farmers must take land issues to a husband, brother, or father; Azong et al. 2018; Fon 2011). One man complained that "instead of putting nkeng, now they run to the police brigade, they run to the divisional officer, they are looking for their judgment, not the truth!" Reviews of Cameroonian land tenure agree that land is an "intense battleground for struggles over ownership, access, and control" (Fisiy 1992, 72) and that "most rural Cameroonians are little more than squatters on their own land" because the state owns all land, making all other rights and claims secondary and deeply contested (Alden Wily 2011, 5). Institutions like the Kwifon are the *de facto* controllers of land in the Grassfields, but the national government has *de jure* rights. Cameroon's traditional land tenure systems have been repeatedly legislated out of existence, yet they continue to ratify claims and allocate land. It depends on who is claiming what resource for which purpose. Colonial and postcolonial land law, enacted to modernize Cameroon through individual property rights without relinquishing central government control, have instead produced corruption, ambiguity, and contradiction about who belongs to what land, and what obligations that "belonging" entails (Awafong 2003; Baye 2008; Geschiere 2004). It was in this disorderly context that ICBP constructed community-based forestry in Oku, and this new institutional layer atop a tenurial muddle explains why the project's gains were so fragmentary and evanescent (Zama 2001).

Oral histories emphasize that dracaena was the technology that formed the Oku social landscape. This territorialization was based on kinship, rank, and boundary plants. Dracaena and fig trees are botanical monuments of a family's land claims. As one ritual expert explained this,

> When a new market or village is founded, the first thing to do is to plant a fig tree [*Ficus* sp.] with some nkeng. That big fig and nkeng there below my house in this little valley, which my whole family lives around, is the center of my family's area. They show the beginning of my family in this place. Each fig tree with nkeng in Oku shows the beginning of some family, village, or set of compounds, or an old homestead foundation site.

Another local historian said that the fon sends a messenger with nkeng and ficus rootstock to establish a new village. Like dracaena, ficus also takes root from cuttings, which makes its vegetative agency particularly effective for emplacing people in a landscape (Gautier 1996). It is also exactly the same combination of plants that Chagga elders recall for village establishment on Kilimanjaro, so the significance of dracaena for place-making in Oku may represent one long-term legacy of the spread of Bantu languages and farming (Sheridan 2008). In brief, ironworking farmers spread throughout Africa over the past 2500 years and produced much of its dazzling cultural and linguistic diversity, as well as its widespread social commonalities. In both Oku and Kilimanjaro, for example, medical specialists are named by the same linguistic root, *ngang*. Oku relies on the Ngang medicine society while the Chagga seek healing from a *mganga,* and both use dracaena. Similarly, widespread concerns with ancestral spirits, witchcraft, polygamy, and autochthony across Africa are evidence for an ancient shared cultural repertoire, much like how Greek and Roman cultural features influenced European and Mediterranean societies (Ehret 1998). In the Grassfields, this historically reproduced cultural repertoire was (and is) "a vocabulary that different people use in different ways to say different things, but that everyone can understand" (Nkwi and Warnier 1982, 54).

Archaeological, genetic, and historical linguistic reconstruction of this deep history suggests that the Nigeria-Cameroon border zone was the cradle of the Bantu language family but does not mean that Bantu speakers moved out of the region in enormous waves of people (de Filippo et al. 2012; Vansina 1995). Instead of big arrows showing migrating groups, the new models of the Bantu expansion show many little arrows, large and small reversals, and slow differentiation (Bastin et al. 1999; Nurse and Philippson 2006). "Shimmering beadwork" and "kaleidoscope" are the major metaphors for the resulting pattern of cultural diversity and repeating elements (Kopytoff 1987, 77; Nurse and Philippson 2006, 5). In my 2008 article, I drew on Kopytoff's model of the African frontier to suggest that *Dracaena fragrans* was part of the continent's vocabulary for generating and creatively composing social organization. I argued that as people settled new land, the first settlers' graves became shrines showing both the power of ancestors and the legitimacy of land rights. Boundary plants such as dracaena marked these as "places of power" (Colson 1997), and the lack of cultivation on those sites allowed secondary forest succession to produce sacred groves (Sheridan and Nyamweru 2008). When population growth compelled some people to move to their frontier, the cycle started over again. This model could explain why the same boundary plant marks property, denotes shrines, and symbolizes peaceful exchange relationships in both Central and East Africa.

Dracaena's place-making role is apparent in the story of Oku's foundation. Archaeological study of the Oku palace shows that the area's first pottery and ironwork date from the 15th to 17th centuries (Oslisly et al. 2015), and oral histories roughly corroborate this settlement history. The ironworking Ntul people were living near the Oku forest when members of the Mbele clan arrived from the northeast and assumed leadership of the area (Gufler 2009-2010; Jeffreys 1961; Koloss 2000, 31).[5] This period's political dynamics are now only vaguely recalled, and undoubtedly viewed through the prism of the present, but the most famous Oku fon was Mkong Moteh. He is now considered Oku's true founder in the early 19th century and revered as one of its gods. According to oral history, he killed most of the Ntul and quickly consolidated power by reorganizing kin relations into a monarchy. As the original inhabitants of the land, the remaining Ntul sacrifice to maintain Oku's prosperity, while the Mbele dynasty retains political and economic control.

The Mbele clan's history describes a Solomonic test of two leaders' vitality (Ndishangong 1984). The elderly fon of Rifum (east of present-day Kumbo) chose Tatah, his wife's oldest son and a hunter, to succeed him. But the people preferred the younger son, a musician. The old man gave each brother a dracaena cutting, saying that the plant that grew best would decide the succession. The hunter was often away from home, so the younger brother's supporters prevented his nkeng from flourishing (probably by twisting the stalk to discourage root growth). The musician brother founded Nso', the most powerful fondom in the region, at Kumbo (Nkwi and Warnier 1982, 132), and Tatah retreated to the Oku forest, befriended the Ntul fon, and eventually succeeded him as fon of Oku. This is why the Oku and Nso' fondoms are considered "brothers" today. Oku people often conflate this story with Mkong Moteh, and my questions about Oku's foundation inevitably led to accounts of how this later fon (Tatah's son, but five fons after him, according to the lists of kings in Koloss 2000, 29 and Ndishangong 1984, 23) negotiated with the Ntul people. As one historian summarized this relationship, "Mkong Moteh brought his nkeng, and the Ntul people had theirs." Another insisted that this fon "used a nkeng stalk like a compass" to found Oku in exactly the right place. The lesson here is that dracaena has long been a strategy for accumulating "wealth in people" in Oku (Guyer and Belinga 1995). The people of Oku celebrate their fondom's establishment annually at Mkong Moteh's old palace at Lumeto, in the Kilum-Ijim forest, where the fon sprinkles the crowd with purifying and vitalizing medicine using a stalk of nkeng (Bah 1996, 14).

What was life like in precolonial Oku? The area's two major ethnographers, Hans-Joachim Koloss and Nicholas Argenti, interpret its society in opposite ways. Koloss, who became a member of the Oku Kwifon, presents the social, political, and symbolic complexity of Oku as a functional system. He argues that its masquerades and symbols express and reinforce social order (2000; 2012). Argenti counters that Oku's masks, dances, and symbols of order are unspeakable "social memories" that hide the catastrophic violence, terror, and disorder of the precolonial slave trade (1999; 2006; 2007; 2011).[6] The broader regional history suggests that the Grassfields was relatively isolated from the Atlantic slave trade because of low population density and the lack of navigable rivers (Nkwi and Warnier 1982). Indigenous Grassfields slavery was much less violent than the inhumane misery of the extractive slave economy bound for New World plantations (Chem-Langhëë 1995; Nkwi 1995). Enslaved people who remained in the Grassfields became members of their owners' families, not commodities and an underclass. The binary of Koloss' functional order and Argenti's repressed violence is a false choice that reduces cultural complexity to competing theoretical paradigms. A Grassfields slave on a fon's diplomatic mission was recognized by his distinctive cap, walking stick, and a dracaena branch. With these tools "he could travel far and wide and even pass through war zones without being harassed" (Fomin and Ndobegang 2006, 640). Was this functional peace-making or hidden violence? Dracaena probably symbolized neither functional structures nor structural violence in the precolonial Grassfields, but instead expressed complex social relations of institutional and structural power. And power, as this book's case studies show repeatedly, is ambivalent and requires cultural and historical contextualization.[7]

Colonialism transformed these relations of power and authority over people and land. The German and British colonial periods (respectively, 1884–1916 and 1919–1961) were quiet in Oku compared to the changes elsewhere in the Grassfields. The German colonial state's demand for plantation workers amounted to imperial slavery (Argenti 2007, 93; Goheen 1996, 65), and Grassfields fons collected tribute and taxes with German military support (Nkwi 1976). The British policy of indirect rule formalized this patron-client system, which then transformed Grassfields social institutions (Geschiere 1993; Rowlands 2005b; Rudin 1938). Colonial rule selected jagged pieces of tradition and assembled them into a façade for the completely alien system, which then became standardized as authentic (Vansina 1992b, 17). As a small and remote source of kola nuts and masks, Oku was much less integrated into the colonial system than its neighbors Nso' and Kom. Its first encounter with the colonial state was in August 1905, when Captain Hans Glauning from the German station at Bamenda traveled to the mountain kingdom to assess its potential for mining and agriculture. Fon Mkong Ndakoh met Glauning disguised as a woman, "with a basket on his back and a hoe, as a sign that he was not going to fight, but wanted to work for the whites" (Glauning 1906, 236). Oku oral histories assert that this was a subtle symbolic display, not a surrender. Elders say that the fon posed with a woman's *kensooi* basket only to prevent the Germans from attacking Oku. Because he needed to signal his real status in Oku, he also carried a stalk of nkeng. The Germans had seen dracaena being used as a peace symbol elsewhere in the Grassfields, so once Glauning identified the cross-dressed fon, he interpreted the subterfuge as peaceful submission. In this cross-cultural miscommunication story, nkeng mediates the political and symbolic limits of local and imperial authority.

The persistence of boundary plant institutions in the Grassfields is partly a result of colonialism, not simple tradition. In 1919, German Kamerun became French Cameroun (most of present-day Cameroon) and the British Cameroons (two discontinuous areas bordering Nigeria). The new colonial powers' different policies continue to shape the region today. Most francophone Grassfielders were governed by the *"indigènat"* legal framework (Fisiy 1992, 33), through which French administrators interpreted customary law. In the British Cameroons, indirect rule empowered and entrenched royal institutions. This contrast is why francophone Cameroonians now look to France for metropolitan cultural capital, such as Bordeaux wine and Parisian fashion, while anglophones tend to focus more on local markers of status and legitimacy like palm wine and masquerades. Indirect rule made many Grassfields fons into land administrators. Rather than ownership, British Cameroonians held 99-year leases under the colonial state's authority but governed by their fons (Nkwi 2010, 73). Royal institutions and symbols like dracaena became mechanisms that mediated how Grassfields fondoms related to the state, not local matters. These practices at the boundaries of colonial authority mystified the contradiction at the core of British colonialism – the contrast between the fons' ritually proclaimed autonomy and their dependence on the central state.

The postcolonial state layered new contradictions atop these rickety foundations (Cheka 2008). French Cameroon became independent in 1960, and the results of a 1961 plebiscite transferred the northern area of British Cameroons to Nigeria, and the southern area (where Oku lies) to Cameroon. To unify an already fractious polity, in 1974 the state enacted land laws to rationalize and modernize tenure. The state became the "guardian of all land" (Awafong 2003, 94), which effectively centralized land tenure by requiring registration and ownership certificates for all occupied or managed land. These laws abolished 99-year leases, specified that fons were custodians instead of landowners, and formally eradicated local land institutions – which blithely continued existing because Cameroonians found them useful and meaningful. Today in Oku only 12% of landholders have title deeds; the rest rely on royal institutions (Nyanchi et al. 2020, 6). Getting certificates involves professional surveys and concrete pillars, and is complex, expensive, and absurd to most Cameroonians (Njoh 1992). Cyprian Fisiy's review of all land registrations in Northwest Province issued from 1976 to 1986 found that 79% were within the Bamenda city limits, and that 83% of the applicants were elite civil servants and businessmen (1992, 91). Only 5% of the certificates went to rural peasants.

The results are that statutory law both exists and is utterly irrelevant for most purposes, most rural Cameroonians are legally landless squatters, and socially recognized means of securing property – like boundary plants – have become increasingly important strategies to negotiate orderly relationships in this institutional disorder. This history shows that the tenurial significance of nkeng in Oku is an ongoing social process, not a static tradition. Much like its neighboring fondom Nso', daily life and land tenure in Oku are increasingly shaped by commoditization, class inequality, and personal autonomy, and less by palatial authority (Goheen 1996, xiii; Keming 2015). Nkeng has surely played a role in these quiet dramas, but like other researchers in Oku, I encountered a resolute insistence that the plant's sacred significance is timeless and separate from petty matters of political economy. My interpretation of this disinterest in history is that never-ending negotiations over status and land rights in the colonial and postcolonial periods require a fixed point of legitimacy (Meyer and Geschiere 1999). Dracaena represents that which does not change in the Grassfields, making it a useful foil for historical change. When confronted with social change, Oku people speak of cultural continuity.[8]

Social organization and boundary plants on patrol

Place-making in Oku is a drama of narrative performance and strategic struggle. Much like masale in Tanzania, nkeng in Oku expresses a specific cultural definition of vital and fertile life force. In the Eblam Ebkwo language, *keyoi kejungha* means "good breath/spirit" and permeates everything alive, from the earth up to the fon's black-and-white tasseled cap. Nkeng directs this vitality, which makes it a tool for displaying and validating power. This boundary plant connects ultimate and sacred cosmological principles to mundane everyday affairs in Oku, and this occurs in two distinct social and spatial spheres; the domestic and the public. Social action often involves moving a particular stalk of nkeng from one domain to another. The usage of nkeng in Oku is an everyday practice of maintaining borders, resolving disputes, asserting peace, and protecting homes, farms, and the fondom from the disorder that threatens to undo them.

In the domestic sphere, nkeng plants occupy the center of most extended family compounds, where it forms the witchcraft-preventing and health-promoting *efam* shrine. As one elder explained, "the nkeng there prays for peace, fertility, and productivity." An Oku family compound (*kebey*) typically has three mud brick buildings around a *kebook* central courtyard, with the open side of the 'U' facing the road or path (Fig. 4.4). A compound cannot have buildings on four sides because, according to a nationally famous Oku healer, "a *kebey* must be for life, and four sides would mean you are preparing a grave!" The highest-ranking person of the patrilineage segment, usually an elderly man, lives in the building to the right of the compound's entrance (1 in Fig. 4.4). Lower-ranking people live in positions 2, 3, and 4. The latrine occupies a corner far away from the entrance. The efam in the middle of the well-swept packed earth courtyard is a cluster of nkeng and few other medicinal herbs. In most compounds, about a dozen stalks, all less

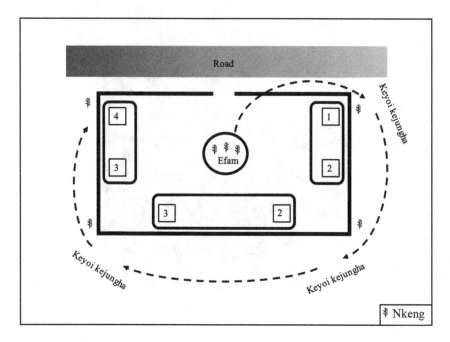

Figure 4.4 Keyoi kejungha moving clockwise around an Oku compound

than two meters high, form the efam. I saw some ancient shrines that were four meters high with dozens of eight-centimeter diameter dracaena stalks atop enormous shaggy rootballs exposed by decades of sweeping and court-yard maintenance. The efam is the last phase of house construction and is always installed by the Ngang medicine society. Once their rituals activate it, the efam "ensures peace and eradicates witchcraft at that compound." The highest-ranking man maintains the efam's protection with daily prayer and periodic sacrifices of *njemte* (ritualized food, usually pounded taro, palm oil, and salt) "if people are not sleeping fine" and at key points of the agricultural cycle. The usual procedure is to fill a small *nchok* calabash with the *njemte*, cork it with a fresh nkeng leaf, and place it in the efam.[9]

The meaning of an efam in Oku corresponds to the social organization of the segmentary kin group, and it territorializes that group's prosperity. It turns the structural power of kinship into a place. It is a domestic landscape feature that links Hirsch's background of cultural potential to the mundane foreground of social action (1995). The efam is the "center of the peace of the family," and anchors the developmental cycle of a domestic group (Fig. 4.5). Food, money, and visitors enter through the gate and should move clockwise, "in the direction of life." Ideally, an Oku compound consists of a married couple and their sons.[10] These men and their families should occupy ranked positions within the *kebey* according to birth order, so that the patriarch

Figure 4.5 The author with a large efam

takes the "warmest corner" at position 1, while the oldest son lives at position 2 on Fig. 4.4, and the youngest at position 4. The residents' relative status therefore declines in a clockwise movement around the compound, signified as the "direction of life." This residential format maintains social order, according to one ritual expert, because person 4 can only know the "secrets of the family" through his relationship to 3, then 2, and finally 1. One purpose of the efam, one elder said, is to keep lower-ranking people in their places, in effect saying "you, 4, don't overtake 3!" And because patriarchs get replaced from cohorts of sons that include labor migrants, university students, and urban elites, there is always some jockeying over succession and authority. Who is the best oldest man to address the gods? Every sacrifice at the efam demonstrates social structure and provides an opportunity to claim higher rank. This structure is also explicitly gendered – a woman in Oku can touch the dracaena in a fence and use its leaves as needed, but she should not interfere with the efam because it is "men's business."

Trying to grasp the efam's meaning, I asked a high-ranking palace official if this shrine was the "heart" of a compound pumping life force through it. He chuckled and said that although the English idiom was appropriate, a better word would be the "navel" of the compound because it is an umbilical cord from the ancestors to the living:

> It is the basic connection site to all and everything in the compound, and the keyoi kejungha flows through it. Keyoi kejungha moves all around the *kebey* so all prospers, and its effects can be seen in coolness and peace, good crop yields, high birth rates, good money from cash crop sales, no war in the country, no accidents, and long lives.

Green and vital nkeng is the conduit for this flow, and the emergent keyoi kejungha moves around a compound clockwise (and downwards through the hierarchy of kinship and gender relations) to ensure fertility and prosperity. The term that I heard repeatedly in Oku for this movement was the efam "going on patrol" to guard the family from illness, malice, and witchcraft. These blessings and protections are transferrable because dracaena's knack for vegetative propagation make it easy to extend its domestic agency to farm plots, some of which are several kilometers away. Farmers plant five nkeng stalks from their home efam on their farms; four on the corners to repel thieves and encroaching neighbors, and one in the center to protect the crops from witches. Anyone can redirect their efam's patrol this way, without needing to consult with (and compensate) the Ngang medicine society that originally set it up. As one healer explained this institutional extension,

> The efam drives away that dry person! It prevents crop theft and witches capturing soil to make witchcraft for damaging that farm's soil fertility. So the efam of a *kebey* works on the farm, no matter how far away, because of that nkeng.

This concern for soil fertility is why efam sacrifices occur during the planting season in March, when the land is bare and most vulnerable to erosion and witchcraft. Oku kinship is not just about social connections, and Oku farm management is not simply about food production. Homes and farms are sites where protective rituals enact struggles over food security, domestic hierarchy, and the metaphysics of life.

This general concern for interwoven human fertility and ecological prosperity is widespread across the Grassfields (Feldman-Savelsberg 1999) and across sub-Saharan Africa (Shipton 1994). In Oku, the public displays of nkeng's power correspond to its domestic role. The Ngang medicine society installs protective and "patrolling" efam on the edges of each Oku village and on the fondom's boundary. The keyoi kejungha circulates, again clockwise, in these larger social domains to maintain prosperity. The center of the entire system is the efam in the Oku palace courtyard between the main gate and the fon's garage. It was once a leafy green clump of ancient nkeng but was replaced with a concrete flagpole base when a local man became Prime Minister. Even without its nkeng, the site remains Oku's *axis mundi*. I watched the powerful Agah juju and its coterie of men from the Ogu military society perform there as part of a funerary celebration (Fig. 4.6). This is an extract from my fieldnotes:

> Agah arrived with its escort, a group of 12 adult men, stripped to the waist and wearing blue and white wraps, necklaces of beads and leopard teeth and black and white caps with dozens of small tassels, all carrying antique handmade iron spears. All around hundreds of excited school kids milled about, agog at the spectacle. Agah wore a blue and

Figure 4.6 Agah arrives at the Oku palace

white robe with a geometric pattern, but its face was invisible within the shadow of a sphere of tightly packed green nkeng leaves and further disguised by a thin cloud of mysterious smoke drifting from the top of its head. It carried a spear in its right hand and a branch of nkeng in its left. Wooden rattles on its feet clattered with each step, and dozens of young Cameroonians held out their cell phones to capture the event on video. The procession entered the palace courtyard and began to chant as they arrived at the palace efam-flagpole base. The men chanted while the green-headed juju whirled around clockwise, flicking its nkeng toward the throne room, toward the king's wives' houses, and toward the princes' compounds. At last the whole group disappeared into the palace complex to conduct the highly secret rituals for purifying the land after a royal death.

Spatial relations of social power and boundary maintenance at Oku's fondom level reflect and recapitulate the cultural capital dynamics of its domestic domain. Because it is the same keyoi kejungha life force flowing through the nkeng in domestic and public shrines, everyday compound life revolves around the same botanical symbol that organizes fondom-wide ritual action. This consonance clothes social relations at both levels with legitimacy and authority. The deeply masculine social order of kin groups and the Oku monarchy flows, clockwise, through boundary plants.

The people of Oku use this symbolic repertoire to generate diverse practices based on the core narrative of fertility, social order, and legitimate property rights. They strategically perform this narrative of vitality for purposes ranging from individual claims to objects and resources to legitimizing institutional controls on behavior. For example, if someone leaves a bundle at the roadside, or simply wants to claim a tree trunk drying on the forest floor, they can place a nkeng leaf to prevent anyone from touching it. When political parties march through Oku on Cameroon's National Day (May 20th), they carry nkeng to demonstrate peacefulness. Oku Catholics form arches of nkeng over a person carrying the bible into a church. When someone reads from the gospel, assistants hold an X of two long nkeng leaves over their head, and Catholic holy water blessings are performed with nkeng.[11] During a family dispute, kin speak their grievances into a calabash full of water, nkeng, and other herbs, and once everyone agrees that the dispute is over, the highest-ranking person present uses the nkeng stalk to sprinkle the water on the disputants. This makes the disputants' bodies into places governed by the same spatial practices of power that underwrite landholding and the management of keyoi kejungha. The most prominent public performances of these place-making and peace-asserting strategic narratives are land disputes. The dispute settlement process varies by the problem's severity, but at any social level nkeng enforces peace.

The Oku Kwifon meets twice during Oku's eight-day week, on market days (*ngokse*) and three days later on *kemeiwiy*. The Kwifon's dispute settlement services are cheap, relatively quick, and culturally embedded compared to the expensive, slow, and abstract justice available from state courts. Statutory legal fees in Oku typically cost at least 10,000 CFA (about USD $18) plus transportation costs to bring court officials up from Kumbo. A land disputant has several increasingly formal options before resorting to statutory law: they can put a nkeng stalk in the disputed area to settle the issue themselves, use the same stalk to request a Kwifon investigation, or they can ask the Kwifon to impose a *kalang*. This is a sharpened stake of dead wood (from the *boboy* plant, species unknown) with its bark removed in alternating 3–5 centimeter stripes. Once the Kwifon hammers the kalang into a disputed area, it cannot be removed until the dispute is settled and the person at fault has paid several chickens to the Kwifon. The kalang is the symbolic opposite of nkeng – its dry inert wood blocks life force. Oku farmers are reluctant to let a kalang remain in the ground long because its lifelessness threatens agricultural and human fertility when the rotting wood "damages the soil" by preventing the flow of keyoi kejungha. Dispute settlement by kalang is therefore a last resort, and generally achieves a speedy resolution.

Land is increasingly disputed in Oku because labor outmigration, agricultural intensification, and population growth have made land claims more necessary, negotiable, and fractious. It is becoming common for neighbors to twist a new nkeng property boundary to keep the stalk from taking root, which may later allow them to move the boundary slightly. Property rights are clearly men's business, but pragmatic day-to-day farm management is in women's hands. Women often borrow fields from one another for seasonal cropping. What happens when a borrower needs to defend her plot from thieves or witches? Planting some nkeng from her home efam at the borrowed plot's center (not the corners) offers protection, but also starts to connote ownership rights. Permanent crops like kola and coffee would make a borrowed plot look like her husband's patrilineage's permanently owned land. Most land quarrels in Oku are tangled matters of long-term land loans gradually solidifying into claims of outright ownership. Many people in Oku are turning to formal contracts between owners and borrowers and writing wills instead of relying on oldest sons to inherit (who then allocate plots to their junior brothers). As these legal documents increasingly replace nkeng, Oku farmers have a cultural dilemma. As one man expressed his misgivings, "paper does not bring fertility to the land the way that nkeng does." Like a kalang, a paper contract is dead wood that does not enhance and direct the flow of life force the way that nkeng does.

As a key symbol with economic, social, and cosmological aspects, dracaena shapes the Oku landscape and fills it with meaning. Every use of nkeng on the land and in society asserts the narrative of peace and prosperity and enacts a strategic struggle for resource access and legitimacy. The issues at play in all of these uses of nkeng in Oku are containing life force and directing the flow of this crucial symbolic resource to produce both places and people. Much of this social labor occurs in masquerade performances.

Masquerades, witchcraft, and life force in Oku

Oku is intensely hierarchical, and most social action expresses fine gradations of status and deference. When you greet an Oku notable with the rank
of *fai* or *shey* (identifiable by their colorful glass trade-bead necklaces), for
example, you must say *"mji"* with your right fist cupped in your left hand in
front of your mouth. The particular color and shape of the knit cap worn
by a man ranking high in a secret society makes this status visible from
a distance. These symbols form a vocabulary of cultural capital, and an
ordinary day in Oku rehearses status ranking repeatedly. The public performances that overflow with status and significance are Oku's juju masquerade dances, which usually accompany rites of passage (especially mortuary
rituals). The orthodox narrative about all Oku jujus is that they are emphatically *not* people wearing masks. Instead, they are wild creatures from the
forested slopes of Mt. Oku, and they carry stalks of nkeng "to make them
tame." Daily life in Oku is full of spectacle because juju performances mark
and formalize social action in institutional arenas ranging from the palace
to recreational groups (like the Kilum Bees Dance Club that I saw parade
through Oku with drummers and an elephant mask), and for processes
from dispute settlement and legitimization to blessings and funerals. Each
secret society has its own masks, medicines, musical instruments, weapons, rituals, and elaborate internal ranking system. The Oku palace is the
axis around which this social action revolves, and itself constitutes a secret
society with its own regalia and functions. Nkeng is part of these secret
societies' shared symbolic vocabulary, and its fresh leaves appear in many
jujus' hands, in the mouths of fierce animal masks, and on the head of Agah
(described above). It represents the authority of secret societies over their
members. One man told me that if he found a single nkeng leaf on his door,
he would know that he had missed a society event and would soon pay a fine.
When the palace's Ntel society reprimands a member for violating a taboo,
its members appear at the offender's house with nkeng stalks, which they use
like arrows to mime shooting a goat (and then confiscate, Koloss 2000, 304).
The Kwifon recruits new members by attaching a single nkeng leaf to an
initiate's clothing at his hip to show their claim on him. The fon sends palace
messengers with calabashes of palm wine, corked with a folded-up nkeng
leaf. Dracaena represents the cultural capital of jujus and secret societies
and expresses and directs that power as social capital.

One juju that exemplifies the power of nkeng is Kheghebchio, whose name
means "full beard." Kheghebchio has a bearded mask of black cloth and a
black gown covered with amulets, talismans, feathers, tiny calabashes, and
animal jawbones all full of medicines. It is one of the most feared and powerful "bad jujus" in Oku because it detects and destroys witches and "bad
medicine" for the Ngang medicine society (Koloss 2000, 292). Kheghebchio
dances with a spear in one hand (topped with a *Solanum* garden egg) and a
branch of nkeng in the other. Fon Sentieh II described its work as "keeping

the peace in" and "forcing the bad out," and explained that someone follows Kheghebchio's dance to sweep its footsteps with another nkeng branch to make them safe for ordinary people to walk upon. The unspoken narrative in these juju performances is that the work necessary to maintain social order is occurring. Without this symbolic labor, Oku society could not exist. These performances assert that a hierarchy of men networked through kinship and status group ranking is the best way to eradicate and prevent the disorder of witchcraft. In this way, each juju performance is also a narrative strategy for reinforcing and legitimizing Oku's orthodoxy. Because the social and spatial organization of domestic compounds mirrors that of the palace, this order is repeated at different social levels.

Order and disorder are ever-present potentialities in human affairs, and social action constantly makes and unmakes both. In Cameroon, this tension of order/disorder appears in the idioms of kinship and witchcraft (Warnier 2007, 142), and in Oku the formula is kinship/monarchy vs. witchcraft. As an icon of order, nkeng is rarely absent from Oku rituals. For example, about two years after a new fon's installation, he blesses a crowd assembled at Mkong Moteh's old residence with a nkeng stalk (Bah 1996; Gufler 2009). The next day, he climbs to Lake Mawes, the volcanic crater lake high in the Kilum-Ijim forest. There he sits on a flat stone covered in nkeng leaves and encircled by nkeng plants. After changing into his ritual clothing (loincloth, jaguar teeth necklaces, and a cap decorated with cowrie shells), and invoking the gods of Oku, the fon walks around the lake clockwise (again, the direction of life and vitality) distributing offerings to the five gods of the lake (among them small calabashes of palm wine with nkeng stoppers). The ritual reenacts a myth of the crater lake's origin, and also legitimizes the new fon's rule by showing him as the conduit for the generalized fertility of keyoi kejungha. There are many layers of symbolism in this ritual, but the importance of nkeng in this case is that, like an insulated electrical cable, it contains power and keeps that power from affecting what it should not. It asserts that ancestral substance and structural power flows from the gods of Oku into the king, who then redistributes this orderly vitality as a public good. In this way, nkeng helps the fon do his work of "keeping the peace in."

But witchcraft (in Eblam Ebkwo, *ebvung*) is the constant chaotic threat to all of this beneficial order, and nkeng is a tool for ridding the Oku landscape, society, and individual bodies of this evil. The power fueling witchcraft in Oku is *keyoi kebang* (literally, "bad breath/spirit"), the exact inverse of the keyoi kejungha that brings peace and prosperity. Rain has keyoi kejungha; lightning has keyoi kebang. The threat is omnipresent, and anything that contradicts fertility and prosperity is likely witchcraft. When strong winds blow down maize fields, farmers fight witchcraft by strengthening those plots with nkeng. An Oku witch has a "bad soul," the *keyus*, that leaves their body and travels to a witches' gathering, an *ekee kevung* ("the fireplace of the witches," Koloss 2000, 323). There the witches offer human sacrifices, usually their close relatives, in exchange for moving up the witch hierarchy.

These gatherings mirror Oku's secret societies, in which men advance to higher rank with payments of food and liquor.

Oku witchcraft is the negative version of its social capital. Trust and intimacy always mean vulnerability to others and their potential for betrayal and disorder (Geschiere 2013). Oku's kin groups and palace institutions work hard to eradicate the existential brittleness of the highly structured social capital in its secret societies, masquerades, and royal institutions. Social life in Oku is about maintaining society against the mundane and terrifying events that might unmake it. The trouble in Cameroon is that witchcraft has increased after social control mechanisms diversified beyond Grassfields monarchies to include the state, international NGOs, and churches (Feldman-Savelsberg 1999, 112). In Oku, medicine societies and the monarchy struggle to maintain order despite this expansion of institutional scale. When I asked one healer how nkeng relates to social change in Oku, he responded by criticizing Protestant denominations for undermining Oku's welfare by refusing to let their members use nkeng to sacrifice or as medicine;

> There are more problems now in Oku because of the new religions. Those churches are witchcraft but nkeng is the opposite. The other church leaders are witches, they have witch in their stomachs. The pastors and deacons all have witchcraft in their stomachs! This means more trouble for the whole country, it destroys everybody, and nkeng attacks witchcraft. Like if a woman goes to ten or eleven months in a pregnancy without birth yet, the nkeng cuts the ropes and cuts the witchcraft to allow the child to come out.

I asked him why people follow the new Christian teachings instead of Oku tradition. He exclaimed "because they get *chop* [in Pidgin English, "food money"]!" He elaborated that Oku Christians are starting to deny their kin, clan, and secret society obligations. This snippet of discourse shows the tight correlations among nkeng, witchcraft, social capital, and fertility in Oku thought. These overlapping domains help to explain the placement of nkeng at particular places in the Oku landscape. For example, the four corners of a footbridge have stalks of nkeng because witches try to enter Oku at stream crossings (Argenti 2011, 280). Dracaena not only marks economic and social boundaries; it encloses orderly social spaces and regulates flows of resources, people, and influences. Calabashes of palm wine in Oku usually have nkeng leaves for stoppers, and landscape, society, and bodies in Oku also have openings that use nkeng to manage their flows of life and vitality.

Life flowing through boundary plants

These layered sociocultural contexts make it clear that nkeng is powerful in Oku. It is a key symbol that summarizes values and significances in multiple and overlapping domains and provides a ritual script for social action.

But what exactly does power mean in Oku? To avoid flattening cultural differences with the preoccupations of current social theory, ethnographic analysis should examine the nature of power, not just who has it or what they do with it. We must investigate the cultural form of power itself, and how it manifests in specific concepts about life and how best to live it. Jean-Pierre Warnier's book on Cameroon's Mankon kingdom (2007) pushes this discussion forward by injecting ideas from Michel Foucault into African studies.

Foucault studied the histories of Western scientific modernism and focused on how Westerners subjectively experience the "governmentality" of particular constellations of knowledge and power (1979). Warnier asks what Foucault might have said about Cameroon and responds that the Grassfields has a "governmentality of containers" (2007, 38). He argues that the people of Mankon govern themselves as "skin-citizens" when they oil and medicate their bodies ("skin-containers"). Their homes have secret pots of medicine buried under the doorways to protect their "house-containers" from threats. Mankon palace and town are also containers with various protective rituals on their boundaries. The Mankon fon therefore has three concentric bodies enclosed by his skin, the palace walls, and the town border. He is responsible for keeping these containers intact and maintaining the flows of food, status, fertility, prosperity, money, and well-being through each body's openings (2007, 159). These concerns for containment and flow explain why Grassfields masked jujus enter houses backwards, why kings spray blessing fluids out of their mouths, and why non-social people like slaves and witches were called "the excrement of the king" (2007, 189). It accounts for Oku healers' practice of rubbing expectant mothers with fresh nkeng leaves to enclose the new life. When the flow of life is closed inappropriately, it affects both human and ecological vitality. One healer explained that

> It is dangerous for a woman to close up a spring while she is farming because then she will also close herself, and so get gynecological problems. Such a spring would have to be re-opened by specialists in order to save both the woman and the land. The water must flow!

The core lesson here is that "the concern with limits, openings, control, storing, and expelling is universally shared in the Grassfields" (Warnier 2007, 201). This symbolic vocabulary illuminates the cultural content of dracaena in Oku because it reorients discussion away from the structural existence of boundaries to the social processes that they regulate and the meanings they mediate.

The previous chapter analyzed closure, power, and vitality in Chagga uses and meanings of dracaena. By applying Knut Myhre's concept of "vectorial personhood" (2017) to the masale plant, Chagga concerns for knotting and enclosing *mbora* life force came into focus. For the Chagga, masale is a vectorial person who monitors appropriate social behavior. Nkeng in Oku also opens, closes, regulates, insulates, and directs the flow of life, but to different ends. The dominant pattern in Oku is centrifugal movement from a center to a periphery and clockwise motion. Just as the nkeng in the efam at the center of a compound protects the household, nkeng also protects farmland from unscrupulous neighbors and witches. The same logic applies to protecting villages and the entire kingdom. Secret societies, especially the Ngang medicine societies, are responsible for arranging and activating boundary plants at the household and village levels of organization, while the Kwifon and fon are responsible for all of Oku.

The dominant metaphor of "patrolling" expresses these centrifugal and clockwise trajectories of contained power. One elderly man in Lui village explained,

> The *ngang* [medicine containing nkeng] patrols the village during various events of the farming calendar. Each ngang from each compound rallies up and patrols the village and does not cross to the next village. The main purpose for this patrol during the harvesting, planting season, wet and dry seasons, and so on is to drive away the evil of keyoi kebang and restore peace.

Applying Warnier's ideas, we can see that the Oku "governmentality of containers" is not only about pot-kings who contain and disperse life and prosperity; the fractal pattern repeats in commoner lives as well.[12] Bodies, compounds, and communities are all containers that require social and symbolic labor to prevent blockages and maintain the flows which, as one healer told me, "secure people's lives and make the peace." Closure and flow are particularly useful concepts for understanding Oku ritual practice. For example, a *nontock* queenmother's death made hats taboo. I learned this when a shopkeeper admonished me to remove my cap because I should not "close myself" during the three weeks of mourning. In the *keman* medicine ritual, Oku men request the ancestors' blessings and forgiveness with a chicken and some nkeng. They form a "ritual chain" (Koloss 2000, 421), each with their hands on the shoulders of the man in front of them so that the request and the blessings flow through their collective body.

This same cultural model of containment and flow applies to Oku's "skin-citizens," who maintain their body-containers by using nkeng to patrol their bodily periphery by washing with water full of the plant's concentrated keyoi kejungha. This can be part of daily bathing or as part of a healer's therapy. Healers often advise barren women to wash with nkeng water. But nkeng bathing is most elaborated and commonplace in Oku childcare practices. Most homes keep nkeng in the kitchen in several calabashes tucked under a shelf near a three-stone fire. These *ebseck feyin* ("calabashes of the gods") ensure that children thrive. One healer explained,

> When a child is born, a calabash like this is prepared. This calabash helps stop a child from nightmares, bad dreams, and the child grows without trouble until he becomes a man. An adult. Myself, I still wash

myself with these calabashes! I bring fresh nkeng for cleansing. I use it to wash the child. You put water in the calabash when you want to wash a child. You add *kefu feyin* [*Basella alba*], the grass of the gods. Life comes from these calabashes!

Such a calabash is also called *feyinawane* in Eblam Ebkwo, "god of child-birth." Many have white dots painted as "eyes" to let them watch over the children. "The *feyinawane* is our mother's womb," one man told me, "we all use it." A well-prepared calabash has three nkeng stalks from the soft green growing ends of plants in the family efam. These nkeng-filled calabashes express parenthood; a parent's death means that the calabashes must be remade to "change the hand" of the caregiver. Even broken *ebseck feyin* calabashes are still useful. Women often store seed for replanting in such a container because, as one elderly woman said, "it still has the peace in it."

The second childcare practice that features nkeng prominently is the treatment of twins, who are *ghon emyin*, "children of the gods" in Oku. Throughout the Grassfields, twins indicate exceptional fertility and prosperity (Bah 2000; Diduk 1993). In Oku they demonstrate an abundance of keyoi kejungha. Many Grassfields palaces celebrate the mothers of locally famous triplets, quadruplets, and quintuplets with life-sized sculptures. Having twins in Oku earns the parents special praise-names, *noghonghene* for mothers and *baghonghe* for fathers, meaning "mother or father of children" (Bah 2000, 27; Wambeng 1993, 112). A twin birth means a party, and the celebrants plant a nkeng stalk at the compound's edge. As it grows, its leaves' green glossiness shows that "the child's future will be bright," one healer said. A twin-mother brings an empty calabash and a nkeng leaf to the palace and receives a mixture of palm oil and salt (*ntoowe*), some water, and a fresh nkeng stalk from the fon. Until her twins are about 10 years old, she wears a garland of the *kefu feyin* vine and carries a single nkeng stalk in a long-necked calabash. When encountering an infertile woman, she can donate some of her surplus fertility by using the stalk to sprinkle water on the less fortunate woman. At home, the twin-mother must never beat the twins with her bare hands or a stick the way she would a single child, but only with a harmless nkeng leaf. These twins demonstrate the fon's success promoting Oku's fertility, and this is why the mural at the Oku palace's entrance shows the fon giving a twin-mother a stalk of nkeng. On special occasions, such as the enthronement of a new fon, all of Oku's twin-mothers bring their twins to the palace and reenact the ntoowe ritual with the new king, who blesses the group of twins with nkeng.

Twins remain significant throughout their lives in Oku. Unlike other children, who get no special wealth, twins have a special bag, the *kebam emyin*, in which the family keeps gifts for the children. The loop that closes the woven jute bag is tied shut with a nkeng leaf. As the babies receive the bag, extended family members give them a few leaves of nkeng, and immediately "beg" for their return. The kinsfolk reciprocate with some coins and cowrie shells, which a healer places in the bag (Argenti 2011, 279). The mother can access the money if she needs something for the twins, but only after asking their permission. Only the children, the mother, and the officiating healer can touch this bag. When I asked why dracaena closes this container instead of a padlock, a ritual expert said that "the nkeng make the twins' hands not be dry." Twins and other infants with signs of being "strong" (such as those born in the breech position, or with an umbilical cord around their neck) need to have generosity (not "dryness") modeled for them because they are likely to become influential and wealthy.[13] Most Oku healers, according to ritual experts, are themselves twins who have "special talents given by God." Twins in Oku share a single soul, and this duality enables them to transform into animals, just as the Oku fon becomes a leopard (Argenti 1999, 35). One elderly twin joked with me during an interview, saying "are you expecting me to turn into a snake or a chameleon right in front of you?"

The theme running through these narratives about pot-kings, calabashes, and twins is that keyoi kejungha requires a technology of institutional and structural power to direct its clockwise centrifugal flow in socially appropriate ways. Sometimes Oku people use nkeng to stop the flow of life force. Oku funerals are occasions for families and secret societies (and their jujus) to ask their deceased members to depart in peace. At the funeral for the queenmother (nontock) that I attended, a somber group of women danced a counterclockwise circle (the direction of death), wearing garlands of the *kefu feyin* plant and holding nkeng in their right hands (Fig. 4.7). "They are

Figure 4.7 Women dancing counterclockwise at a funeral

giving peace to the woman who passed by doing the dance of the gods," one high-ranking woman in the Fembien society told me. Meanwhile a solemn *nchinda* palace attendant played a double gong (doon-DOOOK!) in a slow counter-clockwise circuit around the dead woman's house. Young men crowded around the grave diggers in a patch of taro and sweet potatoes and waited for a series of jujus to witness the interment of the pink, yellow, and white coffin in the notched side of a three-meter-deep grave. I noted the relative lack of nkeng to direct this burial. It appeared only at the end, when an officiant blessed the grave with a nkeng stalk, put it and its calabash on the grave mound, and broke them. The ritual expert hosting me explained that this "closes" the grave so that the deceased may have peace. It would not be appropriate to use nkeng in any other way, he said, because it conducts the flow of life. A grave, of course, is another container. In Oku the cultural focus is on the immanent life force flowing from the ancestors and manifesting in both landscape and society – not the individual occupants of particular graves. Oku's tombstones are in church cemeteries, and most people bury their dead at home without concrete graves. On Kilimanjaro, masale plants embody specific ancestors in the landscape, but in Oku nkeng plants are conduits for collective fertility and vitality and the structural power of collective institutions.

Because the Oku landscape is suffused with structural power, nkeng is an indispensable tool for managing it. Several ritual experts told me independently that it "focuses keyoi kejungha." A senior Kwifon member summarized nkeng's significance by saying that it is "like a pen writing on the land with ink. It hardly ever dies!" The key attribute of nkeng for the people of Oku is that its permanence, intensity, and robustness form a trajectory. These characteristics make it an excellent "ritual attractor" and a "technology of power" that organizes and disciplines human agency. Oku's landscape, social institutions, residential areas, farms, and bodies are all containers through which keyoi kejungha flows, and nkeng opens, closes, and redirects this life force. The uses of nkeng in Oku, from the leaves marking bundles of firewood to cures for infertility, draw on concepts from the Grassfields symbolic repertoire to open and close containers, both physical and symbolic. It is a vegetative institution for managing the flow of resources, people, and meanings, and it both constitutes life and shows how to live it.

Like all technologies and institutions, nkeng's social trajectory reflects the organization of power. Although many of the ritual practices that use nkeng demonstrate the centrifugal distribution of power from a center, it is also a centripetal force creating a particular sort of subjectivity. Nkeng directs people's attention and their movements toward the centers around which they travel. In Oku, these centers are farm plots, family compounds, and the palace itself, so the social life of dracaena constantly refers to centers of power in an agrarian patriarchal monarchy. The posthumanist identification of nkeng as a privileged actant here would not give much insight into its place in a system of privileged access. Because the construction and maintenance of order and prosperity (in matters of land tenure, politics, and ritual practice) are ultimately the exclusively masculine functions of the fon, his palace hierarchy, and patrilineage elders, dracaena is a mechanism in a strongly hierarchical system of governance based on gender, kinship, and status groups. It relates tributary social power, in all three of its manifestations in interpersonal, institutional, and symbolic domains, to landscape formation. It also relates environmental features to an interior landscape of emotion, affect, and meaning in culture-specific ways. In contrast, the plants that Oku farmers produce as commodities, such as coffee, potatoes, and kola nuts, do not produce these richly layered experiences of moral order, social memory, and the feeling of belonging to a place.

Conclusion

If the "fresh" and generally positive feelings that the Oku people have about dracaena relate to its tributary political economy and its broader preoccupation with fertility and the "governmentality of containers," then what does this say about their senses of personhood and place? Dracaena is symbolically overloaded in Oku and pervasive. It asserts order, health, and prosperity, even when society is beset by witchcraft, illness, and poverty. Its hyper-significance is always indexed to social order and the Oku monarchy, as one *fai* clan leader explained;

> Nkeng's greenness and freshness show peace because they show the vitality of its life force. This means that the peace should be long-lasting just like the long life of nkeng, even after it is picked or cut. It has a lot of authority because it appears all over Oku. When a man holds a piece of nkeng, he should hold it with both hands. It represents the authority of the fon and the Kwifon.

My questions about dracaena in Oku repeatedly generated smiles. It delights Oku residents' eyes not just because it is culturally significant, but because it is indexed to an economic, political, and symbolic orthodoxy that contains and directs the flow of life itself. Hernando de Soto (2000) might regard Oku's land as "dead capital" because it cannot be mortgaged, but in Oku land is "lively capital." Nkeng is an institution that converts economic, social, and cultural capitals by channeling life and growth. Its stalks and leaves are "sensuous signs" and "material presences that arrest the senses" (Newell 2018, 10) by connecting generalized vitality to a specific social hierarchy. A person in Oku cannot look upon the nkeng leaf held by a masked dancer or corking a calabash of palm wine without considering how they feel about property rights, their homes, the palace, and their own bodies. But when I asked about plant personhood and whether they considered nkeng to be a sort of person, I got blank stares. Even though it "goes on patrol" and provides social support at every turn, for Oku people it is a plant pulsing with life force, not a social being that happens to exist in the form of a plant. Nkeng expresses "the ideological definition of personhood in terms of rules, roles, and representations" (Jackson and Karp 1990, 15), but it does not have selfhood, individual agency, and its own vegetative gaze like on Kilimanjaro. In Oku, people regularly experience their own self-hood and personhood through nkeng and the life force of keyoi kejungha that it directs on behalf of a hierarchical social order.

This experience, although powerful, is not reducible to an abstract "pleasurable perception of nature's personhood" (Efird 2017), as Oku's ongoing problems with population growth, deforestation, and biodiversity collapse demonstrate (Ingram et al. 2015; Ndenecho 2011). The widespread concern for life and vitality in Oku is only partly located in dracaena's botanical characteristics. Focusing solely on the agency in nkeng, as a plant turn scholar might, would obscure attention to the contradictions in this social-ecological system. Jane Bennett, for example, uses concepts like "distributive agency" (2009, 21) to find causation beyond human hands. Approaching agency this way generates insights about how people interact with non-human material entities, but it does not address a core issue of social science: how social inequality relates to cultural forms. This Oku case study of nkeng shows that material properties of matter (in this case, botanical characteristics) get culturally elaborated into economic, social, and symbolic forms via particular relations of hierarchy. Plant scholars must pay attention to matters of economics and social organization to account for the similarities and differences in the social lives of the same plant on Kilimanjaro and in Oku. At first glance, the social trajectory of dracaena in Oku appears quite similar to the Kilimanjaro case, but a deeper look shows that it is instead an edifice constructed through a different arrangement of similar cultural materials. Dracaena in Oku is indexed to a "vertical" society compared to Kilimanjaro's more "horizontal" social contours. The insight that the same elements can lead to different forms urges us to focus on the cultural construction of society, rather than a static vision of social structure. Nkeng is not a structure that simply functions to reproduce the intricacies of Oku society. Oku's boundary plant operates as a multivocal technology of power in a monarchy based on kinship and tribute, contributes to the construction of culturally distinctive landscape and subjectivity, and ultimately mediates change.

What does the future hold for nkeng in Oku? The ritual experts and social historians that I interviewed agreed that its institutional efficacy and cultural meaning are decreasing as population pressure and the decline of palace institutions erode its complexity. A cultural contradiction is looming for Oku because its social system's legitimacy and authority are predicated on ever-expanding fertility and constantly flowing vitality, but its political economy is becoming increasingly brittle because of rapid population growth, civil war, and commercialization. Whether the Kilimanjaro-style commuter lifestyle can take root in Oku probably depends on transportation infrastructure and the settlement of Cameroon's civil war. For now, the social-ecological system of Oku is becoming tattered and its boundary plant disentangled – but not broken. Compared to ineffective and corrupt state institutions, the entangled governmentality of nkeng is an efficient and emotionally satisfying meshwork for many people in Oku. Dracaena may be trending toward less complexity as a polymarcating boundary plant, but as long as Oku's social and cultural elites find it a useful technology of power, it is likely to remain a green, fresh, and vital way to construct and experience landscape, society, and meaning.

Notes

1. This chapter is based on a brief site visit in 2013 and three weeks of fieldwork in 2014. I conducted 40 interviews in a mixture of English, Pidgin English, and the Eblam Ebkwo language with the assistance of a trilingual research assistant.
2. The less significant species of dracaena in Oku are *kembunglé* (*Dracaena afromontana*, a large shrub or tree with long green leaves) and *nkeng engwaken* (*Dracaena bicolor*, a shrub with white/green or yellow/green striped leaves). *Cordyline fruticosa* is known as "red nkeng" in Oku but is "just a flower."
3. Also known as *Pygeum africanum*, African Cherry, and *eblaa* in Eblam Ebkwo. Europe imports about 3000 tons of *Prunus* bark annually, making it Africa's major medicinal plant in international trade. Two-thirds is from Cameroon (Cunningham et al. 2002, 3).
4. Fon Sentieh II "disappeared" in May 2021 (one never says that a Grassfields fon dies), and his successor, Fon Ngum IV, is encouraging ecotourism and cultural tourism through social media. See https://www.facebook.com/fonofoku
5. This story is a common Grassfields narrative about fondom origins. The Tikar plain in northeast Cameroon has a different cultural repertoire, yet many Grassfields fons claim Tikari roots. The "Tikar problem" in Cameroonian historiography has shifted from explanations of population movements toward viewing "Tikar" as a discourse that asserts and legitimizes power (Fowler and Zeitlyn 1996).
6. A second structurally violent system existed in the precolonial Grassfields. In the 19th century, about half of the adult males in the Grassfields never married or even had sex (Warnier 2007, 36; 2012, 109). These "cadets" remained celibate unless they were lucky enough to be chosen by their kin groups to receive wives and inherit titles. The royal elites were (and are) promiscuously polygamous, often accumulating dozens or hundreds of wives from client patrilineages, and this prevented social mobility for most cadets (Chem-Langhëë 1995).
7. On Koloss and Argenti's theoretical impasse, see Fowler (2011) and Warnier (2014). There is a general lack of data on 19th-century slavery in the Central African interior. Nwokeji and Eltis sketched the general contours of slavery in Cameroon for the 1822–1837 period based on lists of "Liberated Africans" freed from captured ships (2002). The researchers identified 21% of the names as "Tikari" and from the northern Grassfields. If these proportions hold for all of the area's captives for 1751–1840, about 17,466 captives left the area around the time of Oku's foundation, or about 200 per year. Because Oku was a small remote polity, its total annual export of slaves was probably just a few

people (and based on Nwokeji and Eltis' figures, mostly children). This evidence does not support Argenti's model of endemic violence or Koloss' model of social harmony. As Warnier summarizes the middle position, Grassfields slavery involved trading instead of raiding, and was a "quiet violence" (2007, 200).

8. Oku barely appears in the colonial record. In part this is due to administrative neglect; Oku never had a resident colonial officer, and the first postcolonial administrator arrived in 1993 (Ndishangong 1984, 22; Koloss 2000, 40). Fon Sentieh II told me that Oku has no colonial records of any kind.

9. In their Eblam Ebkwo dictionary, Blood and Davis translate *eyfam* as "boundary" (1999, 7). Koloss translates *efam* as "place for medicine" (2000, 459). I use the term "shrine" because this is what Oku ritual experts said.

10. Colonial-era Oku commoners were seldom polygamous, but important men could have four to ten wives (Bah 1998; Koloss 2000, 77). Today polygamy is rare in Oku, except for the fon's approximately 140 wives.

11. The incorporation of Oku symbols into Catholic practice is based on the 1959 Second Vatican Council's effort to "indigenize the liturgy."

12. On this "governmentality of containers" in the Grassfields, see Chilver (1990, 236), Feldman-Savelsberg (1999, 84), Maynard (2004, 60), Price (1985), Quaranta (2010, 177), Rowlands (2005a), and Shanklin (1990).

13. Anyone in the Grassfields with unusual abilities may be a "single twin" who destroyed their sibling in the womb. Because the routes to social and economic success in the Grassfields have diversified since the end of colonial rule, the number of single twins has increased in recent decades. For other examples of African "single twins," see Diduk (1993), Ferme (2001, 213), and Maynard (2004, 51).

5   Papua New Guinea
Embodying Places, Emplacing Bodies

Midima village lies at the side of Papua New Guinea's Highlands Highway in Chimbu Province. My research assistants and I had come to this forested hilltop to study the meanings of the *tanget* plant[1] (*Cordyline fruticosa*) in the area, but first we had to watch the village's famous "skeleton show." I sat on the single wooden bench with Alphonse, the show's director and a local magistrate, while an elderly woman organized tourist knickknacks like pig tusk necklaces, wooden masks, and beaded penis sheaths for my inspection. The men's dance group showed how their ancestors had deceived a hungry monster by painting themselves as skeletons so that the sharp-clawed beast would ignore them. Then the women, adorned with face paint, kina shells, bird-of-paradise feathers, and long green cordyline skirts, performed a quasi-traditional *singsing* for me. The touristy formalities over, we got down to work. The dancers retrieved twelve varieties of cordyline within minutes and described the use and significance of each. Alphonse summarized,

> If a stranger comes to this area and sees a tanget, he always knows it means something – but not what it means specifically. Like you, he has to ask. Throughout the Highlands everyone knows that tanget means boundaries, but others need to ask.

When I asked the group if cordyline gives them a "sense of place," they laughed at the obviousness of my question, saying (in Tok Pisin, Papua New Guinea's *lingua franca*), "*tanget em make ples!*" (cordyline makes a place!).

Anthropology students regularly encounter cordyline when they read the classic texts of Oceanic ethnography. Bronislaw Malinowski, who established participant observation as the discipline's core method through fieldwork in the Trobriand Islands, never mentions cordyline, but his books make multiple references to a "red croton" which was later identified as red cordyline (1961, 35, 370; 2013, 235; Panoff and Panoff 1972, 383). Cordyline became the most famous plant in environmental anthropology through Roy Rappaport's account of how the Tsembaga Maring people use the *rumbim* plant to ritualize cycles of war and peace in the Papua New Guinea highlands (1968). Rappaport argues that cordyline rituals maintain an ecological

DOI: 10.4324/9781003356462-5

balance by regulating the populations of both people and pigs and adjusting the boundaries of competing villages.

Boundary plants in Oceania touch upon core disciplinary concerns. In addition to Malinowski's documentation of large-scale non-capitalist exchange networks and Rappaport's focus on ritual as an adaptive mechanism, these plants have roles in the evolution of sociopolitical complexity (Kirch 1994), cross-cultural encounters and transformations (Sahlins 1985), and non-Western personhood (Strathern 1988). New Guinean Big Men (and, in some areas, Big Women) and Tahitian chiefs alike used cordyline to demonstrate power, and in both areas cordyline established who could do what where. Malinowski asserts that "agriculture and its consequences enter very deeply into the social organization" of Oceanic peoples (2013, ix). These implications could be explored through sweet potatoes, taro, or coconut. This chapter and the next focus on cordyline in order to present case studies for comparison to the African chapters, and to provide background for the tangled story of cordyline in the Caribbean. Cordyline is thickly polymarcating in Oceanic societies because its significances span the cultural distance from the pragmatic (its leaves are excellent for cooking fish) to the sociopolitical (it signifies group identity and territorial claims) to the cosmological (it contains sacred power). The interactions between this complexity and different colonial histories are the reasons that while cordyline demonstrates durable cultural continuity in Papua New Guinea, the plant is more related to cultural change in the eastern Pacific.

Before reviewing the settlement history of Oceania, it is important to know how Oceanic cordyline differs from African dracaena. The trouble is that *Cordyline fruticosa* was long classified as a dracaena. Its colors range from yellowish-green to dark purple, with many green/red combinations in between (Whistler 2000, 148). In this book's other chapters, only a few cultivars are polymarcating boundary plants. My focus groups in Papua New Guinea named dozens of socially significant cordyline varieties. The plant's phenotypic variation generated taxonomic uncertainty until genetics established *Dracaena* and *Cordyline* as separate genera within the *Asparagaceae* family of monocot plants (Ehrlich 1989; Simpson 2000, 18). The plant entered botanical systematics as the genus *Terminalis* with the publication of Georgius Rumphius' botany of Indonesia's Maluku Islands. He specified that the name referred to the plant's use as an agricultural boundary marker (1743, 80). Cordyline became *Dracaena terminalis* in the Linnaean system in 1767 based on a sample from southern China (Simpson 2012). This label stuck, and the first dictionary of the Tahitian language gives *D. terminalis* as the scientific name for cordyline (Davies 1851, 265). Both dracaena and cordyline became popular houseplants and ornamentals with interchangeable names, and inhabitants of glass conservatories in temperate regions. The names for *Cordyline fruticosa* are a confusing mix of dracaena, red dracaena, croton, palm lily, and goodluck plant (Lim 2015, 627). The species' name stabilized as *Cordyline fruticosa* (L.) Chevalier in 1919 (Fosberg 1985).

What makes cordyline differ from dracaena is subtle but important. Both reproduce by vegetative propagation, and both regrow after a fire – properties making them ideal boundary plants for horticultural smallholders. Cordyline's durability makes it an unusual shrub. Planting an inch-long chunk of stem can generate a whole new plant (Griffiths 1992). Cordyline engages more social domains in Oceania than dracaena does in sub-Saharan Africa. Cordyline leaves have fibers radiating from the midrib like a feather, while dracaena leaves' fibers are parallel. Cordyline leaves are durable, pleasant-smelling, free of bad-tasting sap, and resistant to acids (Pétard 1986). Cordyline grows from a large rhizome (up to two meters long) rather than a mesh of roots like dracaena. Cordyline tubers contain about 20% sucrose and need about 20 hours of cooking to soften their tough fibers, convert glucofructofuranan carbohydrates to fructose, and produce a caramelized sugary treat (Hinkle 2005, 9). These multiple uses, from fiber to food, help to explain why cordyline was one of the "canoe plants" that ancient voyagers carried across the Pacific (Whistler 2009).

This chapter begins by sketching the place of cordyline in the deep history of the Pacific. The second section focuses on Papua New Guinea as a social-ecological system in which economics, social relations, and cosmology entangle. Part three turns to the history and politics of land tenure, while part four analyzes the links between social organization and territory. The fifth section explores the aesthetics of cordyline and place-making, and the conclusion summarizes how this boundary plant shapes both bodies and landscapes in New Guinean societies.

The vegecultures of Oceania

The settlement of Oceania is an epic adventure that rivals the American moon shot for the sheer audacity of propelling humans into the vast empty space of the Pacific. Unlike the American astronauts' collection of moon rocks and Cold War bragging rights, Oceanic peoples were colonizing new lands, and not just relying on a temporary life support system. Along with taro, coconut, and bananas, cordyline was a botanical technology for these seafarers. Cordyline is indigenous to Southeast Asia, not Oceania, making its current biogeography its social history (Hinkle 2007). Cordyline plays many roles across Oceania, in societies as different as loosely networked Papua New Guinean horticulturalists and Tahitian centralized chiefdoms. Like the social history of dracaena in Africa, Oceanic cordyline is one element in a variable meshwork of land use, social organization, and cosmological meaning.

This book calls the region Oceania to avoid the pitfalls of older names. The terms Melanesia ("black islands"), Micronesia ("small islands"), and Polynesia ("many islands") were coined by 19th-century Europeans to define geographic areas with cultural characteristics that would allow or prevent settlement and trade (Dumont d'Urville 1832). Melanesia was an area of dangerous black savages, while Polynesia had hospitable brown monarchies (Thomas 1989). Twentieth-century anthropology dropped the racism but kept the evolutionary assumptions. Melanesia is generally egalitarian and characterized by Big Man horticultural economies, while Polynesia is an area of hierarchical agricultural chiefdoms and sociocultural complexity – resulting in a general "upward west to east slope in political development" (Sahlins 1963, 286). The contrast is about social status; in Melanesia people pursue exchange relationships to accumulate and achieve status, while Polynesians have long experienced ranked systems based on ascribed status.

Recent work in biogeography, archaeology, and historical linguistics redrew this portrait in favor of "bumpy regional diversity" rather than a smooth evolutionary sequence (Kirch and Kahn 2007, 213). The biodiversity of terrestrial animals has a sharp break at the Wallace Line (an area of open sea) between Borneo and Sulawesi, with Asiatic species to the west and Australasian ones to the east (Fig. 5.1). Human expansion into the Pacific crossed this line but reached a pragmatic limit in the Solomon Islands about 29,000 years ago. Beyond this point, islands are no longer visible on the eastern horizon (Green 1991, 495). The societies and ecosystems between the Wallace Line and the Solomons have therefore co-evolved for tens of thousands of years, while more isolated islands to the north and east have much shorter social timelines. The older area, which includes Australia, New Guinea, the Bismarck Archipelago, and the Solomon Islands, is now known as Near Oceania. The settlement of Remote Oceania began about 3200 years ago after a "long pause" near Samoa. In effect, the vast region from Palau to Kirabati (the former Micronesia) and the triangle defined by New Zealand, Easter Island, and the Hawaiian Islands (Polynesia) have become Remote Oceania based on its common biogeography and human

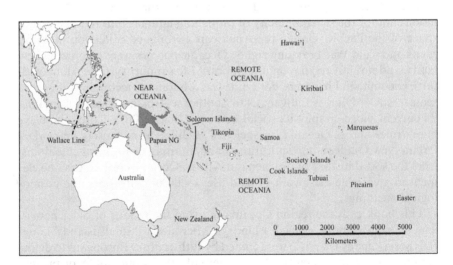

Figure 5.1 Near and Remote Oceania

prehistory, while the old Melanesia has been split down the middle. This new categorization helps to explain the spectacular linguistic diversity of Near Oceania (with about 1300 languages) compared to the roughly 200 relatively homogeneous languages of Remote Oceania (Kirch 2010, 134).

This is the enormous region in which Southeast Asian cordyline became ecologically and socially significant as part of a "transported landscape," alongside coconut, banana, taro, yams, breadfruit, and sugarcane (Merlin 2000). Cordyline is a boundary plant across Oceania, but it shapes social-ecological systems in strikingly different ways. Huw Barton and Tim Denham use the term "vegeculture" to describe how Near Oceanic societies are based in part on plants that reproduce through vegetative propagation, like taro, bananas, and sweet potatoes. "These landscapes," they write, "have been shaped over time through a combination of cosmological belief, social practice and the material properties of plants" (2018, 23). I extend this approach to Remote Oceania to demonstrate how cordyline fits into the vegecultures of Papua New Guinea and the Society Islands in ways that correspond to their different social histories. A dominant model for understanding Papua New Guinean cultures is Marilyn Strathern's relational personhood embedded in decentralized societies (1988), and Lévi-Strauss' concept of ranked "house societies" (1988) is a major model in Oceanic archaeology (1982). Cordyline relates to both dynamics, but it did not simply diffuse across the region as an unchanging cultural trait. Instead, I argue that Oceanic social-ecological systems have always been and continue to be in flux, and that cordyline often mediates different dynamics. The term "custom" obscures and hides the contradictions that beset Oceanic societies (Keesing 1982, 229). Like dracaena's role in African "customary law," the story of cordyline is about continuity and change, even when Oceanic peoples use the plant's indigeneity and rootedness to emphasize only continuity and custom.

No ethnobotanical survey of Oceania is complete without some reference to cordyline. The academic literature insists that the plant's pragmatic and magical usefulness made it part of the regional toolkit (Powell 1976; Leenhardt 1946), and a prime example of plant domestication for purposes beyond food and fiber (Barrau 1965, 289; Sauer 1969, 27). *Cordyline fruticosa* (L.) Chevalier looks different across Oceania. The New Guinean cultivars have diverse colors, sizes, and leaf shapes, while those to the east are mostly large plants with long blade-like green leaves. Recent botanical detective work by Anya Hinkle has shown that the green cultivars of Remote Oceania have sterile pollen and are seedless clones. This indicates how the ancient Austronesian peoples domesticated the plant through vegetative propagation (2004; 2005; 2007), and probably not just for utilitarian reasons. James Fox's idea of "ritual attractors" (2006, 1) provides a more interpretive approach beyond cordyline's pragmatic uses for food and fiber. Fox argues that throughout Oceania, Austronesian houses contain particular architectural elements – often posts – that represent the origins, precedence, legitimacy, and continuity of social groups, and serve as focal points for ritual practice. If we take houses to include yards and gardens, this concept provides a vocabulary for discussing cordyline's importance throughout Oceania and its role in the human settlement of the Pacific.

Cordyline's robusticity made it indispensable in pre-contact Oceania. Cordyline leaves wrap food, line earth ovens, have medical uses, and provide protective clothing, sandals, and thatching (Merlin 1989; Whistler 2009, 89). Cordyline rhizomes are eaten, usually as a secondary or famine food due to their long cooking time, from Anuta (in the Solomon Islands, Yen 1973) to New Zealand (where the failure of tropical cordyline led the ancient Maori to cultivate the endemic *C. australis*; Best 1976c) and Hawai'i (Handy et al. 1972, 224). It also has robust meanings across a vast region. Cordyline stalks and leaves denote property rights in farmers' fields in Indonesia (Erb and Jelahut 2007) and Hawai'i (Menzies 1920, 87), and protects houses from evil spirits in the Philippines (Conklin 1967) and Borneo (Zahorka 2007). These meanings and practices vary across Oceania, and like African dracaena, they form a patchy mosaic with repeating variations on a theme. This chapter and the next form a controlled comparison focused on cordyline's different "ritual attractions" in Papua New Guinea and French Polynesia. The contrast corresponds to the distinction made in the previous chapters about the socially "horizontal" nature of power on Kilimanjaro and its "vertical" dimensions in Oku. The vegecultures of Oceanic boundary plants are rooted in both power and botanical properties.

Papua New Guinea as a social-ecological system

Papua New Guinea is the largest case study in this book in terms of area and population. About 13 million people live on New Guinea's 786,000 square kilometers; about 8 million of them are Papua New Guineans. Because of its relatively late colonization by European powers, the island has the world's greatest cultural and linguistic diversity per square kilometer (Landweer and Unseth 2012). My fieldwork there was a two-week-long transect from the Central Highlands to the coastal city of Madang along the Highlands Highway and through the Ramu Valley (Fig. 5.2). Like a description of life in an enormous house based on brief observations through a few windows, this chapter is a partial account. An in-depth survey of socially significant cordyline would likely find different patterns in the largely patrilineal mainland and the more matrilineal peoples of Papua New Guinea's eastern islands.

The history of New Guinea is a series of ecological revolutions. Foraging peoples crossed the Wallace Line to reach New Guinea about 60,000 years ago, when the area was the northern part of the Sahul continent (Bird et al. 2019). It became an island after the Holocene rise in sea level about 11,000 years ago, although it now seems that a continent's worth of alpine zones, broad mountain valleys, and tropical wetlands fill its spectacularly rugged and diverse topography. A few Southeast Asian mammals (rats, mice,

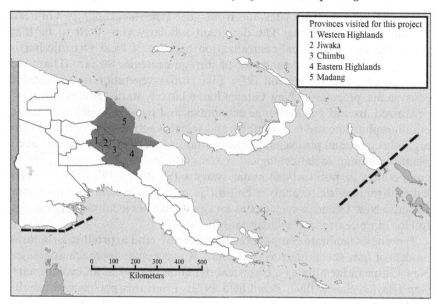

Figure 5.2 Papua New Guinea

and bats) crossed the Wallace Line to join the island's endemic marsupials and its astonishing avian diversity, but pigs arrived with the humans and become a major ecological force (Brown 1978, 17). Around 7000 years ago, Papuans in the central Highlands independently invented agriculture, focusing on taro and bananas (Denham et al. 2003; Haberle et al. 2012). In the malaria-free zone between 1400 and 2100 meters (an area with generally fertile soil, moderate rainfall, and limited weed growth), horticultural societies with wood, stone, bone, shell, and fiber technologies reached high population densities without forming centralized hierarchies. The Austronesian seafarers arrived from Southeast Asia about 4000 years ago, colonized the islands of the Bismarck Archipelago and the northern coast of the New Guinean mainland, and began long-distance trade relations with the Papuan farmers in the island's interior. The arrival of American sweet potatoes (via Austronesian voyaging and trade networks) after 1200 AD revolutionized this farming system yet again because 2200 meters is taro's upper limit, but sweet potatoes grow up to 2800 meters (Allen 2013). The new staple food probably sparked rapid population growth and settlement of areas too high for taro (Denham 2013). It is possible that the Austronesians also introduced cordyline to Papua New Guinea, or the plant may be endemic, or carried there by the ancient foragers (Hinkle 2005, 196, Yen 1987). We can conclude, however, that the social significance of cordyline across New Guinea is not a timeless custom. Instead, this boundary plant's institutionalization was likely the result of the ancient transition to farming lifeways, the regional trading system, and the sweet potato revolution.

This focus on dynamics deviates from older assessments of New Guinean social-ecological systems. The dominant scholarly view used to be that the island lacked political centralization because of Near Oceanic horticulture's ecological limits and lack of storage strategies for taro (Diamond 1999; Hogbin and Wedgwood 1953). Like climax vegetation forming stable ecosystems, precolonial New Guinea had relatively static "climax cultures" organized around gift exchange economies and localized feuding. Classic ethnographies showed how symbols and social practices like cordyline rituals and ceremonial pig exchanges were adaptive mechanisms that balanced plant-pig-human social-ecological systems (Rappaport 1968) and regulated competition in political-economic systems (Strathern 1971). These texts led anthropologists to analyze culture as part of a system, but also constructed New Guinean societies as an egalitarian sociocultural type and a foil for more centralized systems (Sahlins 1963).

Typologies facilitate comparison, but they also tend to produce structural models of how social life stays the same. A processual approach addresses New Guinean farming as a historical dynamic instead of an evolutionary step (Bayliss-Smith 1997; Brookfield 1972). From this perspective, smallholder farming produces food, social relations, and cosmological meanings simultaneously through ongoing experimentation and innovation (LiPuma 1988; Waddell 1972). This view delineates connections between plants and people in New Guinea. Yams, taro, sweet potato, and sago palm all have different planting methods and uses, and this may explain the differences between fiercely independent family farms in the Highlands and more communitarian villages in coastal and island areas. From a processual point of view, the island's social-ecological systems have always been in flux, and the system components (like boundary plants) are changing relationships, not cogs in a machine reproducing social and ecological continuity.

New Guinean kinship and social organization are astonishingly diverse. The key issue for the social history of cordyline is how the island's social groups form and interact. In the 1960s, ethnographers adopted structuralist models of kinship and organization from African studies to analyze Near Oceanic societies with terms like patrilineage, clan, and tribe (Barnes 1962). But New Guineans regularly violated their own rules of kinship and social order. This led 1970s anthropologists to conclude that kinship and descent comprise a folk model of relationships, not actual genealogical structures (Brown 1978, 149).

The shift from structuralism to the processual study of New Guinean institutions culminated in Marilyn Strathern's insight that Western concepts of individuality, personhood, property, and society are wholly inappropriate for understanding the peoples of New Guinea (1988). She replaces these sociological staples with the concept of relational personhood, which defines people as both individual selves and "dividuals" produced by social interaction. Instead of society being the sum of individuals interacting through rules, norms, and institutions, Strathern argues that relations produce people (1988, 173). New Guinean society and culture are, for Strathern, located in human connections rather than pre-existing and determining them, and are continuously reproduced through economic and ritual exchanges. This "relationality" is important for the study of boundary plants in Papua New Guinea because these plants connect to personhood differently than the patterns in the Kilimanjaro and Oku case studies. The relationships that produce people and institutions in Papua New Guinea often begin as human-plant relationships, not human-human relationships that then get applied to plants.

Thinking carefully about personhood and planthood is useful because it can generate new perspectives on classic topics in economic anthropology and gender studies. Marcel Mauss' *The Gift* (1967) was a founding text of economic anthropology. It established reciprocity and exchange relations as core concepts for analyzing non-capitalist economies. Entrepreneurial Big Men and Women in otherwise egalitarian societies demonstrate that gift exchange is as strategic and competitive as Wall Street capitalism, just in different ways (Weiner 1976). Papua New Guinean ethnography revitalized gender studies by moving away from structural accounts of the division of labor by gender to a focus on how gender identities are fluid constructions produced through social action (Herdt 1987). Taken together, these classic texts point toward New Guinean people and plants as processes, not structures – much like understanding time-telling by watching people organize their lives around clocks is far more revealing than analyzing a clock's gears and springs. Cordyline figures prominently in classic texts, like when the people of Mt. Hagen create new ceremonial *moka* gift exchange areas by planting it (Strathern 1971, 37), or when Sambia women use cordyline leaves to protect their initiate sons from a thrashing (Herdt 1987, 137). A relational approach to cordyline and personhood locates this boundary plant at the center of social and symbolic action, even when it lies on the margins. As a hotel-owning Big Man in Mt. Hagen told me, cordyline lies at the core of how property rights connect to Near Oceanic personhood. "You can't have a house without tanget! Otherwise it's just rubbish, and you have nothing to show for your work!"

Most of the ethnographic literature on Papua New Guinea is profoundly localized, and the evidence for significant cordyline reflects the priority of context over comparison. William Clarke, for example, offers a rich description of how cordyline, cultivated crops, and wild plants form gardens, villages, and resource territories for a community of only 154 people (1971). What makes this plant important in contemporary Papua New Guinea is more than the sum of these local significances. As Alphonse said at the beginning of this chapter, the general awareness that a tanget plant must mean *something*, even if its particular significance is unknown, is a major way that Papua New Guineans relate to one another. The potential for meaning is as important as the meaning itself. This boundary plant is increasingly used for intercultural communication because islanders (men in particular) no longer live sharply circumscribed lives in limited territories; instead mobile people are building relational identities in new places with new relationships.

The Tok Pisin term for this pattern is *wantok*, borrowed from the English phrase "one talk" by plantation workers in the mid-20th century to denote the group of people whose shared language and geographic origin give them mutual obligations (Nanau 2011; Schram 2015). It is an idiom for building social networks across differences, not a social unit, and an adaptation to mobility and ethnic pluralism, not an expression of pre-modern ethnic identity. Cordyline relates to wantoks because these boundary plants communicate social significance. They are a pragmatic way to signal "who is your wantok" through shared knowledge about what a particular plant means. In a cemetery at the Ramu sugar plantation, which draws its work force from across the country, graves are marked with a variety of different cordylines. When someone sees the correct tanget variety on a grave, workers told me, he recognizes his wantok.

New stresses in Papua New Guinea are creating new relationships that often involve cordyline as a way of experiencing and reacting to change. New institutions like trans-national corporations and conservation agencies have created new relationships and entanglements for cordyline and society (Jacka 2015; Kirsch 2008). In 2014, the Pacific Mining Partners corporation evaluated the coal mining potential of a coastal community northwest of Madang. Fearing the displacement that other Papua New Guineans have experienced, the Andimarup villagers declared a ban on mining by wrapping a lump of coal with knotted cordyline leaves and placing weapons on top as a promise of violence if the taboo were broken (Tony 2014). When tensions were escalating over the control of conservation and development project benefits in the Crater Mountain Wildlife Management Area, a fight ensued after a landowner claimed land by cutting down the cordyline plants around the project's field office (West 2006, 22). People in Akemaku village near Goroka told me that land management is now disorderly because "our grandfathers knew how to use tanget but anyone under 40 doesn't know much." These anecdotes illustrate some of the ways that Papua New Guineans deal with change with the aid of cordyline plants. The emerging literature on social change in Papua New Guinea points toward messy engagements between the global political economy and localized cultural meanings on topics like religion (Robbins 1995), climate change (Jacka 2009), and the coffee industry (West 2012). Much like the ways that orderly boundary plants have been foils for disorder and change in Tanzania and Cameroon, cordyline is a material and symbolic resource for responding to change in Papua New Guinea's rapidly shifting environment and society.

Cordyline as a *botanica franca*

English terms like land, ownership, and boundary are notoriously difficult concepts to apply to Papua New Guinea (Strathern 2006). Like the other case studies in this book, land claims in Papua New Guinea are never just economic matters. Instead abstractions about ancestors, gender relations, and aesthetics regularly interact with the pragmatics of raising pigs, cooking sweet potatoes, and exporting coffee. Ownership is the wrong question because rights and obligations regarding land are inextricably enmeshed with social relations. The legal scholar Robert Cooter noted, "asking who owns the land [in Papua New Guinea] is like asking which player is the football team" (1991, 769). A second difficulty is that Papua New Guinea's sociocultural diversity means that teams are playing by different rules in the same league. The third difficulty is that social relations regarding land in Papua New Guinea are fuzzy, indeterminate, and in a sort of "kaleidoscopic flux" instead of neat closure (Crocombe and Hide 1987; Sillitoe 1999, 333). Finally, anthropologists have been unable to establish clear causal relations among population density, land management, and land tenure institutions (Brown 1978, 111). The political economy of boundary plants in Papua New Guinea reflects these diverse contexts. Cordyline is a boundary plant in low-density shifting horticultural systems and high-density permanent cropping areas, in both highland and lowland zones. It is significant not only because of its local meaning, but because it represents the general principle of claims-making. Like their national Tok Pisin *lingua franca*, cordyline is a sort of *botanica franca* that allows Papuans to communicate across linguistic and cultural differences. Always pausing before picking a cordyline leaf as a prop for our inquiries, my research assistant looked to a likely landowner, saying, "a man has to *ask* before he breaks a tanget." Papua New Guineans also use topographic features, like rocks, rivers, fences made of wood or reeds, and coral boulders to mark territory, but only cordyline is socially and culturally complex.

Cordyline addresses Papua New Guinean concerns for belonging, containment, and protection. Who belongs to a certain piece of land? What relationships are contained within that space, and which cross its boundary? How can ordinary people prevent invasive threats like witchcraft? Many regional ethnographies mention cordyline as a boundary marker or fence, but the sole text devoted to it is Heinrich Aufenanger's somewhat anecdotal review of land borders and ritual practice (1961).[2] Scholars often mention cordyline (glossed in Tok Pisin as "tanget" without a species designation) on boundaries without describing how different cultivars have specific meanings and uses in different areas. The benchmark text of Papua New Guinean land tenure studies, Harold Brookfield and Paula Brown's *Struggle for Land* (1963; see also Brown et al. 1990), describes Chimbu farmers using lines of cordyline to assert individual ownership in the areas with high soil fertility, but the text does not explore the plant's implications beyond agricultural economics and erosion control. Rappaport's study of ritual and ecology among the Tsembaga Maring describes cordyline as the linchpin holding an entire social-ecological system together, but he only tells us that it is the red-leafed variety of their ten named cultivars (1984, 46, 148). More botanically-oriented ethnographies describe how particular groups use particular cordylines to mark land, direct vitality toward growing crops, and construct dance costumes – and how these uses overlap and intersect (Kocher Schmid 1991; Panoff and Panoff 1972; Sillitoe 1983). This sort of holism is what I heard

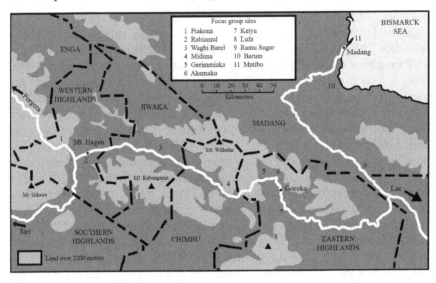

Figure 5.3 My transect along the Highlands Highway

repeatedly in Papua New Guinea when asking about cordyline. Both land and bodies have boundaries, people told me, and the work of cordyline is to enclose, protect, and organize them. Understanding land tenure issues in Papua New Guinea requires attention to both boundary plants and bodies.

I conducted eleven focus group discussions in my journey through Papua New Guinea. Eight were along the Highlands Highway; two were on the coast near Madang (Figs. 5.3 and 5.4).[3] Each followed a similar pattern; after

Figure 5.4 Focus group with a lineup of cordylines

learning what my visit was about, young boys ran to collect sample stalks with three to five leaves. Going down the line of plants, I learned the name and uses of each variety. Usually this led focus groups to discuss and debate the proper names and usage, which I recorded through multiple layers of translation from the local language to Tok Pisin and English. On average, each focus group named ten varieties of cordyline and classified them by usage, leaf size, and color. The resulting data set includes fine points of botanical difference and idiosyncratic social usage, such as "*this* leaf is dark green with a light pink edge, so we wear it for a singsing, and *that* one is dark green with the same pink edge, but it's longer, so we use it to protect sugar cane gardens." Within this thicket of vegetative meaning, I noticed some clear pathways:

1 Utterly pragmatic uses for varieties with stems and leaves well suited for garden fencing, erosion prevention, lining earth ovens (*mumu*), and making packets to cook pig intestines. These varieties usually have long glossy oval dark green leaves.

2 Each site has a particular cultivar called a "land tanget" to mark boundaries between kin groups and villages. Across the Highlands, these are always a uniform bright green, although leaf shape ranged from oval to lanceolate. In two locations, this same cultivar marks garden plots and yards inside the village and even different individuals' places within one garden, but most Highlands groups have different cordylines for these uses. At the coast, a red-leafed cordyline serves this same function. In the Highlands, these cordylines are tightly interwoven with masculine power and kin group identity.

3 Most focus groups use different cultivars for clothing and to institutionalize access to land. Some cordylines are only for ceremonial occasions and some are daily wear. Women's cultivars for singsing skirts and armbands are often variegated green, white, and pink. Men's varieties are usually uniform green, sometimes with thin red edging on the leaves. Costuming varieties are further divided by gender and life-cycle phases; some are only for girls or boys, and others so restricted to adult men that women cannot touch these plants in their own gardens.

4 Only two of the Highlands sites associate green cordylines with death. Except at the coast, red cordyline generally relates to cemeteries, battle sites, and people threatening violence. Dark reddish-purple varieties indicated danger or protection from witchcraft throughout my transect, except at one coastal site where two green varieties (one considered male and the other female) serve these purposes.

5 The cultivars associated with symbolic and aesthetic order beyond witchcraft, like boosting garden fertility, controlling weather, finding thieves, welcoming visitors, and decorating houses vary widely in terms of leaf color and size.

6 Each community has at least two purely decorative and unnamed cordylines.

These meanings and uses overlap in practice. The Chimbu-speaking focus group agreed that the main significance of their bright green *gumane augl* is "men's business and big clan politics" like marking group territories in the forest, and its secondary importance is for lining mumu ovens. But, they said, when women wear it for a singsing (like the skeleton show at the beginning of this chapter), it means "we are all one clan." The point here is not that particular cordylines are meaningful for particular people, but that these boundary plants regularly cross cultural domains. The social field of political economy regularly cross-fertilizes with expressive culture.

The colonial and postcolonial history of cordyline in Papua New Guinea illustrates clashing ideas about land, property, and value. Western notions of property and territoriality are "exclusive" because they focus on people with sole control of something from which others can be legitimately excluded, but Papua New Guineans are more "inclusive." Places and objects are enmeshed in durable relationships among the people in specific exchange histories (Carrier 1998; Weiner 1992). Land and wealth are part of one's self and cannot simply be alienated and separated from a person any more than an arm or a leg can. Instead in Papua New Guinea "the concept of inalienability prevails, prohibiting the sale of land with ownership vested in the clan" (Armitage 2001, 5), and clans exist as social histories of exchanged wealth and people. These pasts have distinctly non-Western meanings, because in Papua New Guinea "the past is commonly understood spatially rather than temporally," as a series of events at specific places (Kahn 2011, 7). The Tok Pisin term *ples* refers to a subjective set of relationships, not the objective location in the English term "place." In Papua New Guinea, cordyline legitimizes some alienations of *ples* despite the overall ban. It bridges the gap between cultural value (the permanence of ancestral territory) and social necessity (new people on old land). When Rappaport's Tsembaga Maring warriors replant the *rumbim* at the end of intervillage hostilities, they confirm a new territorial boundary on alienated land before another cycle of feasting and warfare (1984). Subdividing a farm with a particular variety of cordyline in Geremmiaka village pacifies quarreling brothers, but it alienates each man from the other's land. The meanings of boundaries and the social process of boundary-making are changing in postcolonial Papua New Guinea, and cordyline is becoming a mediator between people, the state, and corporate power. As one focus group expressed it, land falls into three categories; "deeded land in towns and cities, settlements under one traditional landowner in a town or along a main road, and village land organized by tanget."

For most of the 20th century, the Territory of Papua and New Guinea was a *de facto* colony of Australia. Most islanders experienced "government by patrol" rather than direct administration. Australian officers (known as *kiaps*), native police, porters, and translators trekked into remote villages to collect taxes, settle disputes, and recruit labor for coastal plantations

(Dinnen 2019). Believing that customary group tenure prevented cash cropping and economic growth, the Australians copied laws and regulations from colonial Africa to push the colony's flexible collective systems toward English-style individual freehold within clearly delimited village territories (Crocombe and Hide 1987, 348; Lea and Curtin 2011, 22). Much like colonial Africa, these efforts to achieve progress by replacing indigenous complexity with administrative simplicity blurred land issues (Berry 1993; Scott 1998). In practice, *kiaps* relied on indigenous institutions like cordyline to impose administrative legibility onto the Papua New Guinea landscape. Brookfield and Brown report that administrators in the Chimbu area settled conflicts by walking around group territories while locals planted cordyline on their footprints (1963, 146).

Land matters in Papua New Guinea are a mess. The standard figure in the regional literature is that 97% of its land is under "traditional land tenure," but the country's parliament revised the forestry laws to create *de jure* state control over 77% of the country's land area in 2005 (Curtin and Lea 2006). The other 3% of Papua New Guinean land has been "alienated from custom" by title deeds, and is mostly government property, plantations, and urban areas. The Papua New Guinean constitution and 1970s land laws relied on the idea that the customs practiced by territorial groups would determine local matters, but what exactly *is* a custom, and for whom, was never defined. The result is that Papua New Guineans have constructed customs, land groups, and clan histories in response to the demands of ambiguous laws (Filer 2007). These place-making narratives use cordyline to legitimize land claims, making the plant part of governance. After Papua New Guinea's independence in 1975, the new government set up local land tribunals to prevent land conflicts. These administrators rely on oral histories and the presence of boundary plants to show first settlement (Westermark 1997). They often plant cordyline stalks alongside the concrete posts that bank loans require (Benediktsson 2002). This misfit between inclusive, dynamic, and relational land practices and exclusive static institutions generates contradiction and uncertainty for Papua New Guineans.

Is private property the key to Papua New Guinea's economic productivity and development (Chand and Yala 2009)? Or should local land tenure institutions be left alone to "evolve" in response to population pressure on resources? The Papua New Guinean government answered these questions by working with development institutions like the World Bank to write land policies that attract foreign investment. But the Papua New Guinean constitution prohibits customary owners from selling or leasing land except to other citizens (Anderson and Lee 2010). The 1996 Land Act therefore allows the government to circumvent the constitution by serving as the intermediary between foreign commercial developers and local landowners. This is known in Papua New Guinea as the "lease lease-back" system. The state leases the land for 99 years from its customary owners, and then contracts with foreigners in the mining, forestry, and agricultural sectors for resource extraction and production (Cooter 1991). Between 2003 and 2015, about 5.2 million hectares (11% of Papua New Guinea's total land area) was alienated to private companies – often without the landowners' knowledge or permission (Filer 2011; Friends of the Earth 2015).

A 2013 governmental Commission of Inquiry found most of these alienations illegal, and efforts to resolve the situation remain tied up in the country's legal system (Filer and Numapo 2017). This is an excellent example of state territorialization, through which states coercively assert spatial control, produce new social and ecological relationships, redefine terms of resource access, and redistribute control (Vandergeest and Peluso 1995). In Papua New Guinea, the widespread recognition of cordyline as a property institution allows local landholders to challenge state sovereignty and assert different principles of moral economy (Ballard 2013; Curry and Koczberski 2009). Andimarup village's warning to a coal mining company is a good example of vegetative counter-territorialization against state and corporate power. Another example comes from a gold mining company's internal documents (Filer et al. 2000, 82). The company compensated landowners whenever mining operations destroyed gardens, so farmers created fake gardens with a few taro plants and stalks of cordyline.

Cordyline appears in the Papua New Guinean land tenure literature only rarely, which is odd given its broad significance. This conspicuous silence probably reflects the colonial-era division of law into traditional and modern domains. Many Papua New Guinean men told me that they use both tanget and concrete posts to mark land, but that cordyline is more secure because their neighbors recognize it. They were also dismayed that their government regularly ignores boundary plants. When I was traveling along the Highlands Highway, many roadside houses and trees were spray-painted with red Xs, showing that they would be destroyed for a road widening project. Cutting and demolition were just starting when I passed through. "They are even cutting tangets because they don't care!" and "the ones who do not respect tanget are in the government!" were typical laments. Even slum areas organize land with cordyline. I visited a "traditional landowner" near Goroka who earns USD $2400 per month from about 150 tenants living in shacks of plywood and plastic sheeting. The area is technically clan land, but he has effectively privatized it by being the only living man left in his kin group. A chain-link fence surrounds it, and he has "his own tangets" on its corners. He allows renters to plant and harvest seasonal crops in their kitchen gardens, and insisted that the cordylines I saw throughout the settlement were "just flowers for decorating houses," not property markers. Tenants confided that they regarded these as fully legitimate boundary plants, but subsidiary to their verbal contracts for house plots. These examples demonstrate that the Papua New Guinea government contradicts its claim to respect customary land management and landowners redeploy tradition in very uncustomary ways. Like the Africans in Chapters 3 and 4, Papua New Guineans currently practice the governmentality of boundary plants where state institutions are ambiguous or incomplete.

What would Papua New Guinea be like if the social significance of cordyline were recognized in statutory law? The Evolutionary Theory of Land Rights (Platteau 2000) suggests that smallholder farmers' best route to economic development and tenurial security is for states to re-institutionalize indigenous land tenure. Communities would build their own strategies for coping with population pressure, agricultural intensi-fication, and commercialization instead of relying on statutes, cadastral maps, and bureaucracies. In Papua New Guinea, state acknowledgement of boundary plants as land management institutions would encourage cultural diversity and be more consistent with actual land management practices. True, there would be significant transaction costs – such as the need to translate, mediate, and settle disputes among different localized institutions – but this would arguably be better than the current muddle of claims and the injustices of "land grabs" under the lease lease-back sys-tem. Such an enabling environment for boundary plants would also express Papua New Guinean ideas of just and appropriate relations between peo-ple and land. My overall assessment is that islanders have already rede-fined the local meanings of tanget to allow communication across cultural and linguistic differences. They are already evolving new property rights institutions, and this indigenous process is not simply the result of colo-nization or an effect of global capitalism (Brown et al. 1990). Legal rec-ognition of vegetative institutions like tanget in Papua New Guinean law would strengthen smallholder farmers' counter-territorialization efforts against exploitative outsiders.

The downside of this admittedly rosy scenario would be a resurgence of neo-traditional narratives and territorial claims emphasizing already extant social inequalities. The statutory recognition of boundary plants in Papua New Guinea would likely enhance the power of high-status men and entrench male networks, and the *botanica franca* of cordyline would become the language of new struggles instead of the final word on land conflict. The roots of cordyline in Papua New Guinea land matters are entangled with gender, kinship, Big Man status, ethnicity, and ancestors. The next section surveys how cordyline both reflects and contributes to social organization in Papua New Guinea.

Mapping social relations with boundary plants

Marilyn Strathern's argument that Papua New Guineans are both individ-uals and dividuals helps to explore the degree to which plants are personi-fied there, and perhaps even how persons are plant-ified. Being a relational dividual composed of many transactions differs from the Western ontolog-ical construct of individual subjectivity in separate bodies moving through objective space to seek personal advantage. As discussed in the African chapters, personhood is both an experience of selfhood and a culture-specific ideology of moral personhood (Mauss 1938). In the Papua New Guinean context, bodies and places are the always-incomplete products of ongoing interdependencies and transactions, and microcosms of ongoing relations (Strathern 1988, 131). These relational persons create themselves (and places) through reciprocal exchanges, and these often involve cordyline.

On my first day in Mt. Hagen, I experienced a singsing and a meal cooked in an earth oven *mumu* before the Rabiamul village discussion about cordyline. My fieldnote shows how these Melpa-speaking women worked to plantify my person:

> The women had spent 45 minutes at the side of the *moka* grounds preparing for the singsing, their first for a tourist. Each had a collection of bamboo tubes to hold the bird-of-paradise feathers for their headdresses, and they carefully formed each others' costumes of leaves and feathers and painted their faces red with a white mask around the eyes and a white stripe on their chins. They had the skins of the cuscus marsupial possum on their chests and necklaces of shells from a far-off coral reef. Each woman wore a skirt of crinkled and greased green and pink cordyline (*pogla djira*) in front and light green cordyline (*pogla dau*) in back. They swayed and bobbed as they sang a song with lyrics that kept asking me to bring them some water. It took another half hour to store the costume elements in their bamboo tubes and plastic grocery bags; then the mumu began. The women lined a pit with yellowish-green *pogla kia* cordyline leaves, added some hot rocks, then sweet potatoes, more cordyline leaves, breadfruit and pumpkin leaves, more hot rocks and finally a layer of meat. They covered the whole thing with banana leaves and while we waited for it all to cook, they excitedly explained how their crinkled and greased cordyline leaves had invited me to reciprocate with the fat rendering from the ten kilograms of fatty New Zealand "mutton flaps" that my research assistant had bought at the supermarket that morning. This, they said, was the "water" that I was giving them.

I soon recognized these activities as standard practices on my journey from Mt. Hagen to Madang. The cordyline leaves transformed me from stranger to a moral person with whom people could have exchange relationships. Cordyline relates to moral personhood and communal norms for Papua New Guineans, but is also useful, as shown by this singsing and the slumlord in the previous section, for quite selfish reasons. Boundary plants constitute a particularly important component of these tangled relationships because they signify and mediate the overlaps among relational plants, human bodies, and places in Papua New Guinea. Astrid Anderson describes how the people of Wogeo Island (near the island's north coast) talk of leaders and their knowledge as "cordyline plants standing on the beach, in the village, and in the forest" (2011, 12). This botanical expression of rootedness and my singsing experience are good examples of Barton and Denham's vegeculture concept (2018). These anecdotes show how personhood in Papua New Guinea relates to place through boundary plants. Personhood and place are, in Carrier's phrasing (1998), inclusive of one another, not separate experiences (see also Stewart and Strathern 1998).

But if cordylines are relational beings in Papua New Guinea because they contain the social exchanges that involved them, does this make them persons? Not quite. The ethnographic literature suggests that the relations expressed by cordyline are gender and group identities. Boundary plants are, like bird-of-paradise feathers and Manchester United t-shirts, conventionalized ways to objectify personal relations in Papua New Guinea (Strathern 1988, 180). Different cordyline varieties, like the *gini nagatumo* cultivar (narrow leaves striped with green, white, and pink) that women in Geremmiaka village once wore as everyday clothing and the *gini sapu* (oval green leaves with white edges) that the men used, revealed and performed these gender identities. Land matters and identity always have a "family resemblance" in Papua New Guinea because of cordyline. Geremmiaka people look for the bright yellowish-green leaves, up to 60 cm long, of the *gini gonagini* cordyline (*gona* means land, and *gini* is place, so they translated this term for me as "tanget land-place") on village corners. An elder explained the use of *gini gonagini*:

> If one is planted here, the next one can be on a far away hilltop as long as you can see from one to the other. It is the same as a surveyor's peg! We don't grow this one in a garden or use it for fencing or singsing, only for the village boundaries.

Gini inoveya (similar to *gini gonagini*, but with shorter and broader leaves) shows garden boundaries between families within the village. *Gini gundulisa* has lanceolate leaves with a bright green core and wide red stripes on the edges, and marks family territory to show which garden belongs to which brother. Among the ten other tangets in Geremmiaka, two red varieties ritualized gender identities until the 1990s. Dark maroon leaves of *gini maguri*, beautifully greased with pig fat, decorated young women's arms when they emerged from their initiation house after their first menstruation, and male initiates dressed in bright pink *gini sinakona* leaves. Cordyline's bundle of significance links economic, social, and symbolic capital in Geremmiaka and makes land and social identity reflect one another. Cordyline mediates these relationships at the boundaries of bodies, places, and social groups, but is not a moral person in itself. Like elsewhere in Papua New Guinea, boundaries and exchanges form a never-ending tension in Geremmiaka, and never reach closure (Sillitoe 1999, 334). The village's cordylines are a map of its social relations, and like people, every plant has the potential

Figure 5.5 Woman in Geremmiaka holding a *gini nagatumo* cordyline, used for
women's everyday clothing

to enter new relationships. Making a dance costume creates exchange relationships, because villagers always ask before picking someone's cordyline
leaves (Fig. 5.5). No one in Geremmiaka would refuse such a request, the
villagers explained, because those leaves would eventually lead to other rec-
iprocities and favors.

Cordyline is a *sine qua non* of group identity, especially in the Highlands.
When I asked focus groups to identify which cordyline cultivars they considered most powerful, they consistently pointed to the "land tangets" because
they were "about clan politics." Papua New Guinean clans and lineages
are contingent, flexible, and inconsistent groupings, not durable rule-based
structures that generate social action. The narrative performances and strategic struggles embedded in boundary plant rituals construct social groups
that are always becoming. Rappaport says that Tsembaga Maring men
become clansmen by planting a cordyline together on a territorial boundary, not because of kinship (1984, 170). The resulting territorial networks
can determine marriage practices. In Akemaku village near Goroka, elders
explained that the long green lanceolate *gini eyerena* leaves on their clan
boundaries show "where men's business is." When I asked what business

this land tanget did, they said that "it tells the men that they have land, and marks the boundary that women cross to marry!" Papua New Guinean clans and lineages are therefore gendered processes of territorialization and resource access. Men's groupings include ancestors, and cordyline makes the ancestors participants in current social life. The land tangets (*ini makini*) in Keiya village "show where the ancestors lived," and serve as records of past exchanges and alliances. "This one has a very big work!" my focus group declared:

> If a man is jealous or angry and wants to kill someone, a third man will bring a piece of this tanget to the victim, who plants it at home. This prevents the murder and the jealousy is finished because it shows the hands of the ancestors. This prevents conflict and all the compensations that it would involve. If a man sees the land tanget at the house of the person making him angry, he knows that he can do nothing.

This, the elders explained, is how the ancestors limit conflict and keep the territorial group together. Applying Wolf's subdivisions of power (1999) to this statement, these elders are using a boundary plant to convert the structural power of ancestors to institutional power in order to prevent the disruptions of violent interpersonal power.

Social histories of agency, community composition, and accumulated wealth in people are condensed and materialized by these boundary plants. For the Ankave-Anga people on Papua New Guinea's southern coast, ancestors are almost physically present in living cordyline because these plants retain their touch. Pierre Lemmonier stresses that when male initiates have their septums pierced by bone awls, the secret "vegetal theater" for the forest ritual is made from red cordyline stalks (2012, 85). These plants refer to the blood-soaked red clay where an ancestor died, but because cordyline is a vegetatively propagating clone, for the Ankave-Anga these red cordylines are the *same plants* that their ancestors touched, not representations. Marilyn Strathern explains that from a New Guinean perspective, vegetative propagation demonstrates new plantings recycling the life present in the parent plants, and are not "new life" (2020, 65). This vegecultural ontology of plant robusticity means that Ankave-Anga initiates make direct physical contact with their ancestors because their current boundary plants include those older relationships. The cuttings for the ritual come from male elders' gardens, and after the initiation each adult takes home a cordyline for replanting, thereby circulating the physical substance of the ancestors between the ritual and domestic spaces (Bonnemere 1998, 120). These ritual attractors are not abstracted symbols *of* the ancestors; they are tangible relationships *with* the ancestors. At a landscape level, cordyline creates meaningfully thick maps of male territorial groups and their business by embodying the ancestors' touch.

The social relation for which Papua New Guinea is famous in anthropology is Big Men. These are people whose wisdom and skills in oratory, persuasion, and coordination allow them to accumulate networks of followers and clients. They earn "achieved status" in egalitarian societies with few opportunities to inherit it. Big Men compete to organize elaborate gift presentations, known in anthropology as *moka* from Andrew Strathern's ethnography of Melpa economics (1971), and to redistribute the gifts they receive. In the Mt. Hagen area, each grassy park-like moka ground is like a ledger of transactions because its rows of cordyline indicate exchange histories. Robert Foster's description of the Enga people illustrates how Big Men use cordyline to communicate their achieved status. When a man organizes a large gift, he plants a cordyline at the end of the line of pig stakes to show how he has "encompassed" his father's status by giving more pigs than he had achieved. Cordyline fastens and fixes a Big Man's new identity as a bigger man, and the Enga call this "leaping over the father's stakes" (1985, 193), but this status requires reinforcement. Enga Big Men prepare their own graves with cordyline to strengthen their claims of surpassing their elders. More recently, elaborate roadside graves with painted wooden or metal fences and concrete headstones are replacing these cordyline-marked graves in remote uninhabited areas for the Enga (Gibbs 2016). These new and non-vegetative signs of permanent status allow wealthy Big Man families to display cultural capital and proclaim their achieved status publicly.

Big Men also send messages with cordyline. Andrew Strathern provides the text of a moka speech in which the orator stated, "the people around us may think the moka is not yet, but I have sent a knotted cordyline leaf as a message straight to them that it is on now" (1971, 240). The knotted leaves are not as peaceful as Kilimanjaro's similar knots of dracaena. Strathern describes how Big Men magically poison their rivals, and get proof of the assassin's success in a cordyline leaf from the victim's clothing. My Melpa-speaking focus group in Rabiamul agreed, and added that a knotted leaf of *polka kumbulka* (a green cordyline with a reddish tinge on the leaves' midribs) was a Big Man's signal to enemies that he was planning murder. In all of these examples, cordyline communicates social organization like the legend on a map.

Cordyline is much more than a land marker in Papua New Guinea. It is what Latour would call a privileged actant in a network of relational and overlapping dividuals. It affixes those relations onto specially costumed bodies and the landscape of villages and gardens by asserting and solidifying gender/age categories, territorial group identities, ancestral authority, and achieved status. People know who they are, to a degree, by how they relate to these plants, which in turn contain multiple relationships. In the Papua New Guinean relational vegeculture, these boundary plants get personified while people get "plantified," but without plants fully becoming persons or people becoming plants. The next section extends these ideas by examining cordyline's implications for aesthetics, place-making, prosperity, growth, and protection.

Beauty, place, and order

I arrived in Papua New Guinea thinking that the primary use of boundary plants was for property rights, and that their aesthetic and ornamental uses were secondary. I thought that cordyline was about land, and that details like the accordion-folded leaves of a well-made singsing costume were cultural mystifications of resource access. I left convinced that this reductive approach could not account for the complexity and diversity of the Papua New Guinean meanings of cordyline. Aesthetic values are so thoroughly interwoven with political economy and social organization that a purely materialist approach would sandpaper away the texture of cultural difference. I learned from focus group discussions about the beauty of territorial boundaries and the pleasure of decorating oneself with cordyline leaves. In Akemaku village, elders showed me how *gini gutilise*, a red cordyline with two-centimeter-wide leaves, about 30 cm long, with a light green stripe down the center, creates a particularly attractive dance costume. One man explained,

> This beautiful one is used for love magic, the magic man talks to the *gini* before the singsing, and folds the leaves like this to decorate the man before the singsing, and the bouncing *gini* makes the women fight for that man! This is because of the way it bounces, the women, lots of women fight! This leads to conflict among the men and usually some compensation. It is grown in every garden. But it is a dangerous *gini* because men will ask for money and pigs when the women fight for that man. This happens especially when a romantic love match is denied by the arranged marriage of his girlfriend to another man, then he uses this *gini* to get her back for a little bit.

Is my interlocutor talking about beauty, sex, love, social organization, or land use? His narrative is about all of these at once. Cordyline is, like bird-of-paradise feathers and kina shells, part of a Papua New Guinean aesthetic that celebrates the beauty of inclusive relationships that do not reduce to self-interested individuals seeking profit (Sillitoe 1988; Strathern 1988, 341; West 2005). One focus group participant summarized, "the general meaning of tanget is beauty, then organizing boundaries. But a well-organized boundary is beautiful, too, so it's all one thing." The ethnographic challenge of beauty is that it is contextual. There is not a singular unified idea of beauty, botanical or otherwise, across Papua New Guinea. This section therefore surveys the area's mosaic of beauty to argue that cordyline is a thread that binds together a regional aesthetic. The island is full of beautiful places and place-ful beauty, and describing this requires a political ecology of delight.

In Tok Pisin, the most common term for a skirt of cordyline leaves is *arse gras*. For ceremonial occasions, these leaves are folded and greased so that the crinkled leaves glisten with every movement. Ethnographers often side-step the ribald aspects of this term in English by calling it a "rear-covering" (Strathern 1971, 37), "buttocks covering" (Healey 1990, 68), or simply "ornamental" (Rappaport 1984, 186). Understanding what arse gras means requires closer attention to indigenous semiotics. The term for this clothing in some Papuan languages is "tail," specifically a bird's tail (Heider 1970, 243; Sillitoe 1988), which suggests that it signals social intent like a bird's colorful mating display. Christin Kocher Schmid's "botanical ethnography" of the Yopno people in eastern Papua New Guinea offers a deeper interpretation of cordyline's ethno-aesthetics. Kocher Schmid asked what makes particular plants so excellent for decorating bodies and villages. Their key characteristic, she was told, is "swaying" (1991, 256). Plants with long drooping stems that respond to the slightest breeze are able to attract, contain, and redirect a basic force of nature, *gisam*, which involves vitality, well-being, and harmony. My focus groups agreed; cordyline leaves, whether held in the hand or worn as clothing, accentuate a dancer's movements, and not just visually – their swishing rustling sound is also aesthetically pleasing. Life is at its most beautiful when it moves, they asserted (Fig. 5.6). These leaves are not only attractive, they clearly assert boundaries, even when worn as clothing. When I asked one Big Man why these attractive ritual attractors appear on both land and bodies, he grinned and responded that the shared meaning is "do not touch something attractive without permission!" An aesthetically pleasing boundary, swaying with vitality, is also a guarded invitation to exchange relationships for a dividual with relational moral personhood.

As the social histories of African dracaena showed, place-making with boundary plants involves narrative performances and strategic struggles.

Figure 5.6 Dancing with *gini okumaku* cordylines, Amemaku

In Papua New Guinea, these occur on both landscapes and bodies because the body is also a cultivated place, composed of sedimented experiences and relationships. Setha Low uses the term "embodied space" to suggest why walking around a city feels different in Serbia or Costa Rica (2016). In Papua New Guinea, "spatialized bodies" experience overlaps of landscape and personhood when people decorate themselves with the plants that also demonstrate territorial claims. Gender practices and ideologies make bodies spatialized differently, and cordyline primarily spatializes male bodies in Papua New Guinea. In Rappaport's case study, the red cordyline that Tsembaga men plant to end conflict is *yu min rumbim*, the "men's souls cordyline." Women can never touch this boundary plant full of masculine identity and social action (1984, 19, 149). In Paiakona village, I asked who plants the green and pink *polka kaya djira* that women use for singsing costumes. The focus group participants agreed that "it's a man's job, even though it's for women to use." The women in Waghi Barel said that they must ask a man before collecting leaves for a costume. When a resource access institution decorates people, social organization becomes explicit and aesthetically significant. Bodies become gendered places partly through boundary plants, and territorial boundaries reflect a gendered division of labor and access. This is the opposite of symbolic abstraction. Instead of landscape generating intangible ideas like sacredness and beauty, Papua New Guineans use cordyline to make abstractions like gender relations, social history, and cultural values into tangible and visible characteristics of land and bodies (Wagner 1986). A dancer's carefully folded cordyline leaves, glistening with pork fat, are part of a narrative about embodied and emplaced cultural histories. Without this feminist attention to multispecies intersectionality, Papua New Guinean people-plant relationships cannot come into focus (Petitt 2022).

Examples abound in Papua New Guinean mythology and oral history. The *Wantok* newspaper published over one thousand *stori tumbuna* ("ancestor stories") between 1972 and 1997, and Thomas Slone compiled and translated them into English (2001). 36 of these stories mention tanget, and illustrate what wearing this particular plant means in Papua New Guinea. In many stories, cordyline summarizes masculine beauty, prowess, and efficacy. In "Kep killed Kumasi," a cannibalistic wild man dies after his victim's husband uproots the beautiful red tanget growing next to his house. In "A Son was Helped by his Father's Ghost," a man forgot his cordyline buttock covering while bathing at a river, and his enemies tried to kill him by burning the leaves. In two stories, "A Man Lost a Star Woman" and "A Man Grabbed a Ghost Woman from a Hole," women control men traveling far away with magical strings tied to their arse gras cordyline leaves. What these stories about spatialized bodies share is a relationship between a ritual attractor and the seat of masculine agency.

Embodied places in Papua New Guinea use tanget to show this relationship in reverse. Graves turn bodies into places, and the ethnographic literature on Papua New Guinea is full of references to cordyline-marked graves (Busse 2005, 444; Lohmann 2005, 189; Lutkehaus 1995, 97). But what is to be done when there is no body? This is an increasingly important question now that many Papua New Guineans are labor migrants. A woman from Chimbu Province explained one solution to me;

> If a Chimbu person dies outside, the spirit escapes from the body, so when the family gets there to collect the body they call the spirit at the site of the death, then they hear a voice in the bush answer "yes," they cut any sort of tanget near that place to bring the spirit back. If they do not get the tanget together with the body, the parents must go to that location, even to Lae, to collect the spirit, and they collect a tanget at that death place, return to the grave at home in Chimbu and put the tanget at the grave to bring the spirit back to its *tok ples* ["language-place," home area]. This is usually for an accidental death or a murder. The family must transfer a stalk or a whole rooted tanget along with the body if they can, or any other shrubs and soil and stones from that place. This returns the spirit to its home place and peace. If you don't do this at the time of the death, the spirit stays at the place of its accident complaining that no one has come to "pick me up and take me back home" and appearing to various people. Then the people of that place must send a message to the family in Chimbu that the spirit needs collection.

These boundary plants move relationships back where they belong and make them fixed points on a group's territory (see also Aufenanger 1961, 400, 403). Places become bodies and bodies become places through cordyline in Papua New Guinea because the plant's message of economic and social order easily extends to cosmological and metaphysical order and back.

The material, social, and metaphysical entanglements of cordyline in Papua New Guinea exemplify Tim Ingold's vision of life as a tangled "meshwork" composed of the lines and trails along which entities move (2015). In Papua New Guinea these strings are stretchy, and continuously knotted and untangled. Little is permanent and structural in island social life, but cordyline conveys permanence and implies structure. Abstractions about positive and orderly relationships, like protection, prosperity, growth, fertility, and harmony, take material form in this boundary plant, which allows land and life to be manipulated, like when cordyline boosts crop growth (Fortune 1932, 112; Leach 2004, 106). In Midima village, the *gumane kondaugl* cordyline (20 cm leaves, bright pink with dark green at the center) promotes sweet potato production in fields with neat grids of drainage ditches. "We put one stalk for the whole garden, usually at the water outlet. Even now people do this to ensure a big harvest," they said. When I asked why the *gumane* needed to be placed at the drain of the plot, the focus group hypothesized that the plant plugs up the fertility and vitality of the plot while still letting the water flow. In other areas, cordyline marks objects and places that guarantee social growth and prosperity, like the sacred stones that guard community health in New Britain (Panoff and Panoff 1972, 383). The plant can temporarily contain the life-force and vitality of Maring warriors (Rappaport 1984, 132) and ensure that this fertility flows from land to people and back again (LiPuma 1988, 70). It prevents thieves from disrupting these flows (Sillitoe 1983, 128).

The theme that connects these webs of significance is the extension of the human capacity to protect, control, and manage the good things in life. A Goroka man told me that people were stealing from his garden. He hired a ritual specialist to protect it with some "green but tough tangets;"

> We went to my area, and we put them on the four corners with her magic words. Now this protects the farm, because anyone who steals by force, they will experience a problem in life, their pigs die, children sick… so a thief sees the tanget and chooses to steal from a different farm!

The inverse of this logic encourages exchange relationships. In Matibo and Barum villages near Madang, an accordion-folded cordyline leaf hanging from a string is a sign of welcome. The Barum focus group explained that a crinkled maroon-green leaf of *momorr titip* means "you will give me money, you will give me a pig, and it makes the heart of the giver become easy. We welcome you, give us something! If you see it you must give!" My seat for the Barum singsing had a folded leaf hanging over it, and my heart did indeed feel easy.

Cordyline's land protection extends to protection from witchcraft (in Tok Pisin, *sangoma*) in many Papua New Guinea communities. In Waghi Barel, the cordylines (of any sort) at a yard's gate ensure that "anyone who comes must be known, and anyone with bad motives stays out!" In Papua New Guinea the Tok Pisin idiom for these malevolent forces in Tok Pisin is the Poison Man.

> The Poison Man comes to put something in your food and water. Sangoma comes because of bad mind and jealousy. So we plant *gini opu* [a dark purple-black cordyline] with ginger to prevent jealousy and witchcraft by putting it at the gate and around the fence.

One Akemaku villager sought me out after a focus group to elaborate on the offensive and defensive characteristics of his greenish-yellow *gini sapu* cordyline:

> When a person is backbiting me, someone can use *gini sapu* to destroy me. When you have differences with people, or a witch is affecting you, you must say words to the *gini sapu* and plant it around the house.

> After you plant it, you speak at the *gini sapu*, and when the witch comes to destroy you with poison, the *gini sapu* will keep you safe. It will make the power of the witch reflect back to him, and so after all you will see that you are safe from risk. This happened to me! We use this one to protect ourselves from witchcraft or poison. In my house I am safe because of the *gini sapu*. In my life I have experienced this so I know it works, it keeping me safe many times in my life.

In Midima, the purple-black variety prevents the Poison Man from excavating corpses. Other villages used other cordylines, usually red or pink, for various mundane protective purposes, such as marking dangerous snake dens or boulders about to trigger a rockslide. Finally, in Lufa village, purple tangets act as security guards;

> If you have cordyline at your house and the Poison Man comes, they will see the *busa kugupeta* as a human figure standing there so they will not bother you. That's why you put it at your gate or in a good spot where a person would stand outside of the house.

This was the only time I heard Papua New Guineans personify cordyline so clearly. When I described dracaena's vegetative gaze on Kilimanjaro, my focus groups responded that cordyline makes people watch each other's behavior, but the plant does not do the watching.

The colorful variety of cordyline meanings in Papua New Guinea refuses easy categorization as representing aesthetics, gender hierarchies, masculine power, or mystical protection. What connects these meanings into a regional meshwork is the way that cordyline makes bodies belong to places and places parts of bodies. Papua New Guinean cordylines are relations and places through which personhood occurs, not moral persons in themselves. And for many Papua New Guineans, embodied places and emplaced bodies are beautiful.

Conclusion

Summarizing these rich agricultural, social, and metaphysical aspects of vegetative institutions is inevitably reductive because cordyline's significance depends on how its many varieties hang suspended in different webs of cultural meaning at each locality. The same dark red cultivar that connotes danger in the Highlands is a peace symbol at the coast. On Kranget Island, near Madang, a landowner explained that

> When visitors come, you take this red *mazazoz dagan* cordyline, crinkle it with little folds, and hang it by your house from a tree as a flag of welcome. Sometimes you can splash a bit of the white lime powder used for betel nut onto the leaf. It means "you are welcome to come inside, we

have peace." When you have this one in your yard, people say, "he has tanget, he has a proper house, he does not have poison!"

Back in Mt. Hagen, the same leaf would signal violence. The common thread running through this diversity is that cordyline fixes human agency and group identity by materializing them as places on bodies and in landscapes. For example, people in Waghi Barel village reported planting green *go akay* cordylines (also used for land boundaries) when they give up alcohol or tobacco. The boundary plant fixes the power of the ancestors (an ongoing living relationship, not an abstraction) to someone's intentions, making a binding and territorialized promise. Fixed agency and identity explain why Keiya village's foundation in the 1960s required the first settlers to "plant the *ini makini* to make this our place." Throughout my transect, I asked each focus group how the Tok Pisin word *ples* related to cordyline, and if this relationship was missing in the English term "place." All responded with statements like what I heard in Akemaku: that the multivocal relational social life of cordyline connotes *"ples* what it mean to us, not whiteman say place." Much like the ways that dracaena taps into and directs vitality in Oku and Kilimanjaro, in Papua New Guinea cordyline is part of a suite of botanical and ritual technologies of power for manipulating life force. This life force is often conceptualized in Papua New Guinea as a material substance pervading all living, moving things, and is culturally defined as *min* (LiPuma 1988; Rappaport 1984), "grease" (Stewart and Strathern 2001), or *auna* (West 2012, 116). Cordyline contains and directs this life force into social struggles and narrative performances of personal status, group identity, and moral personhood.

The tangled meshwork of cordyline in Papua New Guinea is simplifying and reducing, like in the African case studies of dracaena. Belonging, containment, and protection have new meanings and locations in Papua New Guinea today. One index of this change is the linguistic shift from local names for cordyline to the Tok Pisin term tanget. Older people, especially those born before Papua New Guinea's independence in 1975, know the names and uses of many cordylines. Younger folks just say tanget. Elders agreed that the trend is from cordyline having multiple significances to just two: organizing land tenure and performing group identity. Both meanings relate to struggles for order and structure in a postcolonial tenurial morass and dazzlingly diverse polycultural mixture. This process is partly a matter of the expansion of cultural scale; what had been intensely local and "closed" matters are now national and "open." At the internationally famous Goroka Show (a polyethnic annual singsing), elaborate face-painting, bird-of-paradise headdresses, and tall wigs made of human hair all express external group relations and internal social organization. Traditions and customs are being re-invented in new economic, political, and cultural contexts. Cordyline has, along with kina shells and cuscus skins, become a standardized way to communicate across cultural difference. It is also becoming a marker of cultural authenticity for wholly invented traditions. Cultural tourism is becoming increasingly popular, many Papua New Guinean communities are re-creating customs to form tourist attractions.

Keiya village stopped initiating boys in the 1950s but revived the ritual ordeal in 2012. Before my focus group session, I trekked to a waterfall with several men. We removed our shirts (when female tourists come, one man noted, "we pretend they are men who can wear shirts here") and stood up to our knees in a pool. Our leader thrust long tree thorns up a younger man's sinuses until he bled, and then did his own. Another man stopped taking photos with his cell phone long enough to make a tiny arrow from a shard of freshly broken beer bottle and shoot it with a little bow into our oldest member's tongue four or five times. The three men slowly dripped blood into the clear cool water, holding their chest and arm muscles rigid, while the ritual expert explained that their great-grandfathers had used these ordeals to prepare boys for marriage. The men who bled wore crinkled skirts of dark green *ini akumaku* leaves because cordyline was "needed for this sort of work," and "tourists don't want to take pictures of us in t-shirts and shorts." In this case, cordyline put a patina of legitimacy onto a wholly new purpose for a traditional ritual. When I asked if any of them had endured this rite of passage themselves outside of a tourist context, they exclaimed, "of course not! We're Seventh Day Adventists!" Their experience of the messy ambiguity and contradiction of enacting social change by expressing cultural continuity is dripping with both blood and the structural power of cordyline.

The multilayered significance of cordyline in Papua New Guinea as an intersectional multi-species relationship and a gendered technology of power do not form a coherent and neatly bounded social-ecological system. The overall scenario is an open and variable meshwork of overlapping land-use systems, shreds of social organization, and patches of cultural meaning, with ongoing innovation, experimentation, contradiction, and invention of tradition. Several Highlands focus groups described how new varieties of cordyline had arrived recently from coastal Papua New Guinea and reported that they were studying these new tangets to see what they could do. And much like cordyline is now becoming a popular ornamental plant in both Kilimanjaro and Oku, dracaena is now in Papua New Guinea. It is conspicuous along the fence lines of coffee farms along the Highlands Highway. These all-green and green/white striped exotics are called "flower tangets" to classify them as purely decorative cordylines. In Geremmiaka village, dracaena is "more like a strong flower for your boundary line than a boundary in itself." In other communities, this exotic is acquiring economic meanings. The Midima focus group called its green dracaena "market tanget" and explained that a leaf at the bottom of a basket of vegetables has magic to make the produce look fresher and attract more customers. The people of Akemaku disagreed, citing rumors that the plant steals money, harms a business, and leads to poor financial management. People in Keiya village concurred, saying that dracaena "takes money from your pocket," even though its woody stalk is stronger than cordyline and makes a better fence post. Finally, one participant reported that women plant dracaena to prevent divorce in Goroka, prompting some husbands to uproot the cuttings. Clearly, the cultural elaboration of vegetative propagation is an ongoing process in Papua New Guinea's relational and socially horizontal vegeculture.

The next chapter examines a partially parallel – yet culturally and historically distinctive – process in eastern Remote Oceania. In this strikingly different vertical vegeculture, cordyline relates to rigid and hierarchical precolonial social-ecological systems and postcolonial identities. If the story of cordyline in Papua New Guinea can be summarized as being about continuity and exchange, the corresponding narrative of this boundary plant far to the east is about transformation and inequality.

Notes

1. The term for cordyline in Tok Pisin is sometimes written as *tanket*. When I asked my research assistants whether they wanted to be "Team Tanget" or "Team Tanket," they chose the first. This chapter uses tanget consistently, even when citing sources that use tanket.
2. These texts typically mention cordyline in descriptions of a community's physical layout, like Waddell's Enga houses "concealed behind hedges of cordyline" (1972, 46). For more brief mentions of cordyline, see Crook (1999, 241), Godelier (1986, 5), and Podolefsky (1987, 592).
3. This table summarizes these locations in my transect along the Highlands Highway and near Madang.

Village	Language	Province
Paiakona	Melpa	Western Highlands
Rabiamul	Melpa	Western Highlands
Waghi Barel	Jiwaka	Jiwaka
Midima	Chimbu	Chimbu
Gerimmiaka	Tokano Asaro	Eastern Highlands
Akemaku	Tokano	Eastern Highlands
Keiya	Unggai	Eastern Highlands
Lufa	Huva	Eastern Highlands
Ramu Sugar	Tok Pisin, English	Madang
Barum	Dami	Madang
Matibo	Bel	Madang

Figure 5.7 Focus group languages

6 French Polynesia

Rank and Revitalization in the Society Islands

The village of Papeto'ai is on Moorea Island in French Polynesia, about 16 kilometers west of Tahiti. The cluster of cement-block multi-family homes lies on the narrow coastal strip between the blue waters of Opunohu Bay and the steep slopes of jagged Mount Tautuapae. I met a traditional dance group preparing green *auti* cordyline leaves as costumes for that evening's "spectacle" at the Hilton Hotel. The lead dancer, Stephan, explained that he officiates "traditional Polynesian weddings" for tourists. He puts four auti branches on the beach to create a space full of *mana* (sacred power), binds the newlyweds' hands with an auti leaf, and breaks a coconut over their linked hands. "Anything that is traditional must have auti to create the place," he insisted, "even when Polynesians are in France, we use auti to make a place." One dancer, Tiare, was making a skirt of green auti leaves on a belt of white pandanus strips. She exclaimed, "for me, it's only a costume!" Her grandmother quickly corrected the teenager, saying, "no, it brings peace to both land and body because it chases away bad spirits." Responding to both, Stephan explained that

> The auti is a symbol of mana because during a firewalking ceremony, you must have auti to give the mana, and if you do not follow the instructions you will be burned! I did the firewalk with auti, and the rocks were glowing red!

This chapter explores this discourse and these social lives of cordyline in Remote Oceania. It emphasizes ethnographic material from French Polynesia's Society Islands to show how this boundary plant has mediated ecological, social, and symbolic transformations.[1]

French Polynesian garden centers and roadside plant shops offer a variety of cordylines, but only two are recognized as indigenous: green *auti ma'ohi* and red *auti uteute*. One cultural expert explained that only these two are traditional and significant, while the rest are merely decorative plants

DOI: 10.4324/9781003356462-6

imported from Hawai'i in the 1990s. There are three varieties of auti ma'ohi in the Society Islands:

> *Auti ma'ohi uta* ("cordyline from the mountainous interior"): The 35–55 cm long dark green leaves have a bit of dark greenish-black on the underside of their midribs and a thin red line around their perimeter. The Society Islands' most common cordyline.
>
> *Auti ma'ohi raro* ("leeward or downward cordyline"): Light yellowish green leaves, 30–50 cm long, pointed ends. Associated with the Marquesas Islands.
>
> *Auti ma'ohi tapa* ("bark cloth cordyline"): Green leaves with yellow stripes, grown for use in Marquesas-style dance costumes.

Unless indicated otherwise, all references to auti and auti ma'ohi in this chapter are green auti ma'ohi uta.[2]

This chapter roughly parallels the previous chapter to facilitate comparison. It is based on 24 interviews over three weeks on the islands of Tahiti, Moorea, and Huahine. I begin by revisiting the vegeculture concept before analyzing the precolonial Society Islands as a ranked social-ecological system. The third section examines how ritual and ceremonial spaces mediated land use. The fourth part reviews cordyline's role in European contact and colonization, and the fifth finds this plant marking new boundaries in recent decades. The final section looks at the plant's entanglements with sacred power today. Overall, I argue that in contrast to Near Oceania, cordyline in this part of Remote Oceania has been a mechanism and idiom of change instead of an icon of continuity.

Vegeculture and social ranking in Remote Oceania

Seafaring Austronesian-speakers left Taiwan about 5200 years ago and built large settlements on beaches or with stilt houses over shallow lagoons throughout Near Oceania (Kirch 2000a, 98). Their Lapita pottery, named after an archaeological site in New Caledonia, has distinctive geometric tooth-shaped designs stamped into the clay, and is found from the northern coast of Papua New Guinea to Samoa (Kirch 1997a). The portable Austronesian social-ecological system included mainland Southeast Asian domesticated animals like pigs, chickens, and dogs, and Near Oceanic "canoe plants" like taro, yams, and bananas. The "out-of-Taiwan" settlement model (Bellwood and Dizon 2005) does not mean that a unified racial group left Taiwan en masse, carrying certain cultural traits, and reproduced the same system throughout Oceania, as race-based diffusionist models once argued (Riesenfeld 1950; Smith 1915). Instead the model that best explains the archaeological, linguistic, and genetic data is Roger Green's "triple-I" vision of intrusion, innovation, and integration (2000; Kirch 2010, 137).

This model accounts for the intrusion of Southeast Asian material culture, language, and genes into the Bismarck Archipelago, technological innovations like new types of stone adzes, and the integration of New Guinean crops into the Austronesian farming repertoire between 1500 and 1200 BC. These ancient interactions may explain how Southeast Asian cordyline became so well established in Near Oceania, although the plant may have been indigenous to New Guinea prior to Papuan settlement. Intriguingly, the triple-I model illuminates ancient Austronesian social organization. Genetics show that most Remote Oceanic peoples' mitochondrial DNA is from Asia, and nearly two-thirds of their Y-chromosomes are from New Guinea (Kayser 2010). This suggests that the Austronesians were matrilineal and matrilocal, meaning that settlements were organized around networks of mothers, daughters, and sisters, while men married into these communities and specialized in the long-distance trade of prestige goods like pottery, feathers, and shells (Jordan et al. 2009; Marck 2008). Some of the canoe plants which made further expansion to the east possible probably came from exchanges between pockets of Austronesian-speakers on the northern coast of New Guinea and Papuan islanders through affinal kin linkages during this period (Allen 2013). Cordyline was likely one of these innovations, but the direction of trade is unclear.

In just a few hundred years, the Austronesians expanded from Near Oceania into the western part of Remote Oceania to reach Fiji, Tonga, and Samoa by 1000 BC (Fig. 5.1). Expansion stopped until the first millennium AD, when a new period of exploration and colonization brought their descendants to Hawai'i, New Zealand, and Easter Island. For archaeologists, one hallmark of these arrivals is the presence of charred cordyline roots (Hather 2013; Parkes 1997; Prebble 2008). During the "long pause" near Samoa, these speakers of Proto Eastern Polynesian languages became increasingly sophisticated in long-distance navigation and canoe technology (Kirch 2000a, 232; Kirch and Kahn 2007, 199, Spriggs 2015). Sailing east usually meant sailing into the wind, although El Niño events occasionally reversed this pattern and pushed canoes to the east. Westward voyages also shaped Oceanic settlement patterns and agriculture. Seafarers reached coastal South America and returned west with American sweet potatoes after 1200 AD (Green 2005), allowing the settlement of the New Guinean highlands described in the previous chapter. Some eastward voyages probably failed and resulted in uncontrolled drift voyages back west. This may account for the settlement of New Zealand and the scattering of "Polynesian outliers" (linguistic/cultural/genetic isolates) in western Remote Oceania and Near Oceania, like Nukumanu and Tikopia (Ziegler 2002). Like the Bantu expansion described in Chapter 4, the settlement of the Pacific is best conceptualized as a series of recursive movements instead of a single arrow to the east. The presence of cordyline in the eastern Pacific, even in places as remote as Easter Island (Hinkle 2005, 42), testifies to this great achievement in human migration.

But why would the ancient Austronesians and the precolonial voyagers of Remote Oceania include a marginal food like cordyline in their botanical toolkit? How did this plant merit space in canoes packed with the people and equipment needed to colonize new lands? The material affordances of cordyline, like its leaves' durability and its roots' edibility (if baked long enough), certainly made the plant useful to these ancient mariners, but was it just these physical properties that placed it alongside coconut, breadfruit, taro, banana, sugarcane, and pandanus? We cannot know precisely what motivated the settlers of Remote Oceania to include cordyline, but the plant's regional coherence of economic, social, and cosmological meaning suggests diffusion from Near Oceania, not independent inventions. The primary difference in the social history of cordyline on islands like Tahiti and Hawai'i is that it became an important part of intensely hierarchical social-ecological systems in contrast to the "flatter," less centralized systems of Near Oceania. Applying the Near Oceanic vegeculture concept to Remote Oceanic kingdoms generates new insights. The vegeculture of Papua New Guinea expressed the permanence of the ancestors' social relations through the botanical property of vegetative propagation. The vegeculture of cordyline in Remote Oceania also communicates continuity, but also refers to social rank, elite place-making, and postcolonial identity politics. Cordyline remains a ritual attractor and a "fixed point" (Fox 2006; Meyer and Geschiere 1999) in Remote Oceania, but it attracts different rituals because of the area's trajectory through precolonial, colonial and postcolonial history.[3]

The Society Islands as social-ecological systems

Eighteenth-century Europeans saw Remote Oceanic islands as tropical utopias of ecological abundance, social harmony, and sexual freedom (Salmond 2009). But they were not the pristine Edens demanded by the colonial imagination (Grove 1995). They were delicate and "disharmonic" ecosystems characterized by low plant and animal diversity and high species endemism (Gillespie et al. 2008; Kirch and Kahn 2007, 203). This biogeographic pattern results from the relative youth of many volcanic islands and coral atolls and their isolation from continental landmasses. Human colonization generally caused rapid deforestation and the decimation of bird populations. The subsequent demographic booms induced rapid ecological change, but this varied according to an island's geology and soil type.

Some of the resulting social-ecological systems, like Tikopia, became sustainable as the islanders replaced indigenous vegetation with "food forests" from their stock of canoe plants. Others, like Mangaia, became so environmentally impoverished and socially violent that warfare was for acquiring human bodies to manure taro fields (Kirch 1997b). Remote Oceania's environmental history cautions us against viewing these islands as stable steps on an evolutionary ladder toward social and ecological order. Instead, much like environmental anthropologists and historians have redefined Amazonian forests (Heckenberger et al. 2003) and intensively cultivated African farmlands (Widgren and Sutton 2004) as specialized and nonlinear dynamic systems, Remote Oceanic islands increasingly appear to be the variable and complex results of a standard toolkit of plants, animals, and institutions interacting with differing and isolated ecosystems. The social significance of cordyline throughout the region is part of the flux of both nature and society resulting from these ancient "transported landscapes" (Anderson 2020).

The modern landscapes of French Polynesia reflect both long-term settlement and colonial disruption (Figs. 6.1 and 6.2). In Tahiti, for example, the current distributions of bamboo groves (*Schizostachyum glaucifolium*) and candlenut forests (*Aleurites moluccannus*), correspond to ancient Ma'ohi settlement sites (Larrue et al. 2010). What tourists view as natural landscapes are actually consequences of these precolonial dynamics and the colonial period's demographic collapse. In Tahiti, the part of French Polynesia with the best demographic data, war, disease, and alcohol slashed the population by 85% between 1772 and 1807 (Robineau and Rallu 1990, 82). This catastrophe obscures the precolonial social-ecological system. Many ancient residences, public infrastructure, and sacred sites are now hidden by secondary forest on Huahine (Sinoto 1996) and in the Marquesas (Donaldson 2019). A local historian on Moorea told me that finding precolonial sites in the forest is now difficult because after missionaries forbade the use of cordyline to

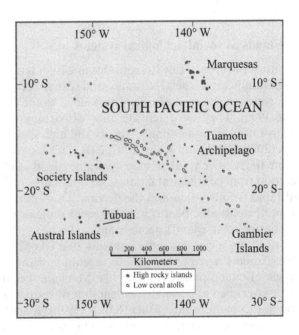

Figure 6.1 French Polynesia

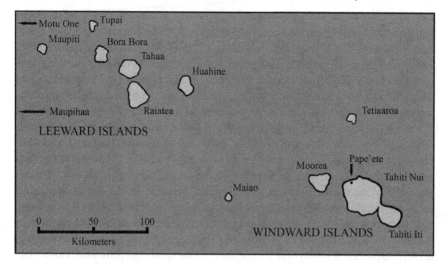

Figure 6.2 The Society Islands

mark those places, the growth of a dense canopy killed these light-loving plants. Understanding cordyline's usage in Ma'ohi agriculture is also difficult because the fields left fallow by the population crisis are precisely those that Europeans made into plantations. Even Charles Darwin mistakenly saw the forested interior of Tahiti as a natural landscape in 1835, when his hiking companions gave him a meal of bananas and baked cordyline tubers that were "as sweet as treacle." He thought that these were wild foods indicating tropical abundance, not the vegetative footprints of a demographic collapse (1915, 405). Recovering cordyline's place in precolonial Remote Oceanic farming systems is a matter of bringing many disparate threads together to re-weave ancient meshworks.

Cordyline may have been the first canoe plant that the navigators of Remote Oceania used to settle new islands. The ethnobotanist Paul Pétard recounts that the seafarers' first action upon making landfall at an uninhabited island was to plant cordyline cuttings to thank Maui, the god of travel, for a successful voyage (1986, 108). These pioneers established agricultural systems based on vegetatively propagated root crops, particularly taro (*Colocasia* spp.) and yams (*Dioscorea* spp.), and largely managed through shifting cultivation. These horticultural systems gradually became intensive in Remote Oceania, with higher labor inputs and increased yields from continuous cropping on modified landscapes. These intensive systems generally combined three strategies; irrigated taro in "pondfields," dryland rotations of taro, yams, sweet potatoes, and bananas, and starch storage systems based on breadfruit (Kirch 1994, 5).

Historical information about the Society Islands' precolonial social-ecological systems is scarce because European explorers and missionaries saw agriculture as unimportant. The farming practices and foodways the least linked to indigenous social organization and religion were those that best survived the European transformation of Ma'ohi society (Lavondès 1990, 72; Oliver 1974, 852). Dana Lepofsky reconstructed this farming system based on historical clues and archaeological remains. It consisted of 47 plant taxa grown in six major subsystems; house gardens, nursery gardens for paper mulberry (*Broussonetia papyrifera*) and kava (*Piper methysticum*), ornamental gardens of sweet-smelling flowers, tree plantations of breadfruit and coconut, dry short-fallow fields of yams and banana, and taro on irrigated terraces or in swampy areas (1999; 2003). Historical records and oral tradition do not particularly associate cordyline with any of these specialized agricultural areas. On Hawai'i, farmers in the Kona area built food-producing windbreaks from lines of stone and cordyline (Kirch 1996, 102; Menzies 1920, 75). We know that cordyline grew around their houses (Henry 1928, 220), and it is possible that pragmatic precolonial Ma'ohi farmers also used cordyline to get food from field margins and yards, but direct evidence is lacking.

Even if we cannot know precisely where, cordyline was a common feature of Remote Oceanic landscapes partly because it was so extraordinarily useful. Hawaiians (and probably precolonial Ma'ohi) made rope, sandals, rain capes, and even whistles from its tough leaves (Abernethy et al. 1983). In the Society Islands, the basic pragmatic value of auti leaves was probably food security. Cordyline or banana leaves lined pits where ripe breadfruit fermented for at least one week. Anaerobic conditions and high fermentation temperatures prevented bacterial growth, and the resulting cheese-like *mahi* paste could be stored in these pits for years (Pollock 1984; Ragone 1991). Cordyline leaves were long-lasting and did not damage the mahi's flavor. A mahi pit 1.2 meters in diameter and 1 meter deep could feed a typical family for an entire year (Lavondès 1990, 60). Baked cordyline tubers were eaten in Hawai'i (Portlock 1789, 38) and "counted very good food" in the Society Islands (Parkinson 1794, 38). It was probably only a secondary food because of the long cooking time required (Wilson and Wilson 1799, 359). Finally, the plant's capacity to vegetatively propagate from cuttings and spread from rhizomes meant that a little labor could turn a gulley into a famine reserve (Hinkle 2005, 15). Cordyline was not just a canoe plant useful for food and fiber; it also reduced the long-term risks of settling new islands.

Ranked chiefdoms grew atop these settlement dynamics and are staple case studies of social complexity in anthropology textbooks. Their elaborate status systems – involving dramatic expressions of cultural capital like the feathered cloaks, royal surfing areas, and prostrating commoners of ancient Hawai'i – are excellent material for comparing non-Western hierarchies to capitalist Euro-American modernity. The question that drives many scholars is what stratified these social systems. Did hierarchies result from systems of symbols that made social differences meaningful and valid? Or did emergent elites manipulate resource allocation, apply force, and only then legitimize their privileges symbolically? Boundary plants stand at the intersection of these symbolic and materialist approaches and can perhaps offer some pointers toward their synthesis.

How do chiefs come to power? The general model of indigenous Remote Oceanic political economies describes wedge-like island segments extending from mountain peaks to coral reefs and controlled by tribute-collecting chiefs (Ward and Kingdon 1995a). But was this a top-down command and control structure? Marshall Sahlins argues that the chiefdoms of Remote Oceania formed around cultural meanings, not political-economic domination (1985). Sahlins suggests that precolonial Hawaiian chiefs were sacred because they were focal points of "mytho-praxis" that shaped how Hawaiian society revolved around cosmology and aesthetics. From this culturalist perspective, boundary plants are components of systems of meaning, and every fluttering cordyline leaf is an emotionally rich performance of cosmic significance. A materialist point of view asserts that universal patterns of political economy drive historical change. This idea propels Timothy Earle and Matthew Spriggs' review of Marxian approaches to Remote Oceanic hierarchies (2015; see also Earle 1997). They argue that leaders gradually accumulated power and authority by controlling flows of materials, commodities, and labor, and that cosmological narratives legitimized the resulting inequalities instead of causing them. A Marxian take on cordyline in Remote Oceania would see it as an institutional bottleneck through which islanders channeled resources, regulated food surpluses, and allocated labor. The story of cordyline in Remote Oceanic hierarchies does not fit neatly into either view, but these approaches provide windows through which we can gaze (however imperfectly) on precolonial phenomena.

In the early 20th century, scholarly accounts of Society Islands social organization measured regional hierarchies with European yardsticks. They described Ma'ohi organization into three economic classes. The *ari'i* were tribute-receiving royals, and subdivided into the *ari'i-maro-'ura*, the ultimate sovereigns who wore *'ura* feather girdles, and the lesser *ari'i-ri'i* nobility. The *ra'atira* were landowning elites without royal blood, and the majority of *manahune* commoners were landless except for tiny kitchen gardens (Handy 1930, 42; Henry 1928, 229). Realizing that these economic strata were also closed marriage networks, subsequent scholars revised this model with a more culturally inflected approach. Douglas Oliver (1974, 749) shows how these rankings depended on symbolic status displays and endogamous marriage rules, not economics. This reconstruction is tenuous because explorers' accounts, oral histories, and missionary documents are often contradictory, ambiguous, and superficial sources. The general picture of Ma'ohi land matters is that the ra'atira gentry determined the manahune commoners' access to land, but that farmers controlled their own labor. The monarchs could only confiscate and reallocate land in times of war. This was a social-ecological system organized by tributary exchange and scarce labor, not one shaped by subsistence and land scarcity

(Coppenrath 2003, 17). Food, tapa cloth, and canoes flowed up the hierarchy, and were quickly redistributed to the chiefs' favorites (Ellis 1829, 374).

Archaeologists are using the "House Society model" to reevaluate the material record of precolonial Remote Oceania (Kahn 2007; Kahn and Kirch 2013).[4] Lévi-Strauss (1982) developed this idea to fill the evolutionary niche between egalitarian kinship-based societies (like those in Papua New Guinea) and hierarchical chiefdoms in the Pacific. He proposed that in a House Society, the House is a corporate moral person with a material and a symbolic estate. The status of an elite group, like the British House of Windsor, replaces kinship or class consciousness as the social glue binding ranked groups together. Subsequent scholars shifted the emphasis from an explanatory social type to an interpretive tool for understanding social processes (Carsten and Hugh-Jones 1995; Gillespie 2000). This approach illuminates the meanings of boundary plants in precolonial Remote Oceania because Ma'ohi physical houses extend beyond domestic architecture to include residential yards and sacred spaces. Homes were 'utuafare, open-air living spaces surrounded by stone walls, wooden fences, and boundary plants, with separate buildings where residential groups ate, cooked, and slept (Orliac 1990, 36). Each House was a group, a space, and a symbolic estate in which physical houses (*fare*) were just components. Gender intersected with these Houses in important ways. Gardening was primarily a male concern, while women produced *tapa* cloth (Lavondès 1990; Lavondès and Charleux 1990). Consumption was also deeply gendered. Men and women maintained separate cooking areas and *umu* earth ovens, young men cooked for older kinsmen, and particular high-status foods (like turtles and tuna) were restricted to male palates. Overall, ranked Ma'ohi Houses were intersectional systems of inequality that shaped the social production of space and the cultural construction of precolonial subjectivity and personhood.

This ranked social-ecological system was never a static structure, and was certainly not functional for the manahune commoners. It was a struggle over landscape, privilege, and meaning within a long-term process of island settlement and demographic expansion. The signs of social stress prior to European contact and disruption include widespread infanticide (about two-thirds of all infants, Oliver 1974, 424), increasing population pressure on resources, widening social stratification, and deepening conflict. A new social category, the *arioi* religious specialists, was spreading from Raiatea to other islands with a message about the war god 'Oro replacing local agriculturally-oriented deities. This consolidated the ari'i nobles' institutional and structural power and allowed non-ari'i to extract and consume agricultural surpluses. European contact led to new inequalities in what was already a complex ranked system. By 1767, six chiefdoms ruled Tahiti through an uneasy balance of power predicated on marriage alliances and 'Oro worship (de Deckker 1990, 25). The arrival of European prestige goods, weapons, and institutions spurred the centralization of power under the Pomare dynasty (1788–1880), and the 19th-century political economy increasingly turned toward plantation agriculture, whaling, and provisioning the emergent European-dominated Pacific trading system (Newell 2010; Toullelan 1990). Christianization, urbanization, education, and wage labor became the major determinants of Ma'ohi society instead of indigenous values and meanings (Toullelan et al. 1990). Anti-nuclear activism, tourism, and a distinctively Ma'ohi politics of indigeneity are the most recent entanglements of a meshwork that was never the static paradise imagined by post-impressionistic painters and honeymooners (Kahn 2011; Saura 2013). Cordyline punctuates this larger story of Ma'ohi continuity and change, but its significance in the Ma'ohi "ranked vegeculture" does not reduce to either cosmological meanings or structures of inequality.

Boundary plants and monuments to hierarchy

If the social significance of cordyline in the Pacific was a simple matter of diffusion from Near Oceania, we could expect this web of meaning to resemble the New Guinean pattern. It would feature networks of relatively equal men negotiating land claims, making space into place, and directing cosmic forces of vitality. Cordyline did all of these things in precolonial Remote Oceania, but in the context of stratified House Societies which expressed status through a particular form of monumental architecture now known as *marae* (Fig. 6.3). The plant defined the perimeters of these ritual attractors and contained the symbolic estate (name and reputation) of Ma'ohi Houses (Henry 1928, 220).

Indigenous land tenure in the Society Islands is now thoroughly obscured by colonial changes in land management and social institutions. The data on

Figure 6.3 Marae Ahu-o-Mahine, Moorea, with auti ma'ohi uta on the left

pre-contact divisions of land and labor are very thin because the European and Ma'ohi elites did not regard gardening and food production as topics for historical records like diaries, ships' logs, and reports (Lepofsky 1994, 4). Much like this book's other case studies, Ma'ohi land tenure was a social process and a web of relationships rather than a structure of rules and roles. Production, exchange, and consumption flowed through a system of tribute. *Manahune* commoners and some *ra'atira* landowners worked the land and presented its produce to upper-level landowners and chiefly *ari'i* families. This tributary political economy did not, as far as the ethnohistorical data reveals, rely on boundary plants to inscribe social boundaries in the agricultural landscape. Boundary markers instead included streams, hills, ridgelines, boulders, cairns, rock walls, and *ti'i* images of stone, wood, or coral – but not cordyline. Boundaries were probably the idiosyncratic knowledge of land managers, not conventionalized symbols (Donaldson 2016, 46; Ellis 1829, 362).[5] The ti'i images were more performative signals of rank than economic institutions, as James Morrison noted in his journal in 1792:

> They have carved wooden images of men that they call *etee* set up as boundaries of their estates ... to remind passengers below and of equal rank to the possessor and owner of that land to strip the clothes off their shoulders and heads as they pass by – in compliment to the owner.
>
> (2010, 182)

Instead of demonstrating individual and group property rights, cordyline marked institutional limits, especially at the marae and dance grounds where groups constructed and performed status. Compared to the direct use of cordyline to etch social relations into the Papua New Guinea landscape, the Ma'ohi pattern was more indirect because it materialized ranked Houses instead of claims to land.

An account from the 1797 visit to Tahiti by the ship *Duff* explains this landscape. James Wilson wanted to estimate the island's total population (but note that at this time it had declined by about 75% since European contact in 1767) and asked a royal servant to help him. The man replied that the best way was to count the number of *matteyna* (*mata'eina'a*) in an area, estimate the number of ti'i for each, and finally reckon the number of people per ti'i. Wilson asked him to define these units. The servant responded that a matteyna was "a principal house, distinguished either by a degree of rank in its ancient or present owner," and that each matteyna

> sets up a *tee* [ti'i] at the *morai* [marae], which entitles it to the liberty of worshipping there; and the other houses in the department of the matteyna claim a part of the same privilege, and are thence called *tees*.
>
> (Wilson and Wilson 1799, 184)

The descriptions of these mata'eina'a neighborhoods in the historical record are ambiguous and variable (Oliver 1974, 969). It would stretch the already thin evidence to declare these residential units the equivalent of the Houses in the House Society model. Wilson's description does, however, suggest that landholding and territory were primarily matters of rank legitimized by access to sacredness, not "property rights" in the Western institutional sense.

The marae of the Society Islands are ceremonial spaces with stone altar platforms (*ahu*) made of limestone or basalt slabs at the end of level rectangular courtyards (*paepae*) paved with coral or rounded stream cobbles, sometimes enclosed by *aua teni* walls of vertical limestone or coral slabs (Emory 1933, 14). Most were "surrounded with the most majestic trees" like *Casuarina equisetifolia* and *Thespesia populnea*, whose whistling leaves were perceived as the voices of the gods (Moerenhout 1993, 236). The relatively uniform marae of the Leeward Islands (Huahine, Raiatea, and Bora Bora) are coastal and are less architecturally complex than in the Windward Islands (Tahiti and Moorea), where marae are more structurally variable, more likely to have wall enclosures, and more often built inland. None date from the first centuries of Ma'ohi settlement (AD 600–1200), and only became prevalent and complex as Ma'ohi society became increasingly hierarchical (Wallin 1993, 105).

Archaeologists look at the emergence of Ma'ohi monumental architecture as signaling the development of social complexity, with parallels from Stonehenge to ancient Mesoamerica. In brief, when archaeologists find large non-residential permanent stone structures, this indicates a complex division of labor within the groups who planned, constructed, and used architecture that was clearly unnecessary for survival. Many marae contained the bones of ancestors. They were sites for prayer, sacrifice, and other interactions with gods, spirits, and ancestors. They expressed territory and group identity indirectly by acting as "guardians of the threshold" at the boundaries of *mata'eina'a* settlements (Garanger 1964, 10).

Teuira Henry was a Tahitian scholar who reconstructed her missionary grandfather's lost manuscript (original ca. 1848) to produce the classic ethnohistorical text *Ancient Tahiti* (1928). She outlines six categories of Ma'ohi marae by degree of social and political complexity (1928, 119). The first three are large-scale and public; the second three are more localized and private:

1 International marae used by chiefdoms from multiple islands
2 National marae signifying the authority of an *ari'i* chief or king over an island
3 Local marae in a district (*fenua*) or valley where residents of a *mata'eina'a* worshipped
4 Family or ancestral marae (*marae tupuna*) for domestic rituals, to which "were attached the hereditary names of the family, without

which they would have no proof of their ownership of the land"
(Henry 1928, 141)

5 Social marae belonging to devotees of particular gods and allowing
  strangers with the same affiliations to build networks
6 Specialist marae for healers, canoe craftsmen, stonemasons, etc.

Marae were often destroyed and rebuilt after a war or changed to reflect new
political alliances. The marae at Papeto'ai at the beginning of this chapter
was once the family marae Tepuatia, but rebuilt as a national marae after
its chief formed an alliance with a leader from Raiatea. Henry's list shows
that marae were indexed to social rank and essentially functioned as political and economic machines for converting economic relations into social
and cultural capital (and vice versa) via ritual performances that legitimized
those accumulations of wealth and people. Henry's account suggests that all
Ma'ohi knew and experienced their social position and rights to resources
via marae rituals, but we know little about non-elite uses and perspectives
on these matters of structural power. It is clear that sacredness (*ra'a* in
most Ma'ohi languages, expressing the flow of *mana*, power and efficacy)
was unequally distributed. Manahune Houses, commoner subjectivity, and
their struggles over rank do not appear in the ethnohistorical record, so it is
unclear exactly how the family marae related to land tenure despite Henry's
description of their close association. Marae were also explicitly gendered
because all but the highest-ranking women were banned from entering
them. The marae did not simply reflect social organization, but were primary mechanisms through which Ma'ohi crafted social and spiritual places
out of physical space. As Robineau and Garanger summarize this struc-
turing dynamic, "marae were by the end of the 18th century the essential
'determinants' of ancient, pre-European Tahitian society" (1990, 72). Marae
legitimized both Houses and inter-island politics in the indigenous dialectic of political economy and meaning. In Eric Hirsch's terms, the marae
were sacred landscape features that forcefully dramatized the insertion of
the cultural background of ranked Houses into the quotidian foreground of
Ma'ohi life (1995). This is why "the greatest pride of an inhabitant of Otahite
is to have a grand Marai" (Banks 1896, 104).

Cordyline enters this story as a boundary plant marking these places
as special, sacred, and unquestionable. Henry's account of establishing a
national marae (1928, 131) explains that the first step of marae construction
was planting a branch of cordyline near a royal house in order to declare that
the district's food production would be devoted to the project. This branch,
the *ti patia* ("cordyline stuck in") indicated that all vegetable produce, livestock, and fish would be devoted to feeding the district's adult male population for the two to three years of construction. Henry does not explain how
Ma'ohi made lower-ranking marae, beyond the fact that all marae began
from a "seed stone" containing the sacred power from another marae. Her
list of the 13 varieties of cordyline in precolonial Tahiti suggests that the

large green variety now called auti ma'ohi uta defined the borders of all marae types: "*Ti-'uti*. A fine variety, formerly regarded as the most sacred of all ti, and so planted chiefly in the marae enclosures for the gods and religious uses" (1928, 37).[6] According to one ethnohistorian on Moorea, this green perimeter of cordyline was called the *mutura'a* and demonstrated the transition from mundane to sacred space. The marae were polymarcating monuments to social rank that used cordyline as an element in a meshwork of economic, social, and cultural capital.

The marae were technologies of power for the House Society complex described above because they materialized the symbolic estates of the groups around them. Foucault might say that the marae were sites of House governmentality where the gaze of socially ranked ancestors disciplined the behavior and shaped the subjectivity of their descendants. This "marae gaze" constructed rights in people, which then determined access to resources through tributary relations. The social and geographic organization of both 'utuafare residential groups and mata'eina'a neighborhoods existed through marae and their ritual performances (Lepofsky 1994, 42; Oliver 1974, 1094; Wallin 1993, 117). At the time of European contact, Ma'ohi House Society, or at least its elite tiers, was organized by the reputation of collective moral persons into a social landscape, not just the economics of property.

Archaeologist Patrick Kirch describes how this pattern emerged. He argues that Remote Oceanic peoples developed marae (and similar structures like Hawaiian *heiau* temple platforms and Easter Island's famous *moai* statues) from a Near Oceanic template (2000b). His reconstructed architectural sequence describes an institutional shift from houses with ritual attractors (like central posts) to burial sites and then formal courtyards with upright stones. Through a series of symbolic and architectural innovations, the houses of the living became dwellings for gods and ancestors, and the ritual focus shifted to these new definitions of Houses as ranked moral persons. This model neatly accounts for the ways that social relations in the tributary ranked chiefdoms of eastern Remote Oceania repeatedly intersect in these monumental ritual attractors. Along with stone pavements and walls, cordyline defined the boundaries of these core institutions of the Ma'ohi House Society and was probably as significant as the monumental stone structures that fascinate tourists today.[7]

This was probably never a stable meshwork. The Society Islands were already convulsing with change at the time of European contact. As Teuira Henry's category of international marae suggests, cultural entrepreneurs were using local symbols of landholding and authority to forge new political bonds across the archipelago. A monotheistic religious and political system, organized around the war god 'Oro, was replacing diverse polytheistic and localized systems in the late protohistoric period (Oliver 1974, 890). As a symbol of tradition, rank, and social boundary maintenance, cordyline became a key element of this struggle over centralization in the decades before and after European contact.

The conjunctures of cordyline and colonialism

On June 19, 1767, the British ship *Dolphin*, captained by Samuel Wallis, found itself surrounded by "some hundreds of canoes" near Matavai Bay on Tahiti's northern coast (Hawkesworth 1789, 189). One of the Tahitians held up a banana plant, delivered a speech, and dropped the "man-long banana" (a sign of peace) into the sea. One brave young man climbed aboard the *Dolphin* and, to his astonishment, was immediately butted by a goat. Once everyone's peaceful intentions were clear, other Tahitians threw more banana shoots up onto the deck and came aboard. In the next days the *Dolphin* crew went ashore to get water and plant a British flag. Several exchanges of rocks and bullets preceded a robust trade of livestock and produce for trinkets and tools. The British departed on July 27, 1767. The Tahitians repurposed the flag as part of a royal *maro 'ura* red feather girdle, later worn by Pomare I as the high king of a newly unified Tahiti and Moorea in 1788 (Beaglehole 1967, 202). The Ma'ohi became globalized as part of the European world-system, and the British flag indigenized as part of the Ma'ohi status-ranking system.

This encounter forms a classic anthropological debate. Marshall Sahlins argues that the particular ways that the Tahitians received Wallis, and the Hawaiians later encountered (and eventually killed) James Cook, were parts of a distinctly Remote Oceanic performance of interactions with the accumulated sacred power (mana) of a god/chief (1985, 74). This "mytho-praxis" meant that, for the Tahitians and Hawaiians, contact with Europe was filtered through the cultural politics of sacred symbols and heroic myths. Sahlins says that Cook had to die because he represented the Hawaiian agricultural god Lono, whose ritual death and replacement by the war god Ku propelled an annual cycle of politicized myth (1985, 104). Gananath Obeyesekere (1992) counters that the idea of Pacific Islanders perceiving Europeans as gods is itself an imperial myth. This scholarly debate raises important issues about the politics and ethics of anthropological interpretation – like the issue of who has the right to speak for the colonized – and the practice of historical anthropology (Borofsky 1997; Sahlins 1995). The core questions are how history happens and how to think about it. Sahlins uses the concept of the "structure of the conjuncture" to analyze how symbolic systems shaped histories of culture contact (1985, xiv), while Obeyesekere takes a Marxian stance that the structure of the global economy (and its contradictions) is the mainspring of history. This section only wades ankle-deep into these turbulent waters to make the point that cordyline was an important component of the European conjuncture with and eventual domination of Remote Oceania.

Cordyline became part of this conjuncture because of its pragmatic values for food and fodder and its social meanings. On James Cook's second Pacific voyage in July 1774, naturalist Johann Forster reported that Vanuatu islanders greeted the sailors by waving branches of "*Dracaena terminalis*" (cordyline) and variegated croton, pouring handfuls of seawater on their heads, and repeating a term for friend (Forster 2000, 480). Cook's crew returned the same gestures the next day, using cordyline branches, and started to exchange gifts.[8] On his next voyage, during which Cook became fatally involved in Hawaiian politics and religion, the *Resolution* and *Discovery* received cooked cordyline tubers in Oahu. Surgeon's Mate William Ellis remarked that the plant grew in "almost every part" of Hawai'i (Ellis 1788, 94). Because sailors made the tubers into alcohol and the leaves were palatable to livestock, Europeans harvested such large quantities of Hawaiian cordyline over the next decades that in some areas the plant was eradicated (Kepler 1998, 216). The most notable conjuncture of cordyline with the colonization of the Pacific, however, was its role in the famous mutiny on the *Bounty*.

The discord between Captain William Bligh and Fletcher Christian is one of the greatest tales of European maritime history. The ship's mission was to acquire breadfruit (*Artocarpus altilis*) from Tahiti, which Bligh had visited with Cook in 1777. This was empire-building botany. The goal was making British sugar colonies in the Caribbean economically sustainable. Trade barriers and hostilities with revolutionary Americans and French colonies had damaged slave plantation profits. Breadfruit is a starchy perennial tree crop and promised abundant calories for enslaved people with minimal labor. The *Bounty* left England in October 1787, and after failing to enter the Pacific via Cape Horn, sailed east and arrived at Tahiti on October 26, 1788. Bligh's name is now synonymous with unnecessary discipline. He strictly supervised food rations and forced his crew to dance for several hours a day for exercise. His sharp tongue, quick temper, and micromanaging leadership style led him to berate his crew nearly constantly (Dening 1992).

The *Bounty* anchored in Tahiti's Matavai Bay for five months to prepare over 1000 breadfruit suckers and other plants (including cordyline, Henry 1928, 26) for transport to the Caribbean. Many of the crew lived ashore and formed relationships with Ma'ohi women. After the *Bounty* sailed west on April 4, 1789, tensions increased on the crowded ship. Matters came to a head on April 27, 1789, when Bligh accused his officers of stealing some of his personal stock of coconuts and put them on half rations. Early the next morning, Fletcher Christian led a mutiny. The mutineers forced Bligh and eighteen loyalists into the *Bounty*'s launch with several days' worth of food and water and basic gear like a compass, canvas, twine, and carpentry tools. Shouting "huzzah for Otaheite!" and dumping the breadfruit into the sea, the mutineers sailed to Tubuai Island (Salmond 2011, 215).

Christian probably thought Bligh would perish at sea. Instead, he accomplished a stunning feat of seamanship. Bligh used memory and guesswork to navigate 6500 kilometers to Kupang, a Dutch trading post in present-day Indonesia, in an open seven-meter-long boat. He lost only one man during the 48-day voyage. After an October 1790 court martial exonerated him, Bligh approached his patron, the famous botanist and Royal Society president Joseph Banks, to help him secure a promotion and receive a new assignment. In 1791, Bligh received new orders and returned to Tahiti with the ships *Providence* and *Assistant* to get breadfruit for the British Caribbean. From April to June 1792, Bligh's crew loaded the ships with thousands of Tahitian plants destined for St. Vincent, Jamaica, and Kew Gardens. The consequences of Bligh's success in St. Vincent are described in the next chapter.

The auti ma'ohi variety of cordyline was a minor but important material element in the structure of this conjuncture. Its sweet tubers were part of the food that Tahitians gave the English (Bligh 1792, 83). Because Tahitian farmers helped to prepare the breadfruit suckers for travel, Ma'ohi technical knowledge about cordyline leaves' usefulness made the breadfruit mission a success. Documentary records are silent on this issue, but oral histories that I collected in Tahiti, Moorea, and Huahine agreed that the fragile breadfruit suckers on both the *Bounty* and the *Providence* were wrapped in cordyline. The ancient Ma'ohi technique to prepare vegetatively propagating plants for long sea voyages was to wrap wet balls of rootstock tightly with cordyline leaves and dry banana leaves. One cultural expert told me,

> You had to be careful to prevent the sap from leaking out of the broken root, so we wrapped them with auti ma'ohi to heal the wound and cool the root and make it stay moist ... When they arrived, the ancestors would remove the dried-out auti leaves and put them in a hole next to the breadfruit rootstock. This was not a fertilizer, but so that its spiritual power would help the breadfruit grow.

Without this indigenous knowledge, Bligh's mission might have failed.

The fate of the mutineers also involved cordyline. The *Bounty* crew returned to Tahiti in June 1789 to get livestock and recruit Tahitian labor for a fortress on Tubuai. The Tubuai islanders repelled the mutineers, so they and their Ma'ohi allies returned to Tahiti on September 22, 1789. Sixteen Englishmen elected to stay on Tahiti, despite the growing skepticism among the Ma'ohi. Bligh's claim to be James Cook's son, and Christian's lie about Cook sending him back to Tahiti to fetch supplies for a settlement in what is now the Cook Islands, were both proven false when other European ships visited Tahiti. Determined to find a remote island where they could escape detection, capture, and trial by the British navy, Christian and a hardcore group of eight mutineers lured a group of Tahitians aboard the *Bounty* for dinner. They cut the anchor cable and sailed away in the night (Salmond 2011, 261), and after allowing eight middle-aged women to leave the ship on Moorea, Christian searched for refuge. Finding a reference to Pitcairn Island in Bligh's books, Christian led the mutineers, some allies from Tubuai, and their Tahitian captives to this remote and harborless island. The presence of marae and cordyline showed that Ma'ohi had once colonized it, but Pitcairn was uninhabited. Christian and his followers built temporary shelters with cordyline leaves, stripped the *Bounty* of useful material, and burned her to the waterline in January 1790 (Salmond 2011, 346). Three years later, conflicts over the Ma'ohi women (who were essentially slaves for sex and domestic labor) led to the murders of five mutineers and all of the men from Tahiti and Tubuai. One of the four surviving mutineers had experience in a distillery and started to produce alcohol from cordyline tubers in 1798 (Bligh and Christian 2001, 252). According to one of the Tahitian women, cordyline alcohol soon led to the deaths of William McCoy and Matthew Quintal, leaving Edward Young and John Adams as the only mutineers to die of natural causes (Te'ehuteatuaonoa 1829). About 50 people, many of them the descendants of the *Bounty* mutineers and their Ma'ohi captives, live on Pitcairn today.

In the subsequent decades, many Pacific islands experienced demographic collapse, social disorder, and rapid cultural change. Contact with European microbes, technology, social institutions, and ideas occurred when the Society Islands were already experiencing political and religious centralization and conflict because of the 'Oro war god cult. When Joseph Banks visited southern Tahiti in 1769, for example, he crossed a field of "numberless human bones, chiefly ribs and vertebrae" (1896, 104) from a December 1768 inter-chiefdom battle. By the 1820s, Ma'ohi social institutions, agriculture, craft production, and belief systems had all but collapsed due to population decline and the region's new status as a source for ships' provisions and a vertex of the Pacific triangular trade of sandalwood to China, tea to Australia, and weapons and alcohol to Oceania (Toullelan 1990, 89). Society Islands politics became centralized under the Pomare dynasty (five generations of monarchs 1788–1880), and Christian missionaries became the powers behind the throne. Jesus replaced 'Oro in a structurally similar formalist and rigid "missionary theocracy" (Robineau et al. 1990, 57), and churches replaced the marae, which the missionaries considered "hideous dens and dungeons of idolatry" (Tyerman and Bennet 1831, 84).

Cordyline exacerbated but did not cause this process. Remote Oceanic peoples had made a mild beer by fermenting its baked and crushed tubers, but it was William Stevenson (an English convict escaped from Australia) who taught Hawaiians how to distill this brew into *'okolehao* alcohol by repurposing the iron pots for rendering whale blubber (Kepler 1998, 214). The technique quickly spread to the Society Islands, where a missionary reported men gathering around stills made of hollowed-out stones, bamboo, and calabashes. These groups drank alcohol directly from the stills before "sinking into a state of indescribable wretchedness and often practicing the most ferocious barbarities" (Ellis 1829, 230). Contemporary observers blamed the population crashes (which had been caused primarily by disease and the disruption of indigenous production systems) on cordyline alcohol (Moerenhout 1993, 507). These evaluations certainly contain more than a bit of missionary moralizing, but they also indicate a reorientation of collective action from integrative marae rituals to less productive pursuits.

The shift from a tributary militarized aristocracy to a centralized state monarchy in the Society Islands in the early 19th century was driven by a dialectic of religion and economics. Pomare I (ruled 1788–1791, regent 1792–1803) unified several islands with European weapons and alliances, but Pomare II (ruled 1803–1821) used Christianity to consolidate institutional power. Representatives of the London Missionary Society (LMS) arrived in Tahiti in 1797 on the ship *Duff*, and in 1812 these missionaries converted Pomare II (Newell 2010, 209; Wilson and Wilson 1799). After the decisive battle of Fei Pi in Tahiti in 1815, Pomare II pardoned his enemies on condition that they convert to Christianity. LMS missionaries advised him on the rituals, behavior, and institutions characteristic of a proper Christian monarch, and on May 16, 1819 they baptized him as king. Pomare II and his LMS advisors produced a document now known as the Pomare Code to outline their plan for the king's absolute authority over a Christian civil society (Gunson 1969). The Code banned infanticide, murder, idolatry, and wandering pigs. It required monogamy and the observance of the sabbath. It also outlawed a long list of anti-social and seditious cultural practices, like acting in a suspicious manner, long hair, and tattooing.

One of the Code's provisions relates to cordyline and explains why many of my informants insisted that the "missionaries made auti forbidden." Article 8, number 24 declares "those who call themselves *rauti*" illegal (Bouge 1952, 18). This refers to oratory warfare specialists, clothed in cordyline leaves (*rauti*) and carrying bunches of auti leaves with stingray tails at their centers, who exhorted warriors to fight (Henry 1928, 299; Oliver 1974, 876). The Code also banned "night-walkers," which probably referred to the *arioi* society within the 'Oro cult. These were highly privileged ritual specialists and "sacred clowns" organized into eight orders and identifiable by their tattoos and costumes of (among other things) shredded auti leaves. During the expansion of the 'Oro religious-political system in the late protohistoric period, they moved from district to district conducting marae rituals and organizing nighttime feasts (Henry 1928, 230). Inverting these practices, Pomare II tasked his followers with destroying all marae in alcohol-fueled groups. These legal and institutional innovations reveal Pomare II as a savvy politician dedicated to undermining competing systems of power and authority. He was also a good businessman, and soon acquired his own ship for exporting salt pork to Australia (Gunson 1969, 69). The relevance of cordyline faded in the Society Islands as the Pomare Code and new markers of value and privilege became increasingly institutionalized.

Pomare II died in 1821, and the child monarch Pomare III in 1827. He was succeeded by his sister (Pomare IV, ruled 1827–1877). Christian theocracy was shaky in Tahiti in the early 1800s, not only because of competing commercial interests in luxury goods and the whaling industry, but because a cultural revitalization movement threatened the entire enterprise (Gunson and de Deckker 1990). Christian acolytes declared themselves *perofeta* (prophets) and asserted that because the millennium predicted in the biblical Book of Revelation had already begun, sin, punishment, and the laws of the Pomare Code were irrelevant. Intense Christian prayer could, they said, confer miraculous powers of healing and walking on water. These visionaries wore long beards and dressed in cordyline leaves and feathers (Gunson 1962). Their followers formed communities glued together by ritualized alcohol use, feasting, and sexuality; a pattern similar to the *arioi* worship of the war god 'Oro described above. At its height, this movement (which the English disparagingly called *Mamaia*, Tahitian for "unripe fruit, good for nothing"), emptied the LMS churches enough to make the missionaries fear state collapse.

The prophets' efforts to revitalize Ma'ohi society through syncretic millenarian Christianity failed in the 1840s because they refused to let the LMS missionaries "pierce them" with smallpox vaccinations. The epidemic spread through the Mamaia network so quickly and thoroughly that "the disease appeared to search them out for its victims" (Pritchard 1842, cited in Gunson 1962, 239). With the flame of resistance snuffed, the structure of the Ma'ohi-European conjuncture became global colonial politics. Territorial struggles between British Protestant and French Catholic missionaries in the Society Islands led to the Franco-Tahitian war (1844–1847) and France's formal annexation in 1888. In its role as a boundary plant delineating sacred spaces and social rank, cordyline generally disappears from the historical and ethnographic record until a second cultural revitalization movement began in the 1960s.

Cordyline had been a ritual attractor organizing sacred space and a technology of power in a ranked House Society, but what replaced it was bureaucratic instead of polymarcating. The missionaries tried to refocus the indigenous land system on patrilineal inheritance, ban land sales and rental to Europeans, and suppress territorial rituals at the marae. After the French government declared Tahiti a protectorate in 1842, a series of land laws gave French settlers property rights and institutionalized a new land tenure system. Each district had a land register book with the names of (mostly male) heads of households, names of the land parcels, and significant boundary marks. The procedure was supervised by a five-member panel, and the *reo Tahiti* term for land certificates, *tomite*, comes from the English word "committee." Claiming these certificates required applicants to demonstrate possession by either living on the land or belonging to the group associated with the marae on that land (Tetiarahi 1987, 54). After Pomare V abdicated the throne in 1880, Tahiti and its island dependencies became a French colony, which then required landowners to register land under French civil law. Few did, and French Polynesian land matters became a dualistic system of registered private lands and joint property lands under the *tomite* system (Coppenrath 2003; Ravault 1982). As one farmer in Moorea told me, "the settlers' pineapple, avocado, and grapefruit farms just erased all of our ways of organizing land." The result is that coastal areas have title deeds and private ownership, while interior lands (and many lagoon and reef areas) are claimed by families according to oral history, genealogy, and the documents left by the 19th-century committees. Each *fenua toto* ("family land") estate exists in a legal state of "indivision" and collective ownership, and intergenerational transfers often get legitimized by cordyline. My Huahine research assistant inherited land from his maternal grandmother, and reported that the first thing he did at the property was planting auti along its boundaries. This method for marking family land is probably more the result of the French colonial *tomite* system creatively repurposing the boundary plant institution than continuity from precolonial landholding by the *ari'i* royalty, *ra'atira* gentry, and *manahune* commoners.

Like the colonial property muddles described in the previous chapters, land matters in French Polynesia are neither indigenously collective nor colonially individualistic, but instead a messy process of claims-making in different institutional forums (Rapaport 1996). Because land sales in French Polynesia often involve getting birth and death certificates from distant family members, they have high transaction costs and are notoriously slow, despite French promotion of land markets. Throughout this convoluted process, cordyline's materialization of House Society institutions faded as statutory law became routinized in Society Islands life. A cultural expert at Fare Pote'e remembered her parents calling her back from a big green auti ma'ohi, saying, "we cannot go over that auti, it is for someone else!" They did not (and, she thinks, could not) instruct their daughter in the finer points of this boundary plant's significance. She knew that it meant property rights, and that it was associated with ancestral areas, but little else:

> When I was playing in the forest with my friends when I was a kid, I was afraid of finding an auti plant because I was afraid I'd find skulls near it! For me, auti meant "something to respect, do not go near!"

What Paul Pétard wrote in 1946 remains true today; "Tahitians are more and more leaving behind the culture of ti and contenting themselves with simply letting the shoots grow spontaneously around their homes" (Pétard et al. 1946, 204). But this disentanglement does not indicate a cultural characteristic crushed under the heel of modernity. Cordyline had been fundamentally a symbol of rank and prestige; it is now a commonplace part of Ma'ohi home ownership. The significant places for these boundary plants are houses and bodies, not marae, with new meanings of emplacement and embodiment. The long-term structure of this conjuncture shows institutional power and cultural capital, once largely the domains of *ari'i-ri'i* nobles and the *ra'atira* gentry, being captured and redeployed by the descendants of the *manahune* commoners. Cordyline is now part of a flexible social-ecological system, not a rigid hierarchical one, and this made the plant ripe source material for new cultural revitalization strategies in postcolonial French Polynesia.

Revitalized boundaries in a new society

Social life in the Society Islands today is very different from the stratified chiefdoms described in the early ethnohistorical accounts, yet concepts of indigenous authenticity continue to shape the archipelago's economy, social organization, and cultural imagination. Salaries, formal schooling, and city life became the dominant forces in the 20th century. The indigenous population was recovering from its 19th-century crash, and immigrants from East Asia and other Pacific islands were joining a French administrative elite to form a complex multiethnic society (Newbury 1980, 258). The colonial state wanted French Polynesia to be the keystone of a global maritime network (facilitated by the Panama Canal, which France began constructing in 1881); instead, the islands became an economic backwater producing copra and phosphate. Military infrastructure later pushed the colonial plantation system toward a service economy. The Americans built an airstrip on Bora Bora in 1943 as a wartime supply base, and in 1960 the French colonial state completed Tahiti's Faa'a airport. France moved its nuclear testing program from the Algerian desert to the Tuamotu Archipelago about 1200 kilometers southeast of Tahiti and completed 192 atmospheric and subterranean nuclear tests before the program ended in 1996 (Kahn 2011, 73). In the resulting economic boom, Ma'ohi islanders became wage laborers in a cash economy, land commodification increased rapidly, and an emergent middle class of *demis* (half-Ma'ohi and half-immigrant French Polynesians) gained control of coastal lands and most commerce. The islands' current economic system of tourism, services, vanilla, and cultured pearls is the context for a Ma'ohi cultural revitalization movement that uses cordyline as a key symbol of neo-traditional place-making and social reconstruction.

For much of the Society Islands' post-contact history, outsiders dominated a lopsided negotiation over place and identity. Paul Gauguin's romantic paintings and the Broadway musical (and later film) *South Pacific* entrenched 19th-century portrayals of exotic, alluring, and innocent mirror images of Euro-American modernity. The reappropriation and revitalization of Ma'ohi symbols, practices, and identities began in the 1960s as responses to the French nuclear program, anticolonial movements in French West Africa, and indigenous networking across Oceania. The leaders of these efforts were returning from education in France with a newfound critical consciousness and were seeking allies among other postcolonial peoples in New Zealand and Hawai'i (Saura 1990, 59). Although French Polynesians often found their friends' non-Christian identity politics confusing, these activists asserted deterritorialized pan-Pacific postcolonial identities instead of localized island-based identities. The term Ma'ohi, with the meanings of genuine and indigenous, found new currency as a form of transnational cultural capital. Cordyline's broad regional significance made it part of these new ocean-spanning webs of meaning. Across the Pacific, social elites and cultural entrepreneurs used ideas about landholding and rootedness to reinvent social order and cultural authenticity, despite their islands' incorporation into the world system in decidedly peripheral roles (Ward and Kingdon 1995b). In Vanuatu, for example, the Nagriamel Movement political party, which campaigns for the return of European-owned land to indigenous Vanuatuans, gets its name from cordyline (*nagaria*) and cycad palms (*namwele*) (Walter and Labot 2007, 67).

Ma'ohi revitalization occurred through formal government structures and diffuse social networks. Nationalists, anti-nuclear activists, and indigenous Protestant pastors were converging on a shared vision of sacred Ma'ohi lands, and the French Polynesian government responded with a series of cultural institutions. The Tahitian Academy (established in 1975, devoted to language and literature), the Polynesian Center for the Humanities (established in 1980, responsible for archaeology, cultural resource management, and museums), and the Territorial Office for Cultural Action (established in 1980 to promote the arts) became the official curators of Ma'ohi cultural heritage within a firmly French framework (Saura 1990). The term Ma'ohi itself took root through these institutions, and it reflects a distinctly Tahitian take on culture and politics. Tahitian ideas about indigeneity inspired a social movement throughout French Polynesia that downplayed the language, history, and culture of peripheral areas like the Marquesas and Gambier archipelagoes. An informal network of archaeologists, poets, artists, dancers and tattooists created new cultural forms in the Society Islands in the 1970s and 1980s. For these innovators, Teuira Henry's *Ancient Tahiti* (1928) was the bible of Ma'ohi tradition, and this is why today's revitalized Ma'ohi cultural forms look so much like Henry's account.

Archaeological reconstructions of Ma'ohi places and practices were enormously influential in the 1970s. Archaeologist Yosihiko Sinoto from Hawaii's Bishop Museum began excavating and restoring marae in Raiatea, Moorea, and Huahine in the 1960s, and planted auti ma'ohi uta cordylines around them just as Henry had described. The marae quickly became tourist magnets, but a different archaeological connection to Hawai'i captured the public imagination in the Society Islands. In 1976 a double-hulled canoe voyage organized by maritime archaeologist Ben Finney arrived after sailing over 4000 kilometers from Hawai'i without instruments. The *Hokule'a* demonstrated and revitalized indigenous navigation in Remote Oceania, but the ship was also buffeted by the region's cultural currents. Finney describes tensions between indigenous Hawaiian and *haole* (white) crewmembers, and how cordyline expressed this social drama:

> [some of the crewmen] took to wearing armbands, anklets and headbands made from the shiny green leaves of the ti plant (*Cordyline*) ... The effect was striking, particularly when they were aboard a similarly decorated *Hokule'a*. One day when I flew out to visit the canoe at an outer island anchorage, I found her festooned with ti leaves, looking

more like a floating Christmas tree than a voyaging canoe. *Hokule'a's mana*, that spiritual force essential to all Polynesian enterprises, must be protected, I was told.

(1979, 34)

When the canoe arrived in Pape'ete on June 4, 1976, thousands of people greeted it by waving branches of auti (Ehrlich 2000, 389). *Hokule'a* proceeded west to Taputapuatea, French Polynesia's preeminent "international" marae, on Raiatea (an island once called Havai'i and the legendary homeland of the Hawaiians). Young Ma'ohi men escorted the crew to a reception at the marae carrying staffs topped by rauti leaves (Finney 1979, 280). The ritual complete, the *Hokule'a* returned to Hawai'i, but the crew later learned that they had entered the Raiatea lagoon through the wrong gap in the reef, instead of the pass that would fulfill an ancient prophecy and lift a curse. The ship returned to French Polynesia in 1985, and after a fire-walking ritual on Moorea (during which they were protected by necklaces of woven cordyline leaves), the crew steered for Raiatea and lifted the *tapu* (Finney 1994, 108). This extraordinary story of maritime prowess exemplifies how cordyline fit into the cultural revitalization of the 1970s and 1980s, as an icon of structural power embedded in a broad regional meshwork of resignified meanings and practices.

The revitalization of Ma'ohi arts and literature shows the general pattern of postcolonial indigenous people accumulating cultural capital in narrow social fields. Henri Hiro (1944–1990) rejected his Christian and colonial upbringing soon after returning from education in France and worked to translate anticolonial ideologies into Ma'ohi terms (Saura 2013, 71). He combined antinuclear activism with Ma'ohi cosmology and language, and organized his poetry, films, and political activism around themes of indigenous landholding (Kahn 2011, 22). Bobby Holcomb (1947–1991), a Hawaiian-American immigrant to Huahine, became a leading French Polynesian painter and musician. He enthusiastically adopted a Ma'ohi identity, wore tapa cloth, tattoos, and a crown of flowers, sang pop music in *reo Tahiti*, and tended the restored marae at Maeva, Huahine (Holcomb and Leimbach 1992; Saura and Levy 2013). Dorothy Levy, Bobby's friend, biographer, and curator of the Fare Pote'e cultural center, told me that Bobby was a *hotu pa'inu*, "a fruit that washes up on the beach and takes root." Both Hiro and Holcomb were inspired by Henry's rich description of Ma'ohi cosmology in *Ancient Tahiti*, both revitalized Ma'ohi symbols in their own lives, and both were buried in graves marked by green auti ma'ohi cordyline plants. In the decades since their deaths, Society Islanders' graves have increasingly been marked by cordyline, usually the red *auti uteute*, because Christian churches no longer oppose the practice of using the plant to "conduct the soul to the spirit world" (Henry 1928, 291).

Dance and tattoo followed a similar pattern of synthetic innovation and postcolonial reinvented tradition. Both were banned by the 1819 Pomare

Code, and both returned in the 1980s as mixtures of symbols and practices. For many decades, the Society Islands' major cultural event was Pape'ete's annual *Tiurai* ("July") parade celebrating Bastille Day. Dance groups had begun performing for tourists in high-end hotels in the 1960s, but tattoos were still considered signs of immoral personhood. Two dancers revived the art of tattoo in the 1980s. The Marquesan dancer "Teve" Tuhipua and Tavana Salmon, a Tahitian-Norwegian *demi* and dance impresario, were inspired by the Samoan dancers in Salmon's group. They studied an illustrated account of Marquesan tattoo (von den Steinen 1925) to design Teve's full-body tattoo. They recruited artists from Samoa, where tattoo was still practiced, and secured support from the Tahitian Academy and other government institutions. By 1983, about 150 Ma'ohi had been tattooed by the Samoan masters (Saura 2013, 93). Teve revealed his completed project in 1984 at the *Tiurai* dance festival, and became the personification of Ma'ohi tattoo in performances, advertising, and films (Kuwahara 2005). They used only traditional tools and techniques, so these innovators probably applied the juice of crushed cordyline leaves to "clear up the skin and bring out the tattoos" (von den Steinen 2007, 27). In 1985 the government separated the *Tiurai* festival from Bastille Day and renamed it the *Heiva i Tahiti*. It is now one of the Pacific's foremost cultural festivals, and includes canoe races, feats of strength and endurance, firewalking rituals, reenactments of marae ceremonies, and dozens of competing local and international dance troupes. Like Tiare, the young dancer from Moorea at the start of this chapter, many of these performers rely on cordyline for durable and culturally significant costuming, as a quick Google of the term "Heiva i Tahiti" reveals (Fig. 6.4). The revitalization of Ma'ohi bodily practices and symbols used institutional power, cultural capital from multiple sites in Oceania, and scholarly texts to reinvent tradition.

One of the most dramatic Oceanic revitalizations is firewalking, known in *reo Tahiti* as *umu-ti* ("cordyline oven").[9] In the precolonial period, Ma'ohi kin and mata'eina'a neighborhoods used communal earth ovens to cook breadfruit and cordyline tubers, but the early missionaries outlawed these feasts as gluttonous debauchery (Pollock 1984, 160). There are no references to Ma'ohi firewalking rituals in the early ethnohistorical literature, and it was likely a late 19th century invented tradition (Oliver 1974, 94). According to Arii-peu, a fifth-generation firewalker from Huahine, the umu-ti ritual was imported from Fiji in the 1850s (Kenn 1949, 25). Colonial administrators and missionaries describe Tahitian priests leading acolytes and tourists across beds of red-hot rounded lava cobbles (Young 1925; Lang 1899; Langley and Lang 1901). What had been an agricultural ritual in western Remote Oceania became a tourist spectacle and a strategy for converting indigenous cultural capital into a marketable commodity (Pigliasco 2010). The colonial government incorporated firewalking into an annual government-sponsored cultural festival in 1958 (Stevenson 1990, 267), and in the 1980s self-taught archaeologist, entrepreneurial "traditional high priest,"

Figure 6.4 Choosing leaves for a dance costume

and cultural expert Raymond Graffe began organizing umu-ti events within the government's neo-traditional institutional framework (Saura 2013, 110).

This revitalization followed the same pathway as navigation and dance. It is a sliver of the precolonial social, political, and religious ranking system that relied on close readings of 19th-century ethnography to determine the usage of cordyline leaves. Teuira Henry's descriptions (1893; 1928, 214) became the source texts for firewalking's reinvention. The umu-ti involves, she says, the goddesses Hina-nui-te-'a'ara and Te-vahine-nui-tahu-ra'i, who wore skirts and garlands of shredded cordyline leaves. Following this account, participants wear long green auti ma'ohi leaves and carry bunches of them across the hot stones. Not just any leaves will do; the priests collect cordyline leaves that seem to "float in the air," put them into a marae to "sleep," and strike the oven with bundles of leaves before leading a procession across (see also Huguenin 1902). One cultural expert in Moorea explained how priests find particularly lively cordyline leaves:

> The priest has to go through the forest talking to the plants, saying, "may you come with me, will you come and bless us?" He has to look

for leaves that move in answer to his question even though there is no wind. Finding such a leaf, he speaks to the plant, "now I am cutting this branch, you are coming with me, you will stand on one corner of my fire pit for this umu-ti." At the firewalking place near the beach, the priest puts his auti at the four corners of the fire pit to contain the gods within the sacred space. When the rocks are hot, the priest takes a branch of auti and sweeps the heat from the fire pit. Once the people cross, both they and their land are purified and the crops will then grow.

After giving a parallel account, another cultural expert remarked that cordyline's heat-controlling ability extends beyond the oven, so that a garden with auti makes a home fireproof.

These examples demonstrate how cordyline became re-entangled in a colonial and postcolonial project that wove together government institutions, cultural entrepreneurs, and old ethnographic documents. This use of cordyline to mark new symbolic boundaries occurred as the Society Islands' political economy was rapidly shifting to a service economy based on tourism. Anthony Cohen argues that "as the structural bases of [a community's] boundary become blurred, so the symbolic bases are strengthened through 'flourishes and decorations,' 'aesthetic fields,' and so forth" (1985, 44). This helps to explain why cordyline, as an ancient icon of structural power, was regularly deployed in the Ma'ohi revitalization. The results are ambiguous. One man on Huahine told me, "we know that auti is important, but we don't know why." This boundary plant had indicated the edges of sacred spaces that expressed social rank, but now cordyline shows the boundary between tradition and modernity in a new polyethnic and postcolonial context. Local meanings have become enmeshed in regional practices that demonstrate a new indigenous moral personhood. Cordyline had been emplaced in the landscape and on Ma'ohi bodies as part of a localized system of rank but is now more an embodied institution that performs a broader pan-Oceanic identity.

Decentralized protection and power

The ethnography of plants and people across the Pacific is full of references to the power of cordyline to manage risky places and help people experiencing difficulty. One elderly woman that Anya Hinkle interviewed in Moorea always puts a few auti leaves in her luggage when traveling to protect her from theft and sorcery (2005, 36). University of Hawai'i football fans wave cordyline leaves to cheer on their Rainbow Warriors (Lewis 2019). In Tikopia, a cordyline leaf necklace provides protection from illness and misfortune (Firth 2012, 182). In New Caledonia, the son of a Protestant missionary was once welcomed with yams and the explanation that "we were in a state of disarray and extreme misery when your father came with the Word. He is the cordyline planted in our country!" (Leenhardt 1946). What unites these twigs of vegetative significance is the regional emphasis on cordyline as a ritual attractor with the power to maintain a particular cultural order. What is striking about the social meanings of cordyline in contemporary French Polynesia is that the plant has moved from collective spaces in a hierarchical social system to become a commonplace component of anyone's well-maintained house and healthy body. It is now a ritual attractor full of cultural capital and accessible to anyone, not a carefully circumscribed privilege. This boundary plant once again marks sacred sites because of the revitalization dynamics described above, but the locations of structural power and symbolic order are now much more decentralized.

Cordyline marks spaces and bodies as places of power. When asked to explain what the plant does at marae, along pathways in an island's forested interior, or in a dancer's costume, my informants responded that it directs and controls mana. During an interview with an elderly woman from Raiatea who had married a Moorea man, I tried to grasp the mana in auti. Holding up a blade of grass and an auti ma'ohi leaf, I asked which had more mana. Her answer showed how mana is a variable and socially active force:

> The grass doesn't have much mana beyond its own life, but the auti has much more. I know it has more mana because I know how to use it and I feel the extra life from the auti in my own life.

Scholars often gloss mana as equivalent to the English term "power." Anthropologists used mana in the sense of "impersonal spiritual power" to construct models of the universal evolution of religion from diffuse animism to institutionalized orthodoxies about specific gods. This decontextualization of mana in service of the comparative method constitutes a "pervasive translation error" because it transformed indigenous verbs about socially potent and efficacious actions into a Western concept of power as a physical property to be generated, transmitted, or used up like electricity (Keesing 1984). Douglas Oliver failed to find many references to mana in his survey of contact period ethnohistorical documents (1974, 68), so it is possible that the term was not commonly used in the 18th and early 19th centuries. And given that recent decades' Ma'ohi cultural revitalization relied on old ethnographic texts, it is possible that current Ma'ohi discourse about cordyline's mana represents reappropriated academic discussion, rather than direct historical continuity. In recent scholarship, mana is being reconceptualized as an ethical and aesthetic relation instead of a jural-political structure or an indigenous intellectual error (Crépeau and Laugrand 2017; Tomlinson and Tengan 2016).

Nevertheless, when my Ma'ohi friends looked at marae surrounded by neat lines of green cordyline, they stressed that the plants "keep the mana of the gods well contained" and protects the people outside. Mana is ambivalent, not good or bad, unlike the unambiguous goodness of Chagga *mbora* and Oku *keyoi kejungha*. Because Ma'ohi consider the marae fearsome ancestral spaces instead of parts of their daily lives, those cordyline are, like the marae themselves, mixed and muted symbols of strength, vitality, conflict, and anxiety (Donaldson 2018; 2019). In this way, the mana of the marae and their green perimeters expresses a complex history of precolonial place-making, protohistorical religious and political violence, colonial repression, and postcolonial revitalization. The Fare Hape cultural center in Tahiti's central Papeno'o Valley is a good example of this dynamic. This was a densely settled area in the precolonial period, but the Tahitians who survived the early 19th-century demographic collapse abandoned their rich farmland because missionaries pressured them to live at the coast. When the government began evaluating the island's hydroelectric potential in the 1980s, Tahitians organized to protect the Anapua marae at the valley center from flooding. The dam project was scrapped, and in 1994 an activist network started the Haururu cultural NGO to reinvent the ancient site as an educational center, with areas for devoted to Ma'ohi medicine, language, environmental knowledge, genealogy, and spiritual practice. When I visited in 2015, the Anapua marae was surrounded by tall green cordylines. I saw a petroglyph-covered boulder ringed with auti ma'ohi and grave sites marked with red auti uteute. Several of the marae's standing stones were wrapped with cords of braided auti and seashell necklaces; one had a pearl necklace as well. "Auti here is a sign of the gods and harmony," explained my guide. Other Tahitians who I met in Pape'ete were less sure about Anapua. "Can a Christian even go to a place with so much mana safely?" one wondered.

My questions about cordyline's mana got clearer and more enthusiastic responses in discussions of how the plant's leaves mark the boundaries of dancing bodies. Several Tahitian dance experts emphasized that the most important part of the cordyline plant in dance costumes is the *aumu*, the tender and soft new shoot of a leaf growing from the end of a stalk. It has the most life and vitality. When I visited the Tiki Village performance center and tourist wedding facility on Moorea, the lead dancer told me that auti ma'ohi leaves "help contain and control the strong mana of the dance." Stephan, the dancer introduced at the beginning of this chapter, suggested that although auti ma'ohi is aesthetically pleasing, "we do not decorate our bodies with auti to look beautiful, we do this to keep the power of the dancing inside!" An elderly lady, also on Moorea, when reminiscing about her favorite cordyline costumes, said, "auti is good, it has mana, but I'm not sure how or why – it just is." The accounts that I heard about the role of auti in dance, other than occasional complaints about how the "spectacles" at the tourist hotels are more like loud and flashy European rock concerts than quiet and subtle Ma'ohi performances, fell between these assessments of cordyline as an explicit container for mana or an ambiguously positive social practice. In examining the long-term transformations of this boundary plant, I conclude that although it remains a ritual attractor in Remote

Figure 6.5 Line of auti ma'ohi on a property line

Oceania, its significance has moved away from sacred spaces and bodies onto secular and decentralized postcolonial ones.

When I asked Ma'ohi about the importance of cordyline in their lives today, the first thing that most people mentioned was its ability to protect houses from spiritual threats (Fig. 6.5). It appears throughout the region as garden hedges (Pétard 1986, 102). After I explained my project at a bed-and-breakfast in Pape'ete, my host exclaimed, "I have auti for protection from evil spirits here in my garden, just like everyone else in Tahiti!" Several homeowners reported placing stalks of auti ma'ohi in the rafters of their houses during construction for "spiritual protection." Auti can also be a completely pragmatic warning signal. Some sharp curves on Tahiti's roads and anywhere that a poisonous *hutu reva* tree (*Cerbera manghas*) grows are marked with stalks of auti ma'ohi uta cordyline. In one interview with a homeowner and her sister-in-law in Moorea, after a brief conferral the pair summarized their use of cordyline for everyday protection and purification:

> The main thing we do is take some leaves and wave them in the air if there is something bad in the house, like an emotional or psychological state, not a bad object. "Something bad" means an emotionally upset feeling, feeling like you're carrying something heavy, that makes you sick. The second thing you do is plant it everywhere you want to get

spiritual protection. A house can have even just one of these protective auti, and sometimes you bring a branch inside the house and put it in a bottle with water on the table to protect the house. Just one is enough, but if you want you can put more around.

This theme of the generalized protective capacity of cordyline runs throughout the regional literature and is a staple of Remote Oceanic ethnography (for examples from Hawai'i, see Emerson 1902, 13; Handy et al. 1972, 222; Pukui et al. 2002, 190). The spirits kept at bay by cordyline in the Society Islands are the *varua 'ino*. These are typically angry ancestral spirits sent by quarreling families to trouble one another. Stories about these spirits describe pressure and stabbing pains in the back or chest, which cordyline and other herbs can alleviate by "cleansing the house." Cordyline is now more an individual choice to decorate or protect than an assertion of collective social rank. For many Society Islanders, these meanings have shifted from where they materialized rank in a House Society at the marae to the bodies and apartment buildings of a postcolonial service- and tourism-based society. The boundary plant that had marked a system of exclusive rights to people and their tribute became an inclusive marker of homeownership, bodily autonomy, and health.

In this book's previous case studies, boundary plants were either moral persons in their own right, or tools through which people act as moral persons. In the African and Papua New Guinean chapters, these meanings intertwined with culture-specific concepts of life force and vitality to produce distinctive meshworks of landscape, social organization, and personhood. The corresponding meshwork in the Society Islands is a bit different; the ethnohistorical evidence suggests that moral personhood for precolonial islanders was within a strongly hierarchical system of ranking. In contrast, the archipelago's cordylines now relate to a revitalized, reinvented, and unranked Ma'ohi indigenous personhood. Sometimes this personhood is broad and shared by all Pacific Islanders, like in the examples of revitalized navigation and tattoo. Sometimes personhood is island-specific. These new sorts of subjectivity hybridize with Christian and French definitions of moral personhood (Saura 2013). A story from a dance instructor in Moorea about a dancer in her company illustrates these entanglements of personhood, indigeneity, and cordyline:

> There was a dancer here who was a victim of sorcery. He was acting oddly, just ignoring people that he knew. I talked to a *ta'oha* priestess about it, and she spoke to the man. He said that he had seen me in a dream, so he knew that she was coming! The priestess organized a healing ceremony. The man was in one house and I was in another several houses away. The priestess had me put four auti ma'ohi leaves and four *noni* [*Morinda citrifolia*] leaves in some water and break them one by one. Each time I broke a leaf, the man felt a stabbing pain! The priestess took the water from me and brought it to the man so he could bathe with it. The priestess was praying this whole time, and I only learned of the stabbing pains

afterwards. The man returned to himself little by little. The priestess saw the man's ancestors coming out of his body. The problem had been that these ancestors had given him a duty, and he had ignored them, so they were angry. The priestess communicated with his ancestors and found out that the man was supposed to be using some particular traditional medicine with mana that he had been instructed about when he was young, but he had given it up. After this healing he got better. The problem is that he was from Wallace and Futuna near Samoa, and working here on Moorea he was forgetting his own traditions. He was focusing too much on Tahitian dance. So he started keeping to his traditions and he got better. He is back home now, and his ancestors are happy.

This narrative of identity and healing shows how moral personhood can be measured and regulated by mana and marked by cordyline. In this case, the problem was that the man's Tahitian social practices had eclipsed those of his home area.

Cordyline had primarily marked places in the Society Islands to which people belonged, but now it is more about protecting places and bodies that belong to people. This new sense of protection is not simply a defensive mechanism to reduce risk in an uncertain social world; it is also aesthetically pleasing to many Ma'ohi. Understanding this demands a political ecology of beauty and grace to account for these aesthetic aspects of access to material, social, and symbolic resources. One woman described how she feels when she looks at a cordyline plant with this evocative sensory description:

> It is a relief, it's peaceful, a release from stress. People say you can take some auti ma'ohi and wash your face with it, or make a *lei* necklace, to refresh yourself. When the leaves are young and fresh, you can crush them to get a refreshing aroma. I made my children smell it when they were little so they would learn the smell of home.

This ethno-aesthetic and ontologically distinctive account is strikingly similar to the delight that the people of Oku find in the green vitality of dracaena and the beauty of bouncing leaves in Papua New Guinea. The green hedges of auti ma'ohi in the Society Islands are not only ways to struggle against unseen dangers, but also narrative performances of indigenous authenticity, legitimate emplacement, and a culturally nuanced sense of how to live properly with beauty and grace.

Conclusion

The European colonization of Remote Oceania was far more disruptive than the other case studies discussed in this book so far. In Tanzania and Cameroon, the institutionalization of dracaena within social systems organized by kinship, gender, and status persisted and interacted with colonial and postcolonial ways of organizing land and people. In Papua New Guinea, cordyline still connotes, albeit in new ways, social continuity, the fixedness of relational personhood, and culturally legitimate means to turn spaces into meaningful places. Social relations in the Society Islands were, in contrast, shattered and reassembled into entirely new practices and institutions. The ways that boundary plants have mediated ecological, socio-political, and symbolic transformations in these regions differ because of both indigenous cultural dynamics and the particular histories of these areas' conjunctures with the world system. Cordyline and dracaena relate to land tenure, social organization, and cosmology in these areas, but in different ways.

Following Barton and Denham's observation that Papua New Guinean people and plants share a vegeculture of entangled material properties of plants, social practice, and cosmology (2018, 23), this contrast suggests that the Society Islands' vegeculture is different for three reasons. First, cordyline and other canoe plants had formed a "transported landscape" with which ancient seafarers recreated older social-ecological practices on new islands. The resulting systems were relatively isolated compared to the interaction spheres in areas like the East African highlands, the Cameroon Grassfields, and New Guinea, at least until the late protohistoric period. Second, cordyline's precolonial significances reflect its context in status hierarchies, not reciprocal exchange networks. In social terms, the power of cordyline was "vertical" in the Society Islands compared to its more "horizontal" power in Papua New Guinea. As Society Islanders became more organized by Christianity, French citizenship, and a service-based economy, the plant's old meanings of rank faded. Third, the specific history of missionary colonialism, French nuclear weapons testing, an economy based on mass tourism, and a new postcolonial politics of indigeneity gave cordyline vastly different tasks in the Society Islands than its *botanica franca* function in largely horticultural postcolonial Papua New Guinea. Cordyline relates to protection and order in both areas because it is a ritual attractor that mediates transformations of economic, social, and cultural capital, but what is being protected and ordered is the result of different histories of cultural construction. The narrative performances and strategic struggles of place-making in Remote Oceania have distinctive social forms and cultural logic, even if the green leaves of a dancer's costume in both areas appear much the same.

This Remote Oceanic case study extends the argument that boundary plants are botanical technologies of power for producing and constructing people and place to new social fields like navigation and body art. Ma'ohi revitalization of social practices and cultural meanings has occurred, in part, through cordyline because it had been a significant marker of Houses as corporate moral persons through marae architecture and rituals. Moral persons in French Polynesia are now complex mixtures of Christian community, French nationalism, and regional indigeneity instead of parts of status-ranked Houses, and the social life of cordyline has followed this shift. As one Huahine woman explained,

> Auti is not in our lives today. Now even people in one family don't know each other. Now people are more individualistic; we used to be all together ... today you can't find many people thinking about the spirits. Even I don't think about these things very much, so the meaning of auti has reduced down to just decoration.

A man at a roadside fruit kiosk in Moorea agreed; "in the past everybody used plants, but now it's all about money instead of meaning and purpose." The colonial unraveling of cordyline's entanglements in the Society Islands has led to this boundary plant's individualization, popularization, and simplification. The difference between current reconstructed Ma'ohi practices and precolonial struggles over rank is that the new people and places are a "thinner" meshwork. As one cultural expert lamented, "now auti ma'ohi is just a flower that people grow around their houses."

The meanings of cordyline and Ma'ohi identity in domains like firewalking and dance are vibrant but narrow. These are what Miriam Kahn calls "counter-spaces" of indigenous authenticity and resistance (2011, 183). From this perspective, every time dancers wear cordyline to perform for tourists, they are engaging in a longstanding contest about personhood and legitimacy, but in a performative space compartmentalized from group status, landholding, and territory. These reinventions, revivals, and transformations suggest that unlike its role in Papua New Guinea, cordyline in the Society Islands is now more an index of change than a marker of continuity and tradition. It had been an institution for social ranking at the boundaries of marae but became a symbol of postcolonial autonomy and identity. What had been a ranked vegeculture is now more a "revitalized vegeculture" that relates more to the contradictions of postcolonial moral personhood than to precolonial social practices.

I went from Papua New Guinea to Tahiti, Moorea, and Huahine to follow the ancient Austronesian navigators from Near to Remote Oceania and construct a regional contrast in the social histories of cordyline – but this account of a boundary plant's journey is incomplete without tracing its resignification in the Eastern Caribbean. William Bligh clearly saw his breadfruit mission as contributing to the British colonial empire, but he could not have imagined the new landscapes and social histories that his introduction of cordyline, one of the Tahitian "curiosity plants" on the *Providence*, would produce in the Caribbean. The next chapter follows Bligh to St. Vincent to argue that British imperial botany interacted with a post-slavery peasantry to resignify and re-entangle cordyline with reinvented African meanings and practices. This book's African and Oceanic case studies have shown how boundary plants were key elements of two of humanity's great migrations. The following case study extends this approach to a third great movement of people and plants, the Atlantic slave trade. It shows how an Oceanic boundary plant became enmeshed in both landholding and religious expression in a postslavery society.

Notes

1. The Tahitian terms for cordyline are *ti* for the roots, *auti* for the whole plant, and *rauti* for the leaves (Merceron 1988, 205). I use the term "auti" and specify roots and leaves as necessary. *Ma'ohi* means "indigenous, native, and authentic" in *reo Tahiti*, the Tahitian language. In this chapter, I follow my informants' usage of "auti ma'ohi" to designate cordylines with large green leaves. I use the term Ma'ohi to refer to the indigenous people of the Society Islands instead of local names like *ta'ata Moorea* for the people of Moorea or the imperial term "French Polynesian." The use of Ma'ohi for all of French Polynesia's indigenous peoples hides linguistic, historical, and sociocultural diversity (Saura 2013).
2. Green dracaena (*viri tinito*, "Chinese chance") grows in many gardens in the archipelago. One cultural expert on Moorea remarked that when you crush dracaena leaves, you can smell the good luck coming.
3. French Polynesia is not politically postcolonial like this book's other case studies because France continues to govern it as a semi-autonomous "overseas country." The French president is the head of state, and France controls currency, university education, justice, and security matters. French Polynesia is culturally postcolonial in the sense that new subjectivities like Ma'ohi identity politics emerged as direct colonial rule lapsed.
4. I capitalize the term "House Society" to distinguish the ranked social unit of the House from domestic architecture.
5. The Tahitian-English dictionary by Andrews and Andrews translates *'ofa'i 'oti'a* as "boundary mark" (1944, 97), a term derived from the noun *'ofa'i*, stone, and the verb *'oti*, cut.
6. James Morrison, one of the sailors caught up in the *Bounty* mutiny, corroborates Henry's account of cordyline around marae. On Tubuai Island, where the mutineers stayed before heading back to Tahiti, the marae were flat pavements "planted with the *tee*, or sweet root, having a long stalk about six feet long and as thick as a man's finger" (2010, 64).
7. For general references on the use of cordyline to enclose sacred sites in the region, see Abbott 1992, 115 and Yen 1987 for Hawai'i, Fison 1885 for Fiji, and Leenhardt 1946 for New Caledonia.
8. For an image of this encounter, featuring a Vanuatan man holding a cordyline stalk and English men striking heroic poses, see William Hodges' painting, "The Landing at Mallicolo [Malakula], one of the New Hebrides," at https://www.rmg.co.uk/collections/objects/rmgc-object-13382
9. For archaeological analyses and ethnoarchaeological reconstructions of precolonial cordyline ovens, see Carson 2002, Cox 1982, and Fankhauser 1987.

7 St. Vincent

Dragons in a Postslavery Peasant Society

The St. Michael Spiritual Baptist Church in Fitz Hughes, on the north Leeward coast of St. Vincent, is a concrete block building with a corrugated metal roof. The missing blocks near the roof's peak form a ventilation hole shaped like a cross. The main hall has a center aisle between rows of wooden pews. Colorful flags hang from the ceiling, the floor is covered with brightly patterned plastic sheets, and the yellow walls have notices like "no eating in the church" and "women must keep their heads covered in the church" alongside biblical verses and a portrait of Jesus. The altar holds sprays of artificial flowers, white candles, a large brass bell, a golden ankh, and several books. The room's focal point is the white concrete pole with a spiraling red stripe standing at the exact center. The pole holds up three flags instead of the roof; a Union Jack, the French tricolor, and one all green. At the base of the pole facing the back of the church, a terra cotta vase holds eight pink and green cordyline leaves, next to several calabash bowls holding white candles, a small brass bell, pigeon peas, and rice. On the altar side of the plinth a calabash bowl contains five cordyline leaves arranged like a star.

I was early for the Sunday morning service, and there were only two people there, a middle-aged lady and a little girl. I explained that the pastor had invited me, and that I was interested in cordyline. "Oh!" the woman exclaimed, gesturing at the pink leaves at the base of the pole, "the red dragon speaks to us on our journeys to other lands!" I asked if it speaks with a human voice, and she replied, "Yes, it appears in front of you in the air," using her palm to mimic a leaf floating an arms-length away from her face,

> and it speaks in ordinary English. You can ask it a question and Jesus answers in the voice of the plant. We use it with the calabash with peas and rice, because these all came from Africa and are the real local things. The red dragon at the bottom of the central pole is for leadership.

DOI: 10.4324/9781003356462-7

The altar's vibrant nylon flowers were for beautification, she said, but the cordyline is spiritually important:

> The red dragon is our guide on the journey in the mourning room vision. We need this to help train our young people in the correct way to live. The red dragon is for boundary in the carnal world, and for finding the way to wisdom in the spiritual world!

Other worshippers had arrived during her explanation, and once eight women, one old man, and six children had gathered, my teacher led us in singing, "Oh that man should praise the Lord, God is goodness" (Psalm 107:31) repeatedly while another woman went around lighting candles.

This chapter fills in the gap between the 18th-century European introduction of cordyline to the Caribbean and its current economic and cultural significance as a property rights institution and a symbolic link to Africa. How did an exotic plant become an icon of authenticity and legitimacy in St. Vincent? How did the short history of plantation agriculture shape this social process, compared to the deep historical and cultural roots of boundary plants in Africa and Oceania? And to what degree do Caribbean boundary plants represent African land management and symbolic practice?

Core issues of Caribbean historiography and ethnography reverberate in these questions. How do scholars explain and interpret the region's particular mix of continuity and change? Anthropologists once confidently identified some Caribbean cultural practices, like the pole at the center of the church at Fitz Hughes, as "African traits" that had survived slavery (Herskovits and Herskovits 1947, 306). To a degree, anthropologists did not know what to make of the Caribbean. Its indigenous people had been all but exterminated, and there were precious few traditions or natives to study. Instead, it is the world's most modern and Westernized region because its landscapes and peoples resulted almost entirely from European colonialism. As Michel-Rolph Trouillot puts it, old-school anthropology preferred pre-contact cultures, and the Caribbean is "nothing but contact" (1992, 22). Sidney Mintz answered this dilemma with a series of pathbreaking works about how the region's plantations shaped Caribbean social institutions and cultural practices (1965; 1974; 1985; Mintz and Price 1976; 1992). His main idea is that enslaved people created new creole institutions within the plantation system but separate from the masters' institutions. This creolization hypothesis is about change, in contrast to the emphasis on continuity in the African traits literature (Chevannes and Besson 1996; Price 2001). More recent work avoids the false choice in these two positions by focusing on dialectics of continuity and change. Enslaved people and their peasant descendants in the Caribbean built new institutions from African cultural materials, but also by appropriating, inverting, and reinventing European institutions (Besson 2002). These creolization and peasantization processes, through which small-scale farmers created semi-autonomous economic systems, social institutions,

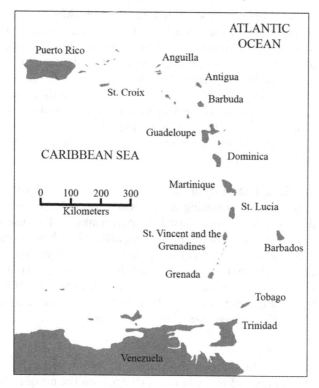

Figure 7.1 The eastern Caribbean

and cultural meanings, was shaped in part by boundary plants. This chapter uses the dialectical approach to continuity and change because although Europeans brought cordyline to the region, its subsequent history and cultural elaboration have been creole innovations and constructions.

Several themes in this scholarship are vital for understanding the social history of boundary plants in the Caribbean (Fig. 7.1). First, Mintz and his successors show that economic institutions, particularly slave plantations, determined the constraints and opportunities for cultural construction long after the end of chattel slavery in the 19th century. Second, Afro-Caribbeans actively resisted systemic injustice and continue to creatively build peasant societies and creole cultures amid desperate circumstances (Besson 1979; 1984; 2018). Third, both slave and peasant societies are better conceptualized as predicaments and dynamic processes than as structural forms (Hauser 2017). Finally, the Caribbean's fulcrums of change were a set of interrelated economic, social, and moral contradictions in landscape and society. Because feeding slaves undermined profitability, many plantations gave laborers land to grow their own food (a pattern dramatically different from food supply and land use on North American plantations). These "provision grounds" were quasi-private property belonging to people who were themselves property. This diluted white power (Mullin 1994, 127) and created a contradictory social situation; entire colonies relied on the partial autonomy and cropping choices of people who were subordinated and controlled. It also generated a moral dilemma because planter elites denied the humanity of the people they depended on (Trouillot 1988, 72).

After emancipation, provision grounds (and the yards around houses) became the economic nuclei around which Caribbean peasantries formed (Carney 2020; Trouillot 2002, 202). New Institutional Economist Douglass North uses these provision grounds as his prime example of how economic institutions form. North writes,

> to get maximum effort from the slave, the owner must devote resources to monitoring and metering a slave's output and critically applying rewards and punishments based on performance. Because there are increasing marginal costs to measuring and policing performance, the master will stop short of perfect policing and will engage instead in policing until the marginal costs equal the additional marginal benefits from such activity... owners are able to enhance the value of their property by granting slaves some rights in exchange for services the owners value more. Hence slaves become owners too.
>
> (1990, 32)

In this amoral and functionalist narrative, provision grounds represent a balance of forces instead of a terrain of struggle on the margins of a structurally violent and racist system. This chapter approaches creole Caribbean institutions as the results of long-term conflict, not short-term function. It argues that cordyline defined provision grounds and yards, and this shaped how a peasant society formed within a deeply asymmetrical system of institutional and structural power.

This chapter begins with a description of plantations as the Caribbean's characteristic social-ecological system. Next, I provide ecological, agricultural, and social background on St. Vincent and summarize its history of land management and social change. The implications of this process for kinship, residence, and social networks constitute the next section. With these contextual foundations established, I turn to the symbolic and religious dimensions of cordyline in St. Vincent, with a particular focus on the Spiritual Baptist Church introduced above. The conclusion suggests that cordyline was a critical technology of power for creole place-making in ways that differ from this book's other case studies because of the specific context of racist oppression.

Boundary plants in the Plantationocene

After geologists began using the term "Anthropocene" to designate a new era of human-dominated planetary history, anthropologists cheered the adoption of their anthro- prefix and promptly set out to redefine the concept.

Their central critique is that our biosphere's existential crisis was not created by all of humanity uniformly, but by privileged groups and nation-states vacuuming up resources and spewing out carbon dioxide. Alternatives like "Capitalocene" (Moore 2016) and "Chthulucene" (Haraway 2016) emerged to refer to, respectively, the domination of a specific economic system and a hoped-for world in which all species co-exist, but Plantationocene is the increasingly popular portmanteau term. Donna Haraway coined it in 2014 (Haraway et al. 2016, 556) to designate the ways that farming and extractive industries rely on "slave labor and other forms of exploited, alienated, and usually spatially transported labor" (Haraway 2016, 206). A plantation is a "system of multispecies forced labor" (Haraway et al. 2019, 5). The argument is that New World slave plantations generated a global system of multi-species exploitation in which plants and animals are now literally enslaved in farms and feedlots. The Plantationocene label elides together the enormous changes caused by the Atlantic slave trade, the introduction of crop and livestock species to new areas, and the corporations behind soybean and oil palm plantations from Brazil to Southeast Asia.

The concept illuminates the human appropriation of biospheric productivity, but it also turns the world system into a monolith. If used uncritically, it also creates a false equivalence between extracting labor from oppressed people and the domestication of honeybees. Indeed, actual plantations are not emphasized in Plantationocene scholarship (Carney 2020, 2). Scholars like Mintz and Trouillot would probably insist that the Plantationocene should be a question about difference and complexity, not a universal explanation. They would warn us against top-down assumptions of homogeneity and uniformity and insist on studying the historically particular (Mintz 1974, 264; Trouillot 2002, 204). My bottom-up account of Caribbean boundary plants follows them and shows how specific contradictions in plantation agriculture fostered the development of Vincentian social relations and cultural forms that were not mandated by a generic Plantationocene.

The spread of Bantu-speaking farming societies across tropical Africa and the Austronesian expansion into Oceania were demographic events that led to new social and cultural forms across much of our planet's surface. The Atlantic slave trade was just as significant a population shift, but was destructive and violent instead of constructive and compositional. Approximately 12.5 million Africans were taken to the New World between 1501 and 1867 and endured dehumanizing conditions and structural violence that persist today as systemic racism. Multiple obstacles face scholars of these people's social history. Captured from culturally different parts of Africa, they became locked into economic, social, political, and cultural meshworks and "total institutions" that oriented entire social-ecological systems toward profit extraction. Yet generalizing about plantations is difficult because they varied by crop, colonial regime, and ecological conditions. Regional historians, archaeologists, and ethnographers therefore construct controlled comparisons by focusing on the dynamics of change. Foremost among these is the improvised invention of culture despite the plantations, not because of them. Alongside the structural violence of slavery and the racist colonial hierarchies that followed it, these scholars focus on the agency, resistance, and negotiation strategies of enslaved peoples and peasants (e.g., Delle 2014; Robotham 2018; Thomas 1984). To find examples of the disempowered exercising power, scholars look to the internal contradictions of plantation systems. The major fault lines were institutions like yards and provision grounds where enslaved peoples had partial autonomy. This chapter examines these "botanical gardens of the dispossessed" (Carney and Rosomoff 2009) in order to locate cordyline as a key part of Caribbean transformations.

In Chapter 6, the *Providence* crew was busy preparing 2126 potted breadfruit suckers wrapped in cordyline leaves. The expedition's botanists, James Wiles and Christopher Smith, also selected 472 food plant specimens and 36 "curious and useful plants" for shipment (Newell 2010, 166; Powell 1977, 393). The ship sailed west from Tahiti on July 18, 1792 and arrived in St. Vincent on January 23, 1793. As the *Providence* approached Kingstown Bay, the two Tahitians on board (Maititi, a royal servant, and Paupo, a stowaway who had assisted the botanists) exclaimed "Otaheite, Otaheite!" thinking that they were back home (Grove 2006, 161). The next day Alexander Anderson, the superintendent of the St. Vincent Botanic Garden, received the ship's smallest and most sickly plants (1983a, 53). Bligh's "floating forest" reached Jamaica on February 5, 1793; after distributing plants to botanical gardens and plantations, the *Providence* sailed for England with a full load of the best plants from each port it had visited. Two years after leaving England, the *Providence* reached the Kentish coast in August 1793 and began transferring the plants to the Royal Botanic Gardens at Kew. This example of imperial botany (Schiebinger 2004) led to institutional innovations in the Caribbean.

Although Bligh introduced cordyline to St. Vincent, the plant had already reached the Caribbean in 1787, when Jamaica's Hinton East imported "*Dracaena ferrea*, the purple dragon tree" from China (Edwards 1793, 479).[1] The first catalogue of plants in East's botanical garden includes an entry for "*Dracaena* nov. spec, [new species], Otaheite, HMS *Providence* 1793" alongside the Chinese *D. ferrea* cited above (Broughton 1794, 9). The historical record does not indicate this Tahitian cordyline's color – was it green *auti ma'ohi uta* or red *auti uteute*? It seems likely that the *Providence* carried both varieties as breadfruit wrappers and for the "curiosity" of a red plant. In any case, Bligh certainly expanded the genetic diversity of cordyline in the Caribbean. In the 1806 catalogue of plants in the St. Vincent Botanic Garden, Bligh's cordyline appears as red "*Dracaena ferrea*" under the category of "exotics, curious or ornamental" (Guilding 1825, 41). By the middle of the 19th century, the terms "dragon-tree" and "dragonsblood" had been established as anglophone names for both cordyline and dracaena (Grisebach 1864). A list of plants introduced to Barbados, for example, includes the "purple dragon tree" (misidentified as *D. ferrea*) and a West African "sweet-scented dragon tree" (*D. fragrans*; Schomburgk 1848, 590). The result of these misnomers is that today Caribbean people commonly speak of "white dragon" for *Dracaena fragrans* (because of the color of its stalk, not the leaves), and "red dragon" or *sang dwagon* for *Cordyline fruticosa*.

This botanical history establishes how an Oceanic plant came to the Caribbean, but how did it escape imperial control and become a vital component of Afro-Caribbean cultural practice? This chapter argues that Vincentians turned a monomarcating colonial boundary plant into a polymarcating institution with roots in both land management and religious practice.[2] Before addressing this transformation, a review of the status of boundary plants in the region shows why St. Vincent is a compelling case study.

In my early work on boundary plants, I found references to dracaena marking property and protecting homes in the New World and hypothesized a connection to West African practices (Sheridan 2008). During a visit to St. Lucia, I found that I was wrong. Dracaena was unimportant, but pink and green *canne wozo* was a boundary marker between unrelated people. Realizing that this was cordyline, the plant I had encountered in Rappaport's work (1984), I surveyed the regional literature for socially significant dracaena and cordyline. Fig. 7.2 shows the results and reveals that cordyline is associated with ex-British colonies (Dominica, Guyana, Jamaica, St. Vincent, Tobago, and Trinidad) and located most strongly in the eastern Caribbean.[3] In 2011, I visited Trinidad, Tobago, St. Vincent, and Dominica to investigate if cordyline was a polymarcating boundary plant in any of these areas. In Trinidad I heard that "people use the metal peg for the government, red *rayo* (cordyline) for their neighbors." I visited the remains of the Arnos Vale sugar plantation on Tobago to see rusting rum production machinery and the red cordylines on the estate's corners. I found long rows of young cordyline plants at a Tobagoan horticultural shop. The gardeners said that they didn't know the plant's history, but that they "sell rayos when people want boun' marks." In Dominica government foresters told me that they recognize old house sites when they find breadfruit, plantain, and the red dragon in the deep forest. Dr. Lennox Honychurch, Dominica's eminent historian and museum director, confirmed this, suggested that these were escaped slaves' homes in the 19th century, and added that the red dragon is a common grave marker on the island.

It was in St. Vincent that I found the most complex example of polymarcation. Cordyline was obviously ubiquitous as boundary markers and signs of human agency. Vincentians told me "everywhere you see it, it is the hand of someone," and "when you speak of the dragon, you know what you mean when you point!" The staff at Montreal Gardens in Mesopotamia Valley told me how the red dragon prevents evil spirits, disease, and theft. Melford Pompey, a bishop of the Spiritual Baptist church, described how cordyline connects worshippers to Africa and forms the center of their

Location	Species	Local name(s)	Significance(s)	Source(s)
Brazil	Both dracaena and cordyline	Peregun or peregum for both	Ritual practice, opens/closes paths, induces trance	Anthony et al. 1995, 114; Behague 2006, 100; Capone 2010, 54; Voeks 1990
Costa Rica	Cordyline	Caña de indio, ti plant	Old boundary marker, in secondary growth forests and abandoned villages	Gargiullo 2008, 89
Dominica	Cordyline	Malvina	Boundary marker	Quinlan and Hagen 2008, 150
Guadeloupe	Cordyline	Sandwagon	Protects houses and yards	Benoît 1990, 328
Guyana	Cordyline	Dragonsblood	Repels witches	Lindsey 1901, 126
Jamaica	Cordyline	Dragon, dragon blood	Land surveyors' marker, boundary marker, prevents menstruating women from harming crops, slave burial ground marker	Brassey 1885, 236; Cassidy and Page 2002, 158; Chevannes 1994, 26; Craton 1977, 274; Hargreaves and Hargreaves 1960, 64; Powell 1972, 23; Rashford 1994, 38
Panama	Cordyline	Nana, flor, red ti plant	Property rights of indigenous Naso people	Paiement 2007, 94
St. Vincent	Cordyline	Dragon	Grave marker	Zane 1999, 48
Suriname	Cordyline	(not given)	Funerary rituals	van Andel et al. 2013, 258
Tobago	Cordyline	Bondé bush	Boundary marker	Bowman and Bowman 1939, 86
Trinidad	Cordyline	Dragon's blood, rheo, rayo, ravenda	Boundary marker, spiritual protection, protects house from lightning, healing, initiation rituals	Brassey 1885, 236; Bryans 1967, 106; Collett and Bowe 1998, 18; Evans 1982; Hargreaves and Hargreaves 1960, 64; Kingsley 1871, 377; Shannon 1956, 130; G. Simpson 1962a, 329; 1962b, 1210; Williams and Williams 1951, 135; Winer 2008, 119, 750

Figure 7.2 Dracaena and cordyline in the Caribbean region

religious practice. But it was Mother Ogaro, a Spiritual Baptist leader in Chateaubelair, who convinced me to focus on St. Vincent. She told me that she "met the red dragon in heaven" in a vision. After introducing her to people who taught her African dances, the "spiritual dragon" instructed her to bring it to the "carnal" world and plant it at home to keep her safe. She sometimes digs it up and carries it to church. Pointing to the four-foot-high stalk topped with scarlet and green variegated leaves, she declared, "this one half boundary mark, half religion!" I knew then that St. Vincent would be this book's final case study.

Cordyline is not the only boundary plant in the Caribbean. Fig. 7.3 summarizes other species and uses, most of which are generally monomarcating and institutionally simple compared to Vincentian cordyline.

Location	Species	Local name(s)	Significance(s)	Source(s)
Barbados	*Opuntia* spp.	Prickly pear	Boundary marker, esp. at roadside	Sheller 2007
Barbados	*Cocos nucifera*	Coconut palm	Boundary marker, esp. on corners	Sheller 2007
Barbuda	*Cajanus cajan*	Pigeon pea	Boundary marker	Berleant-Schiller and Pulsipher 1986, 15
Dominica	*Acalypha hispida*	Chenille	Boundary marker, esp. for a yard	2011 fieldnotes
Grenada	*Artocarpus altilis*	Breadfruit	Boundary marker on property corners	Brierley 1987, 202
	Cajanus cajan	Pigeon pea	Boundary marker, hedge to prevent theft	Brierley 1976, 36
	Cocos nucifera	Coconut palm	Boundary marker on property corners	Brierley 1987, 202
Haiti	*Bromelia pinguin*	Pêgwê	Boundary marker, living fence	Mintz 1962
	Codiaeum variegatum	Krotô	Boundary marker, living fence	Mintz 1962

Figure 7.3 Other boundary plants in the Caribbean *(Continued)*

	Euphorbia lactea	Kâdélab	Boundary marker, living fence	Mintz 1962
	Polyscias spp.	Parése	Boundary marker, living fence	Mintz 1962
Jamaica	*Acalypha hispida*	Cat tail	Boundary marker in yard	Rashford 1994, 38
	Alaypha wilkesiana	Copperleaf	Boundary marker in yard	Rashford 1994, 38
	Bromelia pinguin	Penguin plant, pingwing	Boundary marker, hedges to keep livestock out of gardens, internal field boundaries within a plantation	Hall and Thistlewood 1999, 16, 159; Higman 2001, 214; Rashford 1994, 38
	Codiaeum variegatum	Croton	Boundary marker, gates, keeps spirits in the grave. Often combined with cordyline	Chevannes 1994, 27; Rashford 1989; 1994, 38
	Crescentia cujete	Calabash tree	Grave marker, alongside cordyline	Besson 2016, 212; Rashford 1988, 6
	Euphorbia pucherrima	Poinsettia	Boundary marker in yard	Rashford 1994, 38
	Haematoxylum campechianum	Logwood	Internal field boundaries within a plantation	Higman 2001, 215
	Hibiscus rosa-sinensis	Hibiscus	Boundary marker in yard	Rashford 1994, 38
	Malvaviscus arboreus	Turk's cap	Boundary marker in yard	Rashford 1994, 38
Montserrat	*Cajanus cajan*	Pigeon pea	Boundary marker	Berleant-Schiller and Pulsipher 1986, 15
St. Christopher's	*Citrus* spp.	Lime	Hedges to prevent theft	Grainger 1764, 35
	Citrus limon	Lemon	Hedges to prevent theft	Grainger 1764, 35

Figure 7.3 (Continued)

	Caesalpinia pulcherrima	Flower fence	Hedges to prevent theft	Grainger 1764, 36
	Opuntia spp.	Prickly pear	Hedges to prevent theft	Grainger 1764, 37
	Haematoxylum campechianum	Logwood	Hedges to prevent theft	Grainger 1764, 36
St. Vincent	*Calophyllum antillanum*	Galba tree	Windbreak promoted by colonial agricultural extensionists	Tatham 1911; Wright 1929
	Crescentia cujete	Calabash	Boundary marker, esp. on corners	2015 fieldnotes
	Gliricidia spp.	Gliricidia	Boundary marker, living fence	Glesne 1985, 61; WI Bulletin 1910
	Haematoxylum campechianum	Logwood	Living fences	Anderson 1983a, 4
	Lavandula spp.	Lavender grass	Boundary marker, esp. for a yard	2015 fieldnotes
US Virgin Islands	*Agave* spp.	Century plant	Garden fencing to deter livestock	Olwig 1997, 147

Figure 7.3 (Continued)

This list elides together plantation-era boundary-making, colonial peasant practices, and postcolonial present-day land management, and is in no way exhaustive. Each island has its own historical ecology of boundary plants, which may be more complex than the monomarcation suggested here. Pigeon pea and calabash trees are probably the most promising species for further investigation because of their associations with African ancestral practices. Together, these boundary plants demonstrate how Caribbean yards, gardens, and farms are "subaltern archipelagoes of agrobiodiversity that the Plantationocene spawned" (Carney 2020, 20). The St. Vincent case study that follows is just one example of institutional creativity in the Caribbean region.

St. Vincent as a social-ecological system

The shape of the social-ecological system that grew in the bootprint of St. Vincent's sugar industry was determined by the island's physical geography (Fig. 7.4). In contrast to the rounded hills and sandy white beaches of the

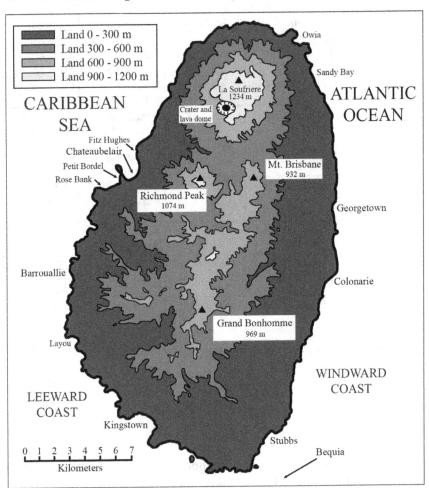

Figure 7.4 St. Vincent

Grenadines archipelago to its south, the main island of St. Vincent is steeply mountainous. 94% of its land has at least a 6° slope, and 80% has a slope greater than 20° (Grossman 1997, 355). From the active Soufrière volcano in the north (elevation 1234 meters), a jagged spine of rugged forested peaks stretches south toward the capital city of Kingstown. Deeply folded ravines of lava rock, known locally as "gutters," emerge laterally from this backbone. The island's eastern Windward side has a moderately sloping coastal plain, but Atlantic waves have left its shore too rocky for shipping. The western Leeward coast, in contrast, has gutters that meet the gentle Caribbean Sea in a series of deep narrow valleys and sheltered bays. This is why the earliest European settlements were on the island's south and west coasts. Its history of vulcanism has blessed the island with moderately fertile soil but cursed it with an ash content so high that the soil holds little water and is prone to erosion. This makes 75% of St. Vincent's land marginal, at best, for agriculture (Richardson 1989, 112; Spinelli 1973, 42). Although most hurricanes pass to its north, St. Vincent has not attracted much investment in tourism because it lacks the broad sandy beaches and bioproductive coral reefs that tourists crave. These physical characteristics make the economic potential of Vincentian agriculture, fisheries, forestry, and tourism severely limited.

These ecological factors determined the island's agricultural history. Its steep interior made it difficult for maroons, squatters, and fugitives to form "runaway peasantries" beyond colonial control (Marshall 2011, 150). Instead, the settlement pattern, social system, and cultural repertoire all reflected the hierarchy of the sugar economy. For the British planters who dominated the 19th-century St. Vincent landscape, "every plant but the sugar cane [was] an eyesore" (Anderson 1983a, 50), and all of the island's moderately sloping land became sugar plantations. This concentrated the population at the coast and made food supply a problem. Should St. Vincent be a "foreign-fed allowance" island where planters supplied rations of expensive imported food, or a "home-fed provisions" colony where the enslaved grew their own food (Stephen 1830, 260)? The planters opted to become a home-fed colony and allocated the enslaved workers provision grounds on estate lands. The sole allowance was three pounds of salt fish per week per worker (Young 1806, 267). Because most provision grounds were remote, farmers concentrated on crops requiring minimal labor, like plantain, banana, yams, sweet potato, and taro (Marshall 1993, 211). The coastal core became a zone of intensive agriculture, monocropping, and social hierarchy, while the interior was a peripheral area of horticulture, agrobiodiversity, and freedom. Yet these strikingly different modes of production formed one system (Berlin and Morgan 1993, 10; United Kingdom House of Commons 1842, 6684). The consequences are still apparent today in the division of agricultural labor among moribund plantations, coastal residential villages with small kitchen gardens, and farms in the roadless mountainous interior. Vincentian farming means climbing steep slopes with hand tools and trekking back down with heavy loads of carbohydrates.

The tempo and location of agricultural labor on sugar plantations shaped St. Vincent's culture. The regional seven-crop complex of intercropped banana, maize, beans, pigeon peas, squash, sweet potatoes, and yams determines what Vincentians consider a good meal (Berleant-Schiller and Pulsipher 1986), but farming is defined as a particularly bad occupation (Glesne 1985; Rubenstein 1987, 139). Vincentians were aghast when I asked why they don't live in the roadless interior, saying incredulously that "you couldn't live in the farmland, you need to be with people!" Plantation agriculture limited how Vincentians constructed social relations and cultural commitments, and each cash crop left an impression. Estate profits declined in the 19th century because the colonial metropole refused to subsidize St. Vincent's sugar, and an emerging class of smallholder peasants tried arrowroot, cocoa, and cotton in the late 19th and early 20th centuries (Handler 1971; Walker 1937). All proved only marginally profitable, and coastal farmers began growing bananas on contract for the British market in the second half of the 20th century (University of the West Indies Development Mission 1969; Grossman 1998). British neoliberal policies allowed Latin American bananas to compete with those from its former colonies, so Vincentian farmers increasingly grew marijuana on their mountain plots (Rubenstein 2006).[4] These boom/bust cycles combined with an underproductive estate system to make the coastal Vincentian agricultural sector stagnant at best, while inland plots are biodiverse yet socially stigmatized "food forests" where people who do not consider themselves farmers grow the crops that feed the island (Glesne 1985, 162; Hills 1988). Increasingly, these cultivators are women because many rural men are labor migrants (Isaacs 2014). Male employment in Vincentian agriculture fell by more than 50% between 1991 and 2001, while female farm employment declined only 6% (Isaacs 2014, 20). Some Vincentian men reject farming because a "garden man" is considered unmasculine and unfriendly (Abrahams 1983, 143).

Today's cultural practices and social forms are the consequences of older struggles between plantations and peasants over land and labor (Berlin and Morgan 1993). Two interlocked cultural paradoxes lie at the heart of many Caribbean societies. First, the institution of corporate "family land" (Besson 1987) coexists with the "overly aggressive individualism" that resulted from a general lack of corporate social institutions beyond plantations (Mintz 1965, 933, Thomas 1984, 20). Second, family lands are often tiny plots barely large enough for a house and a kitchen garden yet have dozens of co-owners. Like the *tomite* lands in French Polynesia, Caribbean family lands are the inalienable property of everyone descended from a landholding ancestor. These kin can activate access rights to grow crops, harvest fruit, or build houses. Three plots of family land in Jamaica's Martha Brae village, for example, had 92 claimants for a single acre (Besson 2002, 292). Family land has the paradoxical quality of unlimited access and severe limits on quantity. All potential claimants cannot use family land at once, making this land tenure institution more symbolic than economic. Token gifts of fruit and vegetables from family land confer membership in an extended kin network, not food security. This makes family land a sort of social capital, not economic capital. Unrestricted access makes family land into the rural poor's flexible safety nets and answers to the plantations' oppressive grip on land (Besson 2002).

The Vincentian forms of these paradoxes result from its structurally violent history of plantation domination. Until government land reform weakened the rigid pyramid of white elites and black upper, middle, and lower classes in the 1980s, 20 large estates had more than half of the island's arable land. The rest of its farms were smallholder plots, usually one or two acres, and either fragmenting into tiny individual parcels or held in common as family lands (John 2006, 102). The next section recounts the history of this long-lasting landholding asymmetry and the role of cordyline. Critics of the family land institution in St. Vincent decry it as "dead capital" and an obstacle to intensifying agricultural production because although this sort of landholding is socially legitimate, it has no formal legal basis and confers no title deed. Family land cannot be mortgaged to give landowners access to credit (de Soto 2000; Graham 2012; Isaacs 2014; Toppin-Allahar 2013). Like the other case studies in this book, St. Vincent's cordyline is part of a process of ambiguous land tenure, frustrated development efforts, and creative place-making.

These land matters have specific social and cultural correlations. Communities in St. Vincent tend to be loose and "open" residential clusters rather than "closed" units with social and symbolic boundaries, partly because labor migration is so prevalent. Social connections are generally dyadic and temporary personalistic ties instead of formal and durable corporate groups. The important social spaces are churches and rum shops, not town halls and public squares. Like elsewhere in the Caribbean, Vincentian culture emphasizes achieved status, individual respectability, and a street-smarts reputation (Rubenstein 1987; Young 1993). This sort of personhood relates to boundary plants because, unlike this book's other case studies, cordyline in St. Vincent is economic and cosmological, but not particularly social. Cordyline's role in the spiritual journeys described in this chapter's introduction is an irreducibly individual experience reflecting these trends. These boundary plants lie at the margins of individuals and society, not between one social group and another.

In April 2021, the Soufrière volcano erupted, prompting the evacuation of the island's northern third. The cement-like mixture of ash and rain that fell will surely become a sedimentary layer atop other crises, like St. Vincent's lack of a coherent land policy, reliance on an illegal cash crop, heavy dependence on labor migrants' remittances, and enduring underemployment and agricultural stagnation (Browne 2016; Kairi Consultants 2008). But like cordyline regrowing after a fire, Vincentians are resilient. They are likely to keep using the red dragon in post-eruption reconstruction because it is cheap, easily visible from a distance, and culturally meaningful. One man explained the meaning of cordyline to me as

> One keep right, one keep left. You keep that side, I keep this side, the right hand is my zone. We keep justice for your neighbors. The government surveyor come to give justice, follow you dragon boundary at them land.

Vincentians are still searching for justice, and I write hoping that cordyline will aid their journey. To evaluate how this boundary plant has guided them thus far, we must return to the historical roots of power by reviewing the political economy of slavery and the formation of the Vincentian peasantry.

Boundary struggles in the provision grounds

When I asked a focus group of elderly Vincentian men about the origins of cordyline on the island, they insisted that "the dragon has nothing to do with recent times. The red dragon came from the old Caribs, who came from Africa before Columbus, they knew there was dragon around here." In contrast, the earliest evidence of cordyline being used beyond the Caribbean's imperial botanical gardens as a social institution is from Trinidad and Jamaica, where there was "handsome copper-coloured dracaena [*sic*]... used everywhere to mark the boundary-lines of estates" on these islands in late 19th century (Brassey 1885, 236). The story of cordyline in St. Vincent is primarily a nineteenth and twentieth century matter, but a brief summary of island history prior to the 1793 arrival of Bligh's "curiosity plants" provides some necessary context.

Europeans colonized St. Vincent relatively late because of steadfast resistance by the so-called Yellow and Black Carib populations. These increasingly integrated peoples, defined in oral history respectively as the descendants of indigenous Kalinago people (mostly on the Leeward side) and shipwrecked or escaped Africans (mostly on the Windward side), were expert fighters who could attack the Europeans and then melt into the mountainous interior. French smallholders from Martinique and Guadeloupe began to establish coffee and tobacco farms on the Leeward coast in the 1720s. Their relations with the Black Caribs were tense but tolerant until the 1760s, while disease decimated the Kalinago population. After France ceded St. Vincent in the 1763 Treaty of Paris, Britain began dividing St. Vincent's land into sugar estates. In the chaos that ensued, estates purchased three times the island's total land area (Taylor 2012, 66). The First Carib War (1769–1773) followed repeated British efforts to survey Black Carib territory and force land sales. A 1773 peace treaty established a clear boundary for the Black Carib land in the north, but geopolitics soon intervened when the French captured most of the island in 1779. France returned St. Vincent to nominal British control in 1783, but long-standing Black Carib grievances over land and French revolutionary support sparked the Second Carib War (1795–1797). A decisive 1796 British victory led to a campaign of ethnic cleansing. The British hunted down and deported 4336 Black Caribs to Balliceaux, a tiny waterless island south of Bequia. Half of them died from disease and starvation (Anderson 1983b, 95) before the British exiled the 2248 survivors to Roatán Island (near present-day Honduras) in March 1797. Many of their Garifuna descendants now live in on the Caribbean coasts of Honduras, Guatemala, and Belize. Some descendants of the Caribs who evaded the British still live in St. Vincent today and self-identify as Yellow and Black Caribs (Gullick 1985; Taylor 2012, 151).[5] The exile of the Caribs is the second reason that although St. Vincent has remote interior regions like Jamaica and Suriname, it did not develop free communities of Maroons (Besson 2016; Price 2002).

This historical review is significant for several reasons. First, it establishes land access as the fundamental struggle in Vincentian history. Second, it explains how sugar plantations came to dominate the island so thoroughly. Finally, it accounts for the fact that cordyline is not significant in Garifuna ethnobotany (Coe and Anderson 1996). When I visited Belize in 2018 to confirm that the plant is important in St. Vincent, but not in its diaspora, Garifuna homeowners in Hopkins and the staff at the Gulisi Garifuna Museum in Dangriga consistently told me that cordyline was "just a flower." This circumstantial evidence implies that cordyline only became a Vincentian cultural commitment in 19th century struggles over land and labor.

Although the early French coffee and tobacco operations relied on about 3400 enslaved workers, it was British merchant ships (primarily from Liverpool) that built St. Vincent's Afro-Caribbean population. They brought approximately 59,000 West Africans to St. Vincent between 1763 and the British slave trade's abolition in 1807. Most captives were from West Africa, and about half were from present-day southeastern Nigeria, Cameroon, Equatorial Guinea, and northern Gabon (Eltis and Richardson 2010, 252). The enslaved population of St. Vincent was 10,391 in 1777, peaked at 24,920 in 1812, and declined to 18,794 by 1833 (Spinelli 1973 72; Taylor 2012, 81). The glaring gap between the total number of enslaved persons imported to the island and its enslaved population is explained by the brutal economics of capitalism. Buying an adult slave was about half the total cost of raising an enslaved baby to begin work at age 14, so planters had little interest in slave families' welfare. The rapid rise and fall of the slave trade to St. Vincent – just 44 years – meant that the island had an unusually high proportion of "saltwater slaves" born in Africa, 38.8% as of 1817 (Young 1993, 46). This demographic pattern raises the tantalizing possibility that former Central-West Africans may have recognized cordyline as similar enough to African dracaena to be culturally significant, but the historical record is resolutely silent on this issue until the late 19th century. The following account of provision grounds connects, albeit loosely and indirectly, the existence of cordyline in the St. Vincent Botanic Garden after 1793 and its uses in the 1880s.

In his history of the British Caribbean, Bryan Edwards argued that the "Negro grounds" constituted a "happy coalition of interests between the master and the slave" (1819, vol. 2, 161). By planting hurricane-proof "ground provisions" like yams and taro along with aboveground crops like maize and beans, enslaved persons got food, acquired an interest in the plantation's success, and earned some money from surplus crops. The plantations avoided expensive imports and benefitted from enslaved peoples' conversion of forest into arable land and rehabilitation of depleted plantation soils. Edwards' rosy and harmonious assessment of injustice and oppression is itself a sort of symbolic violence.

Edwards' generalization obscures the diversity of the enslaved peoples' farms and the specific processes of social and ecological change that occurred in them. These plots varied according to the topography, major crops, and land management practices of each plantation (Carney and Rosomoff 2009, 100; Stephen 1830, 264). We can group them into three types on sugar islands with mountainous interiors: plots near the plantation's core, dooryard or kitchen gardens, and provision grounds on an estate's upland margin (Carney 2020, 6; Pulsipher 1994). The plots assigned by the plantation were typically planted with plantain, taro, and yams, and carefully monitored to ensure that enslaved people spent their half-days of ostensibly free time on Saturdays cultivating there. Kitchen gardens were intensively managed intercropped plots with vegetables, herbs, and medicinal plants (and perhaps a few livestock) in a plantation's "Negro yard" residential area. It was at the mountainous provision grounds, however, that farmers had the most autonomy. They cultivated these areas according to their own needs, design, and labor availability – but only on Sundays during the break mandated by colonial labor laws. Their focus was therefore on ground crops that reproduced through vegetative propagation, demanded only occasional labor, and were safe from casual theft because they were underground.

The terms for these three areas in St. Vincent are "yam piece," "garden," and "mountain," and this system was well established by 1789 (Marshall 1993). A yam piece was about three or four square meters of land marked with wooden stakes and destined for sugar cane (Collins 1803, 103; Warner 1831, 40).[6] It was here that enslaved livestock owners deposited manure to ensure good yam crops – which also helped exhausted soils return to sugar production (Colthurst 1977, 171). The fenced-in gardens were probably about 100 square meters around houses or slave barracks. Enslaved Vincentians worked on the yam pieces and in their gardens during their midday break (Young 1806, 267). The size of the mountain plots varied according to the difficulty of access and labor availability. The boundaries of St. Vincent's provision grounds were marked on only three sides. According to Gertrude Carmichael, a plantation owner and slavery apologist who lived in St. Vincent in 1820-1823, the downhill edge of a mountain plot was "distinctly marked out" to prevent conflict, but the uphill side was left unbounded to let enslaved farmers expand their plots at will (1833 vol. 2, 163). The regional historical record does not reveal much about these markings and the agricultural techniques used at the estates' peripheries (Handler 2002, 133; Marshall 2003, 126). The sole description of cordyline in a Caribbean provision ground comes from a British history professor who visited Trinidad in 1869. After reviewing the food crops and medicinal herbs in a typical garden, he concludes by describing what is unmistakably cordyline:

> ...as a finish to his little paradise, he [a typical estate worker] will have planted at each of its four corners an upright Dragons-blood bush, whose violet and red leaves bedeck our dinner tables in winter; and are here used, from their unlikeness to any other plant in the island, to mark boundaries.
>
> (Kingsley 1871, 377)

As a cheap and lightweight technology with minimal labor requirements, brightly colored cordyline stalks would have been distinctive and efficient, but the historical records to bring these farms into sharper focus remain undiscovered.

The productivity of this farming system was the niche in the racist colonial edifice in which the material and symbolic foundations of the Vincentian peasantry grew. Contemporary observers asserted that enslaved people regarded the mountain plots as personal property even though plantations owned and allocated them. Carmichael states that if a Vincentian planter wanted to revoke an enslaved person's access to provision grounds, six months' notice was necessary, and a new crop had to be ready for harvest elsewhere before the older plot could be fully transferred to other uses (1833, 197). The mountain farms could be inherited, and the "Negro slaves have no idea that any one would or could doubt their legal right to their own property" (1833, 199). In 1807, a Jamaican planter told a British House of Commons committee that his enslaved workers thought of "their houses, their provision grounds, their gardens and orchards... as much their own property as their Master does his Estate" (cited in Anderson 2001, 86). Enslaved people in the British colonies informally assigned their homes and provision grounds to the heirs of their choosing, and there was enough of an informal market for houses and farm plots within estates that some enslaved people accumulated multiple properties (Berlin and Morgan 1993, 37; Brown 2008, 126). Provision grounds were also important sites of contestation and negotiations among enslaved patrons and clients. Established "Creoles" competed fiercely for the privilege of "seasoning" newly arrived saltwater slave children, known as "inmates" in early 19th-century St. Vincent. In return for socializing an enslaved child into plantation labor, feeding them, and teaching them English, the host received a knife, a calabash bowl, and an iron cooking pot (Collins 1803, 76; Young 1806, 267). The competition was because these children supplied two to three years of daily labor on the provision grounds while their hosts worked in the sugar fields. Given the high death rate and short lifespan of enslaved people in the 19th-century Caribbean, these quasi-adoptions were also mechanisms for redistributing homes and mountain provision grounds to non-kin (Mintz 1974, 208).

The mountain provision grounds, and to a lesser degree the kitchen gardens, were places within the plantation system where the enslaved could contest their exploitation with assertions that indirect terms of access were actually primary rights to property. Enslaved people had terms of access to land that almost amounted to formal economic capital, could transfer that quasi-property, and knew that the planters recognized these rights. The meanings of plantation land, labor, and power were negotiable in these peripheral capitalist spaces. Dignity, self-expression, and full personhood were achievable within the structurally violent system, and so it was in these institutional and cultural cracks that social practices grew into distinctively Afro-Caribbean cultural forms (Mintz and Hall 1960; Tomich 1993). Caribbean peasantries formed as this increasingly legitimate economic capital allowed enslaved people to accumulate social and cultural capital. These farms and the marketing system they supported produced enough wealth that some enslaved Vincentians had status symbols, like wine glasses and china, on display in their homes (Carmichael 1833, 130). One Jamaican planter calculated that enslaved people had amassed 16% of that island's money supply (Long 1774, vol. 1, 537) through their own production. This vibrant subsidiary economy within the plantation system probably made plot boundaries important, but it is unclear if cordyline served this purpose in St. Vincent's slavery period.

The full emancipation of the British Caribbean's enslaved people in 1838 was not a watershed moment for St. Vincent. The white "plantocracy" prevented a class of autonomous Black farmers by ensuring "the emancipation of liberated people, but not land, from the St. Vincent planters" (Richardson 1989, 114). The 1834 Abolition of Slavery Act in St. Vincent converted all slaves into "apprenticed labourers" preparing for freedom while working under nearly identical conditions on the same plantations. The apprentices worked 45 hours per week in return for housing, provision grounds, salted fish, clothing, and health care (in Anderson 2001, Appendix I).[7] The issue of which of these terms of access to resources and services were rights contingent on labor, and which were mere "indulgences," quickly became a basic issue of Vincentian politics. In the original formulation of the 1834 Act, field workers would graduate from apprenticeship to freedom in 1840, but skilled non-agricultural laborers would be free two years earlier; the planters therefore defined as many apprentices as possible as "field negroes" (Marshall 1985, 206). The planters identified gaps in the law to reject allowances for the apprentices' free children, customary entitlements like shots of rum after hard labor, and exemption from field labor for pregnant women and nursing mothers. The apprentices responded with violent protest, one work stoppage, and rumors that the government intended to "make them a present of their huts and provision grounds" (Anderson 2001, 86). This transitional period set the course for the 19th century in St. Vincent. After full emancipation in 1838, the white plantocracy kept most of St. Vincent's land and exploited a resentful Black labor force. The former slaves argued that their houses, dooryard gardens, and mountain provision grounds were their own, while the planters insisted that these were rental properties and deductible from their laborers' wages. This lopsided struggle over the meanings of labor and land persisted in St. Vincent until a series of disasters prompted government intervention.

Increasingly idle sugar estates dominated St. Vincent's 1838–1897 post-emancipation period and delayed the formation of an independent peasantry. Grenada provides a useful contrast; over the same period, this British colony developed a relatively prosperous Black peasant class of cocoa farmers on small freehold plots (Shephard 1948). St. Vincent's white plantocracy doggedly persisted, even as sugar profits fell as Europe turned to sugar beets. The case of the Porter family illustrates this pattern particularly well. D.K. Porter and Company became St. Vincent's primary landowner by acquiring other estates at rock-bottom prices through the West Indian Encumbered Estates Acts (1854–1886). In 1882, the company controlled two-thirds of the island (Nanton 1983, 224). As of 1897, Alexander Porter personally owned 25% of St. Vincent's arable land, and another 41% belonged to just nine more landowners (Richardson 1997, 44; West Indies Royal Commission 1897, 119). Most of these plantocrats were absentee owners with attorneys managing their estates (Davy 1854, 184), and more interested in extracting profit than investing in St. Vincent.

The planters used land taxes and strategic inaction to coerce the ex-slaves into low-wage labor contracts and prevent an autonomous landowning peasantry. They refused to fund a survey of St. Vincent's mountainous Crown Lands, which were formally government lands but under plantation control. The survey would have subjected provision grounds to regulation rather than letting each plantation manage their use in exchange for labor (Fraser 1986, 16). Plantations would not buy or process sugar cane from Black smallholder farmers (Shephard 1947). But the tighter the planters' grip on land, the more labor slipped away. The economic logic was apparent soon after Emancipation. In 1842, estate manager Hay McDowell Grant told a British House of Commons committee that Vincentian labor contracts required nine hours of labor per day in exchange for wages and services comprising 1.5 shillings, housing, medicine, two pounds of salted fish, a small kitchen garden next to the house, and a one-acre provision ground (United Kingdom House of Commons 1842, 31). The Black farmers worked on the mountain on weekends and in the kitchen garden during the noontime break from sugar production. Grant estimated that a typical household earned 100 Spanish dollars annually from its own production. This meant that the kitchen and mountain plots gave the same annual return as 277 days of plantation labor. The Vincentians' focus, unsurprisingly, was on their own cultivation instead of plantation crops. Most lived on white-owned estates and got access to housing and provision grounds as contracted plantation laborers, but some became arrowroot-growing squatters in the mountains above the estates – until the plantations got into the arrowroot business themselves. By the end of the century, many Vincentians had become labor migrants, and many plantations were idle. In the 1880s, estates began charging rent for provision grounds to squeeze some income from their labor force, even though these areas were technically government property. The 1897 West India Royal Commission suggested that the long struggle over land and labor had reached a stalemate:

There are… round the sea-coast, thousands of acres of fertile land in the hands of private owners, uncultivated and likely to remain so. The holders of these lands appear to be unwilling to sell them in small lots or at a reasonable price, and are unable to cultivate them.

(WIRC 1897, 48)

The Caribbean's most severe restrictions on Black land rights had failed to coerce labor, so in effect the sugar plantocracy strangled itself instead of reforming its racist system (Gearing 1988; Hall 1978; Momsen 1987).

This period of underdevelopment and oligarchy is when St. Vincent's boundary plants come into focus. Many elderly Vincentians recall that rental plots on estates were marked with bright red cordylines and that farmers "carried dragons up to mountain" to establish provision grounds. But only one man, Joseph Finley from Fitz Hughes, had a detailed description of cordyline in the late 19th century. His father had been an estate manager in the 1890s. He showed workers the boundaries of the yam pieces and kitchen gardens specified in their labor contracts. The estates used green dracaena on corners and filled in the sides of plots with red cordyline, and Joseph's father pointed to these boundary plants when "supervising people to follow the right thing." Summarizing his father's use of these plants, Joseph said that cordyline works to "give justice to the manager, give justice to your neighbor." This account shows that boundary plants had been institutionalized on Vincentian plantations by the late 19th century, but what had cordyline been doing since its establishment in the St. Vincent Botanic Garden in 1793? I think that Tahiti's bright red *auti uteute* cultivar was selected by the planters as a cheap and obvious way to institutionalize the terms of access to estate lands, and that as farmers carried this monomarcating boundary plant into St. Vincent's mountainous interior, it became increasingly polymarcating and culturally complex. Because the scanty descriptions available for 19th century provision grounds focus on food crops instead of social institutions and property rights, it is unclear if this territorialization occurred during slavery, the Apprenticeship period, or the decades-long deadlock between Black laborers and white plantocrats.

Three interlocked disasters – the plantocracy's slow violence, a hurricane, and a volcanic eruption – catalyzed the colonial government to modernize St. Vincent as a nation of smallholder peasants. The overall disaster recovery plan was to settle Black farmers onto marginal or underutilized lands. The first attempt was in 1891 when the colonial government began selling 5-acre plots of Crown Lands above the 300-meter contour line at the practical upper limit of estate lands. The Crown Lands Scheme failed because settlers in these remote and roadless mountains still needed estate work to pay installments on their land. At this time, the declining estates were paying wages below the level set at Emancipation in 1838 (John 2006, 54; Shephard 1945, 16), a strategy for preventing Black farmers from amassing enough cash to purchase land (Fraser 1986, 103). The Great Hurricane of September 1898 killed about 200 Vincentians, destroyed crops, and left half of the island's population homeless. The governor responded by following the 1897 West India Royal Commission's core recommendation – let the sugar industry die and a smallholder farming society take its place (Richardson 1997, 89). The government tasked the St. Vincent Botanic

Garden with providing "economic plants" like cocoa and Sea Island cotton to the planned peasantry, and the Agricultural Department began advising Black farmers (DeFreitas 1965; Howard 1954). The government began buying derelict estates in 1899 and subdividing them for sale to Black farmers who could afford the 25% deposit and complete the purchase by installments within six years (John 2006, 64). After the Soufrière volcano erupted in May 1902, killed approximately 1600 Vincentians, and covered the northern half of the island with ash, displaced families expected to be resettled on prime agricultural land (Richardson 1997, 206). This was a slow process. By 1938, the government had created about 800 plots on one-fifth of the plantations' 19th century acreage, but this was at best a half-hearted reform effort.[8] Most of the settlements and allotments had poor soil and infrastructure because they were the least productive lands of the least productive estates. As of 1938, only one-third of settlers actually lived on their plots (John 2006, 63). In the 1930s global depression, visitors noted that "the abject poverty of the great part of the negro population outside of the capital town Kingstown" (Walker 1937, 222) was due to persistent inequity. A 1935 riot over import duties on alcohol, tobacco, and matches led to another Royal Commission in 1939, which triggered another series of modernist settlements as the solution to St. Vincent's enduring land-holding problem.

The institutionalization of boundary plants in St. Vincent becomes fully apparent in the colonial records and oral histories of this period. The 1917 Boundaries Settlement Act established that surveyed lands have iron posts or concrete pillars on each polygon's corners, and that "line marks shall be cuttings of immortelle trees [*Erythrina* spp.], dragons blood or any growing fence or live hedge" (St. Vincent 1927, vol. 1, 665). The 1944 Forest Policy specifies that Forest Reserves should have "clearly visible" boundaries with concrete markers and "dracaena" planted at intervals (B. Gibbs 1947, 310, 323). It is unclear whether these laws refer to green dracaena or red cordyline, but the accounts of Vincentians who acquired land assert that both expressed ownership at this time. One man showed me photos of the Crown Land plot that his family purchased in 1905, and proudly noted that it has the original red dragons on its corners, showing that it had never been subdivided. Joseph Finley explained that

> Up on the mountain the government take Crown Land and give it out as Buy Land, so people would show that by clearing it and planting red dragon up on the mountain. They would take some cuttings from the estate boundary and carry them up on the mountain Buy Land, they use same dragons as on the estate. They get from estate, carry up to mountain, because there was no dragon up there already.

Another man reported that his family "got Buy Land up on mountain, them carry dragons and they plant it up there, especially at that place the estate land crash to them land up at that place we call farm." This was similar to the process on resettlement estates. One elderly man remembered his mother asking an estate manager, "I cut a kitchen garden?" and then securing the plot with stalks of dracaena and cordyline. Vincentian elders call this "making a straight," and said that it could involve corner posts of the "white dragon" (dracaena) and the red dragon at regular intervals, or it could be all cordyline. One 97-year-old woman showed me how to make a straight by miming the motion of walking in a line and inserting stakes every two meters. "You plant the dragon up at the head and you walk down to the foot," she explained. It is likely that the 1917 law formalized what was already a well-established land management practice from the 19th century.

In 1945, the British government's publication of the 1939 West Indies Royal Commission report (known the Moyne report) shifted the terms of debate on conditions in its Caribbean colonies. The report concluded that most of the colonies' Black subjects lived in conditions "little, if at all, superior to those prevailing under slavery" (1945, 30). The answers were increased development aid and resettlement schemes based on the classic smallholder farmer model of nuclear families living on 3.5-acre homesteads with a mix of tree crops, livestock, fodder grass, and food crops. The colonial government doubted that freehold tenure would ensure efficiency and prevent erosion. Instead the government used 25-year leases and a new Land Settlement and Development Board to control the settlers (John 2006, 72; Shephard 1947). The government acquired 4694 acres from 14 estates for this leasehold system between 1946 and 1961, but it recapitulated the plantocracy land use pattern instead of changing it. The resettlement schemes became nationalized plantations, mostly growing coconuts, arrowroot, and cotton, with their workers living in rented houses with access to kitchen gardens and upland provision grounds. Karl John, the Director of Planning in the St. Vincent Ministry of Finance and Planning in the 1980s, argues that resettlement schemes were a political patronage system instead of the route to a stable landholding peasantry (2006, 82).

By 1967, the skewed land ownership pattern had barely budged since the Moyne report, and St. Vincent had the second lowest per capita income in the Western Hemisphere (University of West Indies Development Mission 1969, 4). From 1946 to 1972, plantations (including resettlement estates) continued to own more than half of all cultivated land (John 2006, 102; Young 1993, 63). Throughout this period of path-dependent institutionalized racism, Vincentians continued to use boundary plants to legitimize landholding (Fig. 7.5). One focus group in the Petit Bordel resettlement area summarized that in the 1960s,

> most people would clear and plant as much land as they could themselves to show the size of the plot they could handle, then they paid the government money and the plot was allocated. Once you got your plot,

Figure 7.5 Petit Bordel resident showing a red dragon that her grandmother planted in 1963

you put either the red dragon or the green one [dracaena] on its corners to show it is yours. Then the concrete markers came later.

All of my interviews pointed to the same pattern: the first step in claiming land was informal but socially legitimate demarcation with boundary plants, followed by formal bureaucratically recognized concrete pillars. As one elderly woman expressed the late colonial situation, "we used both dragons, but the government wanted to use stones, not dragons."

These land use patterns continued after St. Vincent's independence in 1979. My informants who built their homes after the 1970s stressed that no matter if one inherited land, bought it, or got it from a resettlement program, a landowner's first act was planting cordyline; the house itself came second and a garden third. They also reported that the white dragon declined in popularity after independence, and dracaena is now used primarily for purely pragmatic purposes like fencing and erosion control. During this period St. Vincent was experiencing a banana boom (1985–1990) based on contract farming for the British fruit market. Incomes rose, freehold private property gradually replaced the leasehold system, smallholder farming was viable, and architecture became concrete block "wall houses" instead of "trash" (wattle and daub) and "board" (lumber) houses (Grossman 1998, 48). From 1985 to 2000, a land reform program mostly achieved the objectives that had eluded the colonial system. The newly independent government converted 4400 acres of estate land into 1231 smallholder farms, while at the same time tourism and associated service occupations overtook agriculture as St. Vincent's major income earners (Isaacs 2014; John 2006, 228). The big estates now control just 10% of the island's cultivated area, Vincentian landholding is now firmly based on the family land system, and many houses have either hedges of the red dragon or stalks of cordyline at a plot's uphill "head" and its downhill "foot."

This historical account of Vincentian land matters shows the enduring struggle over labor and land as the driver of social change on the island. The core question was always who had which rights to what land, and the evidence suggests that much of this territorialization contest was fought with boundary plants in interstitial spaces beyond the power of the plantations, like yam pieces, kitchen gardens, and mountain provision grounds. This review has emphasized how cordyline (and, to a degree, dracaena) was a monomarcating institution for legitimate land access in a sharply unequal, racist, and structurally violent system. This struggle was about both land and meaning. A few stalks of the red dragon could show an enslaved person, an apprentice, or a free laborer where they had short-term use rights or could demonstrate long-term rights of control. This was a useful ambiguity. The plantocracy lost this long slow battle because of formal government action and informal peasant creativity and hybridization. Michel-Rolph Trouillot identifies four aspects of creolization in the Caribbean; frontier, plantation, enclave, and modernist contexts (2002, 197). The story of boundary plants on St. Vincent shows three of these creolizations; first on a frontier of the British Empire, then within (and against) the violence of plantation agriculture, and finally as part of modernist nation-building. The enclave context of peasant autonomy was never possible because of Vincentian topography and the planters' stubborn racism. The next section examines the social and cultural consequences of this creole territorialization and the Vincentian peasants' new polymarcating meanings for the red dragon.

The social organization of dragons

Compared to this book's other boundary plant stories, St. Vincent stands out for its different polymarcation pattern. Cordyline in St. Vincent is not indexed to corporate groups like lineages and royal monarchies. When Vincentians speak of the red dragon, they say that it denotes "my boundary," not "our boundary." This narrow usage of boundary plants aligns with a truism of Caribbean history and ethnography – that social relations in the region's postslavery societies tend to be dyadic relationships between individuals (like friendship and patron-clientage) instead of larger durable institutionalized groups. The bonds of kinship are relatively weak in St. Vincent, and relationships tend to result from interaction instead of prescribed roles and responsibilities (Rubenstein 1987, 226). Vincentian households are fluid networks rather than decision-making groups (Glesne 1985; Young 1993, 133). The roots of this cultural pattern lie in the radical individualization of chattel slavery, the plantocracy's use of individual plots to entice and fragment the post-emancipation labor force, and high rates of labor migration (Thomas 1984). Access to land was fundamentally a patron-client relationship between estate managers and workers, family land was more safety net than economic asset, and mobile laborers were flexible instead of being attached to places. These observations illuminate how and why Caribbean boundary plants differ from those in Africa and Oceania.

My search for intersections between boundary plants and permanent corporate institutions in St. Vincent was mostly fruitless. The exception was a story repeated by four elderly men about their experiences in October 1979, when St. Vincent celebrated its independence. These men told me that Vincentians marched in the streets of every town waving flags and holding branches of the red dragon. One man remembered being in Kingstown that day, where he received a flag and a branch of cordyline from a government truck:

> I carried one of each! The dragon meant "St. Vincent belongs to us now!" The red dragon is good for many things, so we used it to celebrate and to keep away evil things. The dragon is both land and politics and spiritual, all one thing! We used it to show it is our land, no more slavery on the estates, no more work for nothing! We wave the dragon on that day! All the people! Everyone from the Prime Minister to the lowest man carried the red dragon and flags!

This patriotic display was a strategy to construct a corporate national identity for St. Vincent's new citizens, but it did not form enduring groups.

This is not to say that Vincentians are selfish, do not cooperate, or lack institutions. Their generosity, solidarity, and institutional know-how have all been apparent in their recovery from the April 2021 volcanic eruption. My point is that boundary plants in St. Vincent delimit different relationships between people and land than this book's other case studies. Much of Vincentian social life happens in the yard around a home, and this area is typically defined by hedges, living fences, concrete block walls, or chain-link metal fencing. A gate separates the yard from the street, and is usually marked by cordyline, just like the yard's uphill head and downhill foot. Lower-class Vincentians use their well-swept yards as areas for growing and processing food, laundry, and bathing; for middle- and upper-class people these activities move indoors, and the yard becomes a status display area (Gearing 1988, 348). Many yards feature a shaded wooden bench or a covered porch, and it is here that visitors (like inquisitive anthropologists) are received. These yards are sites for institutional, intergenerational, and interpersonal social negotiation, not just units of real estate. Many Vincentians

Figure 7.6 Chateaubelair resident showing his hedge of red dragons

described their yards by reference to the cordylines that they had planted (Fig. 7.6). Others pointed out which red dragons showed the limits of family land and which represented subdivisions of that land. Yard cordylines can be part of the "crab antics" through which individualistic Afro-Caribbeans compete for reputation, advantage, and resources (Wilson 1973). One woman showed me how her kin had "taken her boundary" and captured a few square meters of her yard by building a concrete block wall on her side of a red dragon. "They say it [the cordyline] good to be in your yard," she said, "but I hope it knock down that wall!"

Like in other Caribbean societies, Vincentian yards form a nucleus of social action and sites for symbolic expression (Mintz 1974, 232). Vincentians use cordyline to create yards under different tenurial conditions, including private "buy land," family land, rental agreements, resettlement schemes, and squatting. These boundary plants mediate flexible relations instead of firmly demonstrating status. Vincentian social organization is a three-dimensional system, like what Jean Besson describes for Jamaica (2002, 277). First, Vincentian marriages take multiple forms, such as "visiting unions," common-law arrangements, and serial polygamy (Rubenstein 1987, 249), none of which necessarily include co-residence or prevent emotional closeness. One Vincentian told me how she treats her father's former girlfriend, with whom he had two children, as her beloved and respected "half-mother." Second, families are loose networks on both sides of one's parentage. These ego-focused bilateral networks can be activated for support and mutual aid. Third, the institution of family land creates ever-growing groups of descendants of original landowners. These ancestor-focused and unrestricted cognatic descent groups exist through symbols of shared family land rather than through residence and resource use. This descent group is a set of potential relationships awaiting activation, not a durable sociological unit. These three intersecting frameworks maximize the number of social ties that a person can activate by flexibly linking the life cycle of domestic groups, exchange relationships, and landholding (Besson 2002, 281). Afro-Caribbean people developed this system in response to the structural violence of slavery, the scarcity of land due to the post-emancipation plantocracy, and the increasing reliance on labor migration. The cordyline next to the gate of a typical Vincentian yard sits at the intersection of these three systems for building and using relationships.

The significance of yards follows from these organizational principles. Vincentian yards are strongly associated with women's work maintaining order in this fluid system, while men are more often found "on the street" and "at the crossroads" (Abrahams 1983). This does not mean that women are group-oriented, passive, and solely focused on "respectability" in contrast to male individualism, action, and quests for "reputation" (Besson 1993; Wilson 1973). Instead, both men and women draw on these contrasting value systems. As we will see in the next sections, cordyline confers order and protection at home, in the street, and even at the edge of heaven. Unlike this book's other case studies, cordyline is not a particularly gendered institution in St. Vincent. This is a consequence of how it is enfolded within the flexible social arrangements of marriage, bilateral networks, and family land. Most Vincentians use government cemeteries, but still regard the graves as family land like many yards. "We put the red dragon at the head of the grave just like we put the dragon at the head of a house plot, men and women all the same" one old woman explained. The orderly domestic domain of the yard extends into the cemetery but does not spatialize gendered resource access like in Kilimanjaro and Papua New Guinea.[9]

A concept from the discussion of Oceanic boundary plants helps explore the symbolic dimensions of yards and family land in St. Vincent. In French Polynesia, the House Society model points to shifting meanings of social ranking and moral personhood and illuminates linkages between precolonial sacred sites and postcolonial homeowners' yards. Translating this idea for St. Vincent leads to the concept of a Yard Society, in which the space around a home serves as the material and ideological context for the objectification of flexible, tangled, and contingent social relations. This is not a type of society or a unit of analysis; Yard Society is an interpretive device for investigating what boundary plants do in a highly individuated society without the ranked corporate groups found in House Societies. Cordyline fills the same symbolic and historical niche as a ritual attractor in St. Vincent that it does in Remote Oceanic yards and gardens, and not simply because it demarcates land. The plant's cultural significance is located in the particular historical and social contexts of structural violence, racial ranking, and labor migration. Emancipation, peasantization, and migration all created new forms of legitimate person-hood (Thomas-Hope 1995). The moral personhood that the cordylines around a yard assert is open rather than exclusive because of these larger contexts of flexibility and mobility. The yard is open to half-mothers, distant cousins, and labor migrants alike, and its cordyline draws people in instead of keeping them out. This identification of a Yard Society allows us to ask what the red dragon does for Vincentians beyond its economic property-marking role. How does it relate to safety, protection, and social order? How do Vincentians use the plants in their yards to materialize and territorialize social relations? The next section addresses these questions by examining cordyline's movements beyond Vincentian yards.

The red dragon is the guide

In Africa and Oceania, dracaena and cordyline at graves demonstrated the emplacement of ancestors into landscapes, from which property rights, political authority, and moral order flowed. For Vincentians, the symbolic side of cordyline regulates spatial freedom, not fixed attachment to place. This ontological perspective suggests that cordyline is a sort of spiritual check valve that keeps relationships moving in the correct direction. This role is most apparent in the management of spirits and in spiritual journeys to places far beyond St. Vincent.

According to my interviews, Vincentians use red dragons to manage relationships with two sorts of spirits: deceased members of their own bilateral kin networks and malicious unknown spirits. Relations with familiar spirits occur primarily at gravesites, which are often marked with stalks of cordyline and croton. The "head" of a grave usually lies to the west, so that the deceased is ready to arise facing Jerusalem at the second coming of Christ. The cordyline that grows upon these graves, usually at the head, controls mobile spirits. The red plant not only makes the grave easy to find, it "stakes the spirit" to keep a ghost contained or to release it. Another stalk of the red dragon at home completes the system. An elderly woman from Chateaubelair explained:

> We just put a bouquet with the red dragon and remove it from the grave when we want the spirit to move around. The spirits just drift around, so they are free, but if you keep some red dragon in the house she will just pass and she let someone sleep. If you stake the spirit in the grave with the red dragon, it would not allow a helpful spirit, like someone from your family, to move about. So we don't put the dragon on the grave if you want your mother's spirit to come visit you in dreams. Don't plant the dragon if you want to get the messages. If you have the dragon in your yard on in a vase in the house, it means that the spirit can come close enough to give you the dream, but they

do not pass inside. Sometimes when a spirit passes you feel a tickling shivery cold on you, that is the spirit of a good family person visiting and guiding you.

I heard similar accounts about the spatial management of spirits in many interviews, usually phrased as "fixing the spirit," "staking a spirit," or "sticking a branch for put that spirit there" as idioms for preventing spiritual trouble and managing intimate relationships. One woman said that her aunt planted a red dragon on her husband's grave, but "she take it out to let him visit her." I could not determine the depth of this practice in Vincentian history, and elders repeatedly said "I met it." Archaeological and ethno-historical investigation might reveal whether enslaved, apprentice, and proto-peasant Vincentians used boundary plants to mark graves in the 19th century, and would illuminate how symbolic landscapes and intimacy were, like in Jamaica, ways to negotiate land access and social status (Besson 2000; Brown 2008).

Unknown spirits are another matter. Vincentians call these malicious beings "jumbies" or demons, and cordyline in a yard protects against them. This extends inside the home and out into the street. When I started inter-views by asking what cordyline does, many people began by saying that the red dragon is "for boundaries, peace, and protection." They contrasted the way that concrete pillars, metal pegs, croton bushes, or gates prevent quar-rels with neighbors with the red dragon's extra ability to prevent evil, jeal-ousy, and the spiteful gossip of "bad mouth." As Mother Ogaro explained, "the red dragon does a lot of work. If a jumbie come, the red dragon prevent it, it get no chance. Any evil thing not come in!" Bringing this protective structural power inside is easy; one simply puts a stalk in water. Typical places for vases and bottles with protective dragons are kitchen tables and the corners of children's bedrooms. Christians also emplace peace and pro-tection by forming a cross from the two halves of a torn cordyline leaf, or by using a leaf as a bookmark to a relevant bible passage. A baby refusing to nurse may "have a jumbie playing with it" and needs a bath of water and red dragon leaves. These examples show how security in land connects to spiritual security through movement from a yard's periphery to the center of the domestic space. Protection at the periphery stretches beyond the yard. Some Vincentians keep a potted cordyline by the front door and carry it to church or parade with it on government holidays. Several old men reported that young islanders used to tuck a cordyline leaf into the back of their trousers and avoid arguments with troublesome people by pointing to it and exclaiming, "talk to the dragon, not to me!"

The best example of cordyline's ontological relation to mobility is the Spiritual Baptist community, and it was their use of cordyline that drew me to St. Vincent (Fig. 7.7). I interviewed at least 25 Spiritual Baptists (some people were reticent about their religious affiliation) and attended four services at two different churches near Chateaubelair. About 10% of

Figure 7.7 St. Michael's Spiritual Baptist Church, Fitz Hughes

the population of St. Vincent are Spiritual Baptists (Kairi Consultants 2008, 26). As specialists in "shamanistic Christianity" (Zane 1999, 4), they also provide services like herbal therapy, dream interpretation, and psychological counseling to Vincentians of other faiths. Many are relatively poor Vincentians who rely on seasonal work, fishing, and gardening. This group was once pejoratively known as the "Shakers" for their ecstatic religious experiences, but here I follow their preferred name as much as possible.[10] It was at times difficult to investigate cordyline's significance with them because my questions often triggered hymns and trembling.

The social history of the Spiritual Baptists dovetails neatly with Vincentian struggles over land and labor. The sect developed from the work of Wesleyan Methodist missionaries to convert St. Vincent's slaves to Christianity in the early 19th century. In 1846, soon after the end of the apprenticeship period and in the early years of the contest between Black laborers and the white plantocracy, a new religious group, the Wilderness People, formed on the Calder estate in southeastern St. Vincent (Boa 2001, 191). Many were former Methodists who could not afford membership dues and Sunday clothing, or who had been expelled for "wild enthusiasms" (Fraser 1986, 251). Without access to churches, they met in yards. In 1862, an estate manager responded to a strike over rum, sugar, and molasses allowances by seizing crops from the workers' provision grounds. Ten days of marching, looting, and arson followed. A Methodist missionary and the attorney general claimed that Wilderness People led this protest (Marshall 1983).

By the 1880s, the white elites were denouncing the new religion as Shakerism and lamenting its spread among the working-class Black population. In the early 20th century, this anxiety grew into an all-out moral panic among the Vincentian middle and upper classes. Amid the dislocations of the 1898 hurricane and 1902 volcanic eruption, Anglican ministers became concerned enough to call for government suppression of Shakerism, particularly on the new government-run resettlements. In October 1912, the government passed the "Shakerism" Prohibition Ordinance, making religious practice punishable by fines or imprisonment. Landowners were subject to the same penalties for allowing the sect on rental land, so Spiritual Baptists hid or fled the plantocrats' estates and resettlement programs (Cox 1994, 221). Throughout this period of establishment and persecution, access to land and labor conditions cannot have been far from the minds of working-class Spiritual Baptists. These details do not prove that cordyline became part of this group's symbolic repertoire during this period, that the new religion expressed class consciousness, or that the institution of the red dragon migrated from the plantation political economy to a religious domain; instead, these correlations form a patchwork of circumstantial evidence.

The hidden Spiritual Baptists remained active throughout the socially rigid late colonial era and became a major political issue. The 1935 riots, during which 4 people died and 36 were wounded when white police fired into Black crowds, were led by pharmacist George McIntosh. He entered electoral politics, and from his position on the St. Vincent Legislative Council advocated religious freedom (Fraser 2011). The Council voted to repeal the "Shakerism" ban but was toothless against the governor's veto. McIntosh and his Labour Party kept up the political pressure until Ebenezer Joshua's Peoples Political Party successfully repealed the prohibition in 1965. Noting that the ban never defined Shakerism, during this period religious activists cleverly renamed themselves the Spiritual Baptists and proclaimed their intent to eradicate Shakerism. In the 1980s, the Spiritual Baptists were somewhat tolerated in St. Vincent (Young 1993, 163) and are now cautiously accepted by mainstream Christians. The Spiritual Baptists were only hidden from culturally unaware eyes during the 1912–1965 ban. Their services continued, quietly, behind closed doors and shuttered windows. According to a retired policeman who once broke up their meetings, Spiritual Baptists could signal their affiliation by keeping the red dragon in their yards, which after all looked like a secular boundary marker. "I believe that the popularity of the red dragon today came from them, because after [the ban] they could use it freely!" he guessed. My Spiritual Baptist interviewees agreed, saying that "we couldn't work the spirit hard, we had to be quiet, but now we are open with the dragon." One man explained that "we had the red dragon at home, but we could not show it on the street or at the baptism." The social boundary of the hidden sect was, in effect, in plain view all along because as one Deaconess told me, "the government did not know the dragon was a spiritual thing."

The social organization of the Spiritual Baptists reflects the individualism and focus on achieved status described for Vincentian kin groups. A high-ranking man can become a Pastor by preaching or a Pointer by guiding the mourning rituals described below; a high-ranking woman is a Mother and can, if necessary, also serve as Pointer. Many Pointers carry leather belts as symbols of authority and discipline (Zane 1999, 42). Subordinate positions, like Nurse, Inspector, Captain, Watchman, African Warrior, and Surveyor are determined by the skills acquired during out-of-body travel experiences. Many of these ritual specializations are precisely the middle-class occupations that lower-class Vincentians are unable to enter. A Nurse, for example, learns to care for patients during her own spiritual journey, and after relating her experience to the congregation, she displays her accumulated cultural capital. She makes a blue Nurse uniform with a white turban-like headtie to wear at church events and acquires a blue flag to carry in parades. The African Warrior (wearing red) dances around the church to keep away malignant spirits. A Surveyor in yellow rings a bell and consecrates the corners, doorway, and center post of the church with holy water to "mark the lines" of the sacred space. The flags in the rafters of the St. Michael church mentioned in this chapter's introduction correspond to these uniforms and represent particular congregants' journeys and social positions.

The scholarly literature (Henney 1973; 1974; Zane 1999) and my own experiences agree that most Spiritual Baptist congregations have a largely female membership and male leadership. In the services that I attended, there were four to five older men scattered among dozens of women and children. Many Spiritual Baptist men build vertical social hierarchies while women focus more on horizontal networking to accumulate social capital. Both strategies make status negotiable and attained through individual spiritual experience – and many of these experiences are guided by the red dragon. I begin this account of the role of cordyline in Spiritual Baptist ritual practice with a description of a service before following their spiritual journeys.

The cordyline leaves next to a white and red pole define the center of the St. Michael Spiritual Baptist church in Fitz Hughes. A typical Sunday service follows the Methodist Order of Worship, but becomes increasingly improvisational as participants "feel the spirit." Early in the service, congregants sing clearly, but deeper into the ritual, many replace the words in hymns with sounds and humming. Services follow a basic structure of welcome, hymns, a sermon, more hymns, and a closing, but this contains much individual variation and improvisation. The Pastor, Mother, or Pointer leading the service can start a hymn by singing a few words or tapping out a rhythm, or a hymn can be signaled from the pews. At any point a congregant may perform a "spiritual task" like preaching or seizing the bell from the center pole and ringing it toward the church's corners, doors, the pole, or the red dragon. Preaching and singing become an increasingly seamless perfor-mance, with speakers sliding from prose to poetry in sentences that become musical phrases. The entire church gradually becomes a huge percussion instrument, with many people pounding their bibles (specifically chosen for their tonal properties) with one hand while they sing. This pattern continues into the sermon, during which participants hum or quietly sing completely different hymns. A good speaker knows how to improvise musical flourishes within the sermon and gently guide the congregation's individual hymns, claps, tapping, and movements to intersect and build into a polyrhythmic *a cappella* symphony that makes the air shake and one's chest vibrate. The verbal cues that I heard repeatedly were phrases about the day's events, which preachers used to syncopate the people humming a gentle "mmmmm ooh oooh mmmmm" with the rhythm section worshippers singing "bum bah bum bah bum." In morning services, I heard "I suffer many things in my sleep this MORNING!" and "a stranger in our midst this MORNING!" (the latter undoubtedly referring to me). The sung responses of "amen!" unified the congregants' diverse forms of participation. In an evening ser-vice I heard the phrase "we ask you to bless our boundaries and center this EVENING!" used the same way. Listening to this gorgeous antiphonal music is the sonic equivalent of watching a flock of birds becoming one vast creature wheeling in the sky.

During these services, many Spiritual Baptists experience the religious ecstasy of a *doption*.[11] Doptions are the percussion section in the symphony described above. They are variable and often idiosyncratic but composed of the same elements. Like the hymns, the doptions make the air throb with intensity and purpose. The performers are improvisational soloists who gradually become a coherent chorus (Henney 1973, 234). The dop-tions that I saw corresponded to Wallace Zane's "Number One doption," in which an "individual stamps on one foot while bending from the waist, then breathing sharply while straightening up" (1999, 101). A doption begins when a Spiritual Baptist concentrates while grunting or groaning in response to a sermon or a song. Trembling, the person steps forward on one foot, the shoulders in line with the knee, while extending the corre-sponding arm forward. In a single motion, the person jerks the foot, knee, shoulders, and arm back while vocalizing. In several of the services that I attended, one woman performed this doption while uttering her distinc-tive "ayyyy-YUP;" some gave a guttural "uhhh!" and others "woooh!" or "oh Jesus!" The Pastor monitors the idiosyncratic twirling, stomping, and bell-ringing participants, and gently steers them toward a syncopated cli-max by modulating the sermon's content, interspersing it with fragments of hymns, stomping, or tapping a shepherd's crook. The preacher signals closure by reaching the end of a familiar biblical story or by interjecting a hymn with a completely different beat than the music and movement being performed. Soon the choral symphony breaks up into individualized solo performances, the stomps become quieter, and the intervals between gasps longer. The Pastor signals a congregant to lead a final hymn, and after a pause, makes a few announcements before the Spiritual Baptists spill out into the street.[12]

Along with music and the central pole, the red dragon is a ritual attractor through which strongly individuated Spiritual Baptists achieve what anthropologists call "communitas" (Turner 1969), a ritualized experience of intense solidarity and community that is usually lacking in Caribbean societies. The attention to the church's corners and the Surveyor's work reflect access to provision grounds and house plots on estates. The images that appear in Spiritual Baptist practices, like ships, captains, whips, keys, locks, uniforms, and flags are clear legacies of slavery and post-emancipation nation-building. These services are embodied colonial history, like Kilimanjaro's ancestral sacrifices, Oku's masquerades, Papua New Guinea's singsings, and French Polynesia's firewalking. In each of these embodied practices, boundary plants connect material matters of claiming and managing land to symbolic actions of creating and maintaining social relations. For the Spiritual Baptists, the red dragon constructs place and personhood most powerfully in their spiritual journeys far beyond St. Vincent.

For Spiritual Baptists, *mourning* (pronounced "mooning;" the term comes from Matthew 5:4 "blessed are those who mourn") is a central religious experience and the way they attain social status and roles in the church. A detailed account is available in Zane's ethnography (1999, 107), so here I provide a bare-bones summary to contextualize the accounts that follow. Each church has at least one small room with a thin mattress on a concrete floor for mourning. The mourner, who is called a "pilgrim" in reference to Bunyan's allegorical *Pilgrim's Progress* (1987),[13] spends eight to ten days there praying intensely while partly blindfolded with tightly tied cloth bands sealed with candle wax. A Nurse provides simple food and water and escorts the pilgrim for bathroom breaks. The Pointer and a Mother visit the pilgrim to read bible passages and pray, but she is mostly alone until she experiences a spiritual journey and receives spiritual gifts. It is here in the mourning room that the red dragon interacts with Spiritual Baptists and becomes a particular sort of moral person and flight instructor.[14] A Deaconess summarized how the plant regulates spiritual movement; "the red dragon is the guide, and it takes us to every land in the spiritual world." One Mother generalized about her mourning experiences, saying

> When you are starting your journey in the mourning room, they give you a leaf of the red dragon. You hold it, only in your right hand, and you look at it as you pray [she mimed pumping her right arm up and down vigorously while holding a leaf in her fist]. Then the red dragon takes you to the cities of Africaland, India, China, and Jerusalem. It is the guide for all journeys. It appears right in front of you and leads you on up the Zion Hill to the boundary of heaven.

Spiritual Baptists refer to these destinations for soul travel as "over yonder." The destinations are consistently parts of the colonial British Empire, but only rarely places in the Caribbean, Canada, New York, or the United Kingdom where labor migrants go. In these "cities" the pilgrim acquires spiritual gifts, like songs, dances, and language, or occupational skills like nursing or baking bread, from other travelers. Mother Ogaro received the cordyline on her doorstep as a gift from the Africans over yonder along with specific instructions for how to place the red dragon amid the other "carnal" plants in her garden.

The most important location for the red dragon in the spiritual lands is at the boundary of life and death. Cordyline guides and halts the pilgrim's movement. In the geography of over yonder, a hedge of the red dragon separates the spiritual lands from heaven itself. One man explained that

> In the spiritual world, the red dragon means "stop there!" on Zion Hill. If you meet it you stop. It is the end of your journey. You need to find that one, if you don't, your mourning continues until you find the red dragon. If you go beyond that red dragon, you will die and not come back. You die in the spirit or you get lost and not come back, so you don't wake up. The red dragon says, "you stop right there!"

He continued with his own experience:

> One time I went to mourn, a day like today. I meet a big water [gestures at the sea in Chateaubelair Bay, visible from his porch], it was very pretty, it was beautiful. On the other side of the water there was a red dragon hedge and a white man, he said "come, come." I stand up, I put my left foot forward, but he said no. I put my right foot in the water, and then a hand grab me by the head and pull me right back. Voice said, "my son if you cross that water you will be in trouble." The red dragon there was the boundary for Zion.

Finding this hedge is how a pilgrim knows that the journey is over and time to turn around. "We gather at the red dragon in the spiritual land at the boundary," a male elder told me. "You meet both strangers and people you know there at the dragon," and it is from these people that mourners receive gifts.

Sometimes things can go very wrong, however. One man said that the Nurse tending him while mourning almost let him "go over yonder" because she was inexperienced:

> I was in the spirit land, and the people there were questioning me. After a while, I find myself down from the height I was at the city on the hill, and there was a lot of red dragon in a line. Those people, they put it there as a boundary. When I mourn, I was up there, I see a gentleman

up there in charge of the place, like comes a multitude of people, they pick you out. These ones, they belong in heaven, put you to that side, to hell that side. So there were three sorts of people, those for heaven, for hell, and those like me told to return. That Nurse did not know how to keep my bands tight on my head and bring me back, but I find my way!

For the Spiritual Baptists, cordyline is unlike secular boundary plants in the Vincentian landscape such as *Gliricidia* and lavender grass because, as one Mother put it, with the red dragon, "there is power, just like a stop sign shows the power of the government, and the red dragon shows you what to do, even from a distance." Unlike the vegetative gaze of dracaena that disciplines people on Kilimanjaro, for the Spiritual Baptists the red dragon guides them to freedom. In St. Vincent, an Oceanic boundary plant determines the vectors of freedom for members of a severely limited post-slavery peasantry.

This red dragon continues to shape a mourner's experience after Zion Hill. Pilgrims fly back to St. Vincent by following cordyline. The leaf commands the traveler to "follow me!" back across the Atlantic. The vase of cordyline leaves and the white and red striped pole at the church's center (which resembles some 19th-century British colonial lighthouses) are navigational beacons for this flying pilgrim to find their body back in the church's mourning room. If the pilgrim receives a cordyline leaf in the spirit lands, its color shows their gift and determines their uniform and flag color. A red leaf is the gift of power, so that mourner has the gift of power, wears red to church, and parades with a red flag as an African Warrior. The Inspector and the Watchman get green leaves of peace, and a yellowish-white leaf shows the love that Surveyors mark with bells and holy water. Variegated cordyline leaves represent Joseph's "coat of many colors" for many Spiritual Baptists, and mourning entitles a woman to wear the correctly colored leaf of the red dragon tucked into the left side of her headtie as a "memorial" of her spiritual journey. After the pilgrim arrives back in her body, she must perform a "shouting" to prove her success over yonder. At the shouting I attended, we sang "All the way from Africa-land, coming HOME to hear them singing" over and over as the pilgrim arrived at the church hall accompanied by ten older women wearing white with yellow crosses on their headties. The pilgrim wore a white robe with red, yellow, white, and green cloth bands around her forehead and eyes. She held a white candle with leaves of the red dragon around it, and the hot wax slowly dripped onto her hand as she stood in the doorway, waiting for us to complete the hymn. She slowly entered the church with her entourage, and after one of her male kin gave a biblical reading about moral order, joined the exultant congregation in singing "then was the baby boor-oor-ORN!" repeatedly while we all danced counter-clockwise eight times around the church. I was pressed into the line as the third person behind the pilgrim. After a few more hymns, the Pointer called for her to explain her journey, singing, "you went to see-ee-EE himmm!" Unfortunately, I did not hear her testimony. Her male kin were giving lengthy explanations of her journey's significance, and I had to leave at 2 am. I like to think that she became a Nurse with a reddish-purple cordyline leaf bobbing proudly on her headtie.

Although many Spiritual Baptists insist that the red dragon "came from Africa" and guides them to "African wisdom," this web of significance was probably woven anew in the Caribbean by poor Vincentians. In St. Vincent, a vase of red leaves and a humble potted plant next to a doorway are cosmological signposts and embodied histories of colonial workers yearning for escape, land security, and meaning in the face of a white supremacist plantocracy. Most of the scholarship on the Spiritual Baptists focuses on their beliefs, symbols, and ritual practices, but their flights to the boundary of heaven with cordyline leaves suggests a material logic behind these symbols. It is clear that Spiritual Baptists and other Afro-Caribbean religions express both African symbolic continuities and creolized Christian institutions. Spiritual Baptist practices are, however, likely rooted in the material experiences of an emerging peasantry struggling against an oppressive and racist system of land ownership and plantation labor. The structural power of the red dragon for the Spiritual Baptists answers the institutional power of the plantation in Vincentian history. Their spiritual journey narratives are, like labor migration, extraterritorial adventures that shed the shackles of political economy and allow travelers to achieve new sorts of personhood and accumulate cultural capital (Wardle 1999). In this process, boundary plants relate to movement, not fixedness. Compared to the vectorial, relational, and revitalized sorts of personhood that this book has explored, cordyline and its Vincentian pilgrims have a "mobile personhood."

Cordyline is a polymarcating boundary plant in St. Vincent that links political economy to social organization and personhood, but in strikingly different ways than the plant's Oceanic biographies. Its vegetative agency has been elaborated into an idiom of human agency in material and spiritual landscapes, but without a focus on collective well-being. The Spiritual Baptists encounter the red dragon in solitary religious experiences that confer personal distinction, not communal experiences. Did cordyline start as a mechanism for access to land and only later become culturally elaborated as a symbol of movement and freedom? I asked many Spiritual Baptists several versions of this question. Most found my search for causation misplaced, admonishing me that "only one big work of the red dragon is for bound mark, but for religion it do many things!"

Conclusion

Superficially, the trajectory of cordyline (and, to a lesser degree, dracaena) in the Caribbean looks similar to this book's other case studies. The red dragon appears on the peripheries and at the center of socially significant units of land. It has materialized and objectified human agency, and shaped the embodiment of history, even under viciously structurally violent conditions. It signifies protection and social order. It creates a triangle of political economy, social organization, and personhood – but in a distinctly creole and Caribbean postcolonial manner. This cultural construction occurred very quickly compared to the millennia-long timelines of African and Oceanic boundary plants. Cordyline's entangled meanings and practices do not reduce to continuity from West Africa, as if enslaved people and their descendants simply recognized cordyline as being the similar enough to African dracaena. Such an approach to identifying cultural commitments as static traits ignores history and the processes that make it. The "how and why" of Caribbean boundary plants matters more than "what and where," and the patchy evidence suggests that cordyline was Africanized in St. Vincent through creole place-making as an oppressed peasantry fought for rights and authenticity in provision grounds and residential yards. This was a different sort of colonialism compared to Africa and Oceania, and the biography of its boundary plants follows these historical contours. This particularistic local history of strategic struggles, narrative performances, vegetative agency, and moral personhood shows that the power of boundary plants is contextual rather than intrinsic. The evidence suggests that sugar estates acquired red cordyline from the St. Vincent Botanic Garden as a cheap, convenient, and colorful land marker, that this practice spread beyond the estates' fields to the semi-autonomous zones of slaves' homes, kitchen gardens, and provision grounds, and then became increasingly elaborated as a Yard Society institution and symbol of freedom and status.

The nuances of Caribbean boundary plants also demonstrate the originality of their creole institutionalization. In the African and Oceanic case studies, many people reported declining, simplifying, and disentangling boundary plant institutions in their societies. They lamented the loss of collective meanings and group organization as land privatization and individualism became increasingly influential. The story of cordyline in St. Vincent shows the opposite trend. Vincentians do sometimes regard individualism as a problem. Elders miss calling "good morning" back and forth across hillsides as neighbors met the new day. The red dragon has become somewhat group-oriented in yards, mountain plots, and churches since its 1793 introduction by British botanical imperialism but is still strongly individuated. Unlike this book's other case studies, Vincentian cordyline does not concern a cosmic life force that humans can tap into, affix to land, and direct to foster social order, fertility, and prosperity. This Caribbean technology of power focuses on individual access to land, personal status, and mobile personhood instead of groups and fixedness, and this individualization is rooted in 19th-century struggles over land and labor rather than recent social change. In Besson and Mintz' terms, boundary plants were creative creole responses to the contradictions of the plantation mode of production, and a medium for resistance to the predicaments of slavery (Besson 2018), not jagged fragments of ancient African institutions surviving the Middle Passage and modernity.

St. Vincent's red dragons demonstrate boundary plants' vitality during the colonial annihilation of social forms and cultural meanings. This thoroughly modern creole institution is an example of resistance at the heart of the Plantationocene. Looking forward, cordyline can contribute to St. Vincent's search for a more just post-plantation economy and recovery from the 2021 volcanic eruption. Only half of the smallholders and farmers on St. Vincent have secure title (Isaacs 2014, 29), and formal recognition of their already robust property rights institution would make the law more congruent with actual social practices. When I visited the St. Vincent Lands and Surveys Department offices to examine old plantation maps, the staff didn't know much about dragons. One surveyor recalled using dracaena to locate iron pegs and concrete pillars, but no one recognized cordyline as a boundary marker. A focus group of rural Vincentians in Petit Bordel responded to my account of that day's work by saying that "sometimes the survey people pull up the plant for the concrete post, but then you can plant your bound mark next to it because concrete is for the government but the red dragon is for the people." Recognition of cordyline as a legitimate part of St. Vincent's social landscape and cultural heritage could, I hope, begin a new phase of creole institution-building.

Notes

1. According to the first *Hortus Kewensis* catalogue (Aiton vol. 1 1789, 454), Benjamin Torin, a British East India Company administrator in Calcutta, donated a *"Dracaena ferrea"* (red cordyline) to Kew Gardens in 1771 from a source in China (Bretschneider 2011, 149).
2. Although cordyline is an ornamental plant in the Grenadines archipelago, it is most socially complex and polymarcating on the main island of St. Vincent. On Bequia, the northernmost of the Grenadines, cordyline is "just a flower." This chapter's references to St. Vincent mean the island, not the nation-state of St. Vincent and the Grenadines.
3. Some of these authors mistakenly refer to "red dracaena." I have corrected these to cordyline. I omit references which lack enough description to identify the species, like Herskovitz and Herskovitz's comment that "dragons blood" protects Trinidadian farms from the "evil eye" (1976, 249).
4. One researcher estimates that marijuana production and distribution involve about 40% of St. Vincent's population and constitute about 30% of St. Vincent's GNP (Edmonds 2014).
5. On the origins of the Black Caribs, see Davidson (1787) for the shipwreck theory and Kim (2013) for the drift voyage from Barbados theory. Crawford (1983) determined that St. Vincent Black Caribs' gene pool is approximately 50% Amerindian. The classic ethnography of the Garifuna is by Gonzáles (1988).
6. Barry Higman (personal communication) suggests that bright red cordyline stalks would have been excellent for marking a day's work, like by showing the area for a Trinidadian slave to dig 350–400 sugar cane holes (Higman 1984, 180).

7. The British government compensated planters for their loss of property rights in people. The St. Vincent planters received £550,777, which was moderate in regional terms. It was about half the amount per capita paid for enslaved people in British Honduras and twice that paid in Bermuda (Levy 1980, 55).

8. In this resettlement scheme's early years, many Vincentian men worked on the American Panama Canal project (1904–1914) to earn cash to purchase these plots (Smith 1915, 4), which deprived their families of adult male labor.

9. The major exception I heard was from several elderly women who said that cordyline relieves menstrual cramps. "Dragon is for making tea, when you get the really red leaves at the right time to make full moon tea," they said. "That's why we say 'it red'em, boil dem bush!' Also, for get them baby!" Lans and Georges report similar medicinal uses for cordyline in Trinidad and Tobago (2011).

10. The corresponding sect in Trinidad was called the "Shouters" (Forde 2019; Glazier 1983; Lum 2000).

11. Some Spiritual Baptists refer to Romans 8:22–23 to explain this term. The Apostle Paul writes that "the whole creation groans and travails in pain" and people "groan within ourselves, waiting for the adoption, the redemption of our body." Paul uses the Roman legal process of slaves or indentured servants being adopted as sons as a metaphor for becoming "sons of God." The Spiritual Baptists use doption to describe the physical experience of ecstasy.

12. Doptions are not the only ecstatic performances in a Spiritual Baptist service, just the most common. I witnessed a dance-like "spiritual swordfight." Two men were singing with bibles in their left hands and striking the books like drums. As the hymn's intensity built, the men met at the front of the church, smiling. Without missing a beat, they mimed a fencing match, using their flattened right hands like swords. They crossed their raised forearms, THUMPED the bible, another cross, THUMP, and a handshake, grinning all the time. See also Zane (1999, 93).

13. Church leaders advised me to read this 1678 text to understand Spiritual Baptist journeys. Bunyan's story is a dream narrative about a journey from the earthly City of Destruction to the heavenly Celestial City atop Zion Hill. This book is the mythic charter for Spiritual Baptist journeys and a source of symbols and images.

14. For a review of the motif of human flight in the mythology of the Black Atlantic, see McDaniel (1990).

8 Conclusion

Beyond Boundaries

I live in rural Vermont, far from this book's tropical landscapes. Still, I am surrounded by boundary plants and vegetative boundaries. Like many Americans, my social status depends on my ability to keep the lawn mown, as neighbors have reminded me from time to time. I planted lines of day lilies (*Hemerocallis* sp.) along the borders of my land and around the mailbox, mostly to define decent-looking edges for my mowing. Some of the trees in my yard are memorials to nonhuman kin because my family has interred the ashes of the dogs we've lost over the years beneath crabapple (*Malus* sp.), *Hydrangea*, and horse chestnut (*Aesculus hippocastanum*) trees. In the woods behind the house, there are stone walls and split-rail fences from the 19th-century heyday of sheep farming here in the Champlain Valley. Inside, I have four dracaena and two cordyline plants that I have grown from cuttings or received as gifts. I received one of the dracaenas, a *D. bicolor* that climbs the stairwell to the second floor, at my mother's funeral in 1995. The bright pink and red cordyline in my sunroom has accompanied me to political ecology lectures. All of these plants represent the boundaries of social relations (of neighborliness, of companion species, of landscape history, of kinship, of professional authority), but none are as complex and polymarcating as what I encountered on my journey to this book. One of the convictions that propelled me through this project was the notion that stories about tropical land managers and their boundary plants can inspire others to live gracefully in meaningful landscapes, and to view farms, gardens, and yards as social and cultural processes instead of static backdrops.

The first part of this concluding chapter revisits the introduction's themes, with particular attention to the methodological and analytical lessons of boundary plants. The second section summarizes and compares the deep histories of boundary plants as components of social-ecological systems. Next, I discuss the shifting nature of territorialization in the five case studies. The final section extrapolates the potential of boundary plant political ecology for building a more sustainable, just, and culturally diverse world system.

DOI: 10.4324/9781003356462-8

Methods revisited

The introduction defined this book as a multi-sited ethnographic political ecology of ethnobotanical institutions. Each boundary plant story built upon the methodological strategies and analytical insights encountered in the others. Together, they form an ethnography of "lateral comparison" (Candea 2018, 16) across five case studies. One of anthropology's classic tropes is that ethnographers present their subjects as culturally different from themselves. Matei Candea calls this "frontal comparison" that homogenizes both "other" and "us." This contrast recalls other hoary analytical binaries like traditional/modern, simple/complex, and primitive/developed. Lateral comparison also has conceptual pitfalls. Comparative data risks becoming like a seashell collection – lifeless structural forms without much information about the creatures' lives. Anthropology benefits from comparative methods that stay in the "just right" Goldilocks zone of not-too-broad and not-too-narrow (Borofsky et al. 2019). But this ethnographic sweet spot varies from project to project and has no foolproof recipe. I oriented this book by focusing narrowly on two plants and making each case study a broad account of political economy, social relations, and cultural meaning. This strategy for multilateral controlled comparison has revealed some patterns, but it does not necessarily explain their specific contents or interpret how and why boundary plants formed separate but similarly tangled meshworks. These tasks require attention to the unique historical sequences of events through which economy, society, and culture co-emerge.

Tim Ingold's lines are threads of movement through space and tracks left on surfaces (2016). The boundary plants in this book form both kinds of living connections. This book is itself a line that braids separate threads of people and boundary plants into one narrative rope. I began this project expecting to walk this tightrope using political ecology as a stable balancing pole. Instead, I found my balance shifting repeatedly as I contextualized each set of field data. I kept finding that each case study's concepts sparked questions for others. Did a variation of concept A apply to another case study, or did I need concept B? The concept of ritual attractors, for example, comes from studies of sacred space in Oceanic House Societies (Fox 2006), but informed the Kilimanjaro chapter's analysis of Chagga homesteads. I examined vegeculture in Papua New Guinea (Barton and Denham 2018), but then had to ask what it might mean in Remote Oceania. This conceptual de-centeredness fostered creative juxtapositions in the multi-sited data set (Falzon 2009). The multiple fieldsites taught me the value of strategic defamiliarization. Just because the same plant was doing similar things in different places, like the cordylines marking graves in Papua New Guinea and the eastern Caribbean, did not necessarily mean that the phenomenon was the same social relationship and institution.

This book's case studies contextualize and historicize boundary plants in social-ecological systems. The structure of each chapter moves from political economy to social organization and only then ideologies and symbols. Each describes the linkages between these analytical levels as strategic struggles over resources and narrative performances that construct social relations and legitimize moral personhood. This book draws on both structuralist and poststructuralist political ecology to examine power. My approach starts with Alf Hornborg's structuralist definition of power as a "social relation built on an asymmetrical distribution of resources and risks" (2001, 1). I then use Eric Wolf's four modalities of power (1999) to analyze how interpersonal and institutional power drive some social relations, while cosmology and structural power shape others. The more I did this, the more I concluded that Wolf's three-part typology of social labor (1982) was the best way to frame the contrasts among the case studies. The roots of power are, as Henri Lefebvre would predict, different asymmetrical social relations in societies organized by kinship and gender than those constructed around tribute or racialized capitalism (1991).

But structure is a blunt tool for interpreting how boundary plants relate to cultural revitalization in French Polynesia, beauty in Papua New Guinea, and the delight that Oku people take in fertility. My political ecology of the good (Ortner 2016) demanded that I reject a hardline materialist position that ideas and emotions are side effects of material phenomena and mystifications of political economy. Symbols, discourses, and subjectivities are important components of social-ecological systems because they legitimize access to material and social resources, direct their flows, and make accumulations of economic, social, and cultural capital meaningful. Thinking of a boundary plant as of person is neither an intellectual error nor an ontological reality; it is instead an interpretive tool for investigating how meanings are embedded in specific political economic and social contexts (Nadasdy 2021). Dracaena's vegetative gaze on Mt. Kilimanjaro, for example, is significant for linking Chagga social organization to landholding and moral order. An analysis that stopped at plant agency and a vegetative point of view might miss what makes dracaena so important in Chagga thought. Approaching Chagga assertions that "dracaena has eyes" as metaphors for deeper structures sidesteps subjective Chagga experiences of the plant as a moral (male) person who constructs order by opening and closing relationships. A poststructuralist and feminist political ecology of boundary plants requires attention to ways that resources, relations, and persons are themselves constituted by the nature of power in a particular society.

Environmental anthropology must be pluralistic and open to differing connections among agency, political economy, social organization, and cultural meaning. Each social-ecological system is an entangled multi-species meshwork, and it is the task of political ecology to cut these knots with hatchets and plant seeds of hope. Knowing where to cut and how to plant is a matter of prioritizing particular tangles for analysis and intervention. My method for doing political ecology in this book knots together material struggles, cultural narratives, and indigenous points of view. There is an inevitable tension between my materialist exposition and my interpretive goal. In my experience of the classic anthropological dilemma of outsider and insider points of view, boundary plants taught me that these can be mutually constituted. Ma'ohi used cordyline to critique the French Polynesian social order and revive indigenous practices, but based this on old ethnographic texts, not living traditions. The lesson for multi-sited political ecology is that there may be no primary causes of continuity and change, like global neoliberal capitalism or local cultural distinctiveness. Boundary plants point to mutual causation among political economy, social organization, and cultural meaning, each of which constitute and substantiate the others. These imperfect abstractions are necessary to construct comparisons and produce knowledge. The roots of boundary plants are tangled meshworks (not ontologically separate worlds), but their ethnography must form clear narrative threads. The map is not the territory, and the ethnography is not the social and cultural context – but both can guide some understanding.

The two plants described in this book share the botanical property of vegetative propagation. Stalks thrust into the ground often take root, making it easy to inscribe lines on a landscape. This biological feature predisposes social engagement, like when British people share cuttings of willows, aloes, and African violets so that recipients can grow their own houseplants (Ellen and Komáromi 2013). Vegetative propagation is socially elaborated into gift exchange, but seeds are not. To a degree, vegetative propagation invites symbolic construction. Mary Douglas argued that the pangolin, a scaly African mammal, is a "natural symbol" of mixed categories (1973). A cutting is a natural symbol of sameness and continuity because the new plant is a clone of its originator, but seeds naturally symbolize children and generational succession. A plant turn scholar might undermine anthropocentric assumptions that only people construct natural symbols by asserting that vegetatively propagating plants are agents constructing meaning. This is an interesting idea, but the examination of distributed agency does not explain when, how, and why plant agency affects social and landscape histories. From Kilimanjaro to St. Vincent, this book's case studies show that boundary plants are sites of struggle and creativity, and that these thoroughly human dynamics have explanatory priority over plant agency. The plant turn introduces new conceptual tools, but its focus on agency is, in my opinion, unproductive without attention to structure (Giddens 1993). Plant agency is, like the ontological turn, more method than truth (Holbraad and Pedersen 2017).[1] Ontological and posthumanist approaches to vegetative propagation and the emergence of polymarcating complexity in more-than-human meshworks ask good questions about the phenomenon of boundary plants, but political ecology provides better answers.

Understanding boundary plants requires more attention to moral personhood than to their agency and selfhood because they are often conceptualized as persons with particular social roles. On Kilimanjaro, dracaena is a witness; in Papua New Guinea, cordyline is a watchman; in St. Vincent the red dragon is a guide. Personhood is a collective representation and a cultural construction (Mauss 1938), and these examples demonstrate that the personification of these boundary plants reflects the different organizations of social labor in these places. Plant personhood is different in a relatively "horizontal society" based on kinship and gender compared to more "vertical societies" rooted in tributary and capitalist hierarchies. In both the Oku monarchy and precolonial Ma'ohi chiefdoms, boundary plants are elite tools for disciplining society and regulating subjectivity, not autonomous persons. Interpretive devices for culturally nuanced models of personhood, like this book's vectorial, container, relational, revitalized, and mobile persons, offer plant turn scholars ways to move beyond plant agency. Ontological questions about indigenous concepts for boundary plants' power and vitality, like Oku's *keyoi kejungha* and Remote Oceania's *mana*, further challenge ethnobotanists to consider how cultural difference shapes their accounts of plant-human systems. The methodological lesson here is that studies of plant personhood and meaning should contextualize plant-human relationships according to a specific political economy and a culturally meaningful articulation of structural power.

This book examines boundary plants first as resource access institutions and only then examines their implications for social organization and the cultural construction of place, personhood, and meaning. This has six implications for New Institutional Economics (Galiani and Sened 2014). First, the case studies support a generally substantivist view of economics as being so embedded in social relations that a strictly formalist view of property rights would miss what makes boundary plants significant. Their purpose is moral and social order, not economic efficiency. In direct opposition to Hernando de Soto's insistence that property without formal title is "dead capital" because it cannot be invested (2000), these boundary plant stories show how people accumulate and invest "lively capital" in material, social, and symbolic forms. Second, these plants demonstrate that boundaries cannot be understood simply as things because they are social and cultural processes. Boundary plants are verbs, not nouns. They are inclusive and connective conjunctions as much as they are exclusive separations (Barth 2000). Third, dracaena and cordyline break down New Institutional Economics' distinction between formal and informal institutions that constrain behavior (North 1990). These plants often convey unwritten and informal rules, but in context, they are usually wholly formal contracts and commitments. Instead of constraint, these living institutions foster negotiation, flexibility, and change. Fourth, the ways that these institutions minimize transaction costs, define incentives, and reduce uncertainty only make sense in specific contexts. Often it is the misfit between boundary plant institutions and statutory law that produces friction, perverse incentives, and pervasive doubt, and these contradictions are historical, not evolutionary. Fifth, viewing these vegetative property rights institutions as "bundles of power" conferring access to resources (Ribot and Peluso 2003) and mediating economic, social, and cultural capital accumulation (Bourdieu 1990) accounts for the polymarcating complexity of boundary plants better than simply identifying them as "rules of the game" (North 1990, 3). Finally, the path dependencies of boundary plants are not simply internal characteristics of isolated economic systems. Instead, each case study showed that property institutions emerged and transformed through the interactions of indigenous practices with colonial rule, usually by getting reified as decontextualized customary law.

Boundary plants provide methodological lessons for students of anthropology, political ecology, ethnobotany, and institutional economics. This book's multi-sited ethnographic political ecology of ethnobotanical institutions focuses on just two plants, but I hope that the topic grows into a multistoried and multispecies garden. The comparative questions that remain are why such similar institutions, social roles, and meanings have sprouted from these two particular plants in five sites, why these stories feature similar dynamics of entanglement and disentanglement, and why boundary plants matter for sustainable rural livelihoods. The next three sections attempt to answer these questions by reviewing the past, present, and future of boundary plants.

Boundaries and routes of power in the past

Each chapter described the history of a property rights institution, its associated social forms, and its cultural elaborations. These trajectories demonstrate continuity, change, and hybridization. The social forms include patrilineages, ranked clans, flexible territorial groups, postcolonial indigeneity movements, and enslaved classes. The cultural elaborations show less diversity and instead cohere loosely around issues of order, morality, protection, vitality, fertility, beauty, autonomy, and peace. Why, then, have the same two plants been significant in culturally similar ways in such diverse social contexts? Why do these case studies show so many "family resemblances"?

Explanations fall into three conceptual boxes. The first, a diffusion model, is simplest. Once a plant was institutionalized as a boundary marker and culturally constructed as legitimate and meaningful, it became a social commitment and a path dependency. As people moved around a landscape or a seascape, the familiar institution helped them recreate society on new frontiers. The social production of space and the cultural construction of place continued after these institutional seeds had been planted, and this explains the coexistence of minor variations and shared regional themes. This scenario fits the patchy evidence for dracaena's part in the Bantu expansion and for cordyline's role in the Austronesian expansion, although both need further archaeobotanical evidence. It does not, however, account for the similarities between the clearly separate trajectories of boundary plants in Africa and Oceania, and diffusion is an anemic theory for understanding boundary plants in the colonial Caribbean.

The second conceptual box is the convergent evolution model. Pyramids existed in both ancient Egypt and Mesoamerica not because of diffusion, but because of limits on ways to stack up rocks. This approach suggests that because there are only so many ways to institutionalize resource access, cheap and efficient boundary plant institutions have been independently invented by agrarian societies. Incentivized by population pressure on resources and market opportunities, smallholder farmers secure access and intensify production by establishing "perimetrics" on their land and investing in the social relations that legitimize those markings (Netting 1993; Sheridan 2008; 2014; Stone 1994). This approach articulates the economic logic of boundary plants but does not address how and why different social relations of access lead to similar meanings.

The third model starts from Wolf's tripartite scheme for the organization of social labor through kinship and gender, tributary, and capitalist modes of production. Each social type has a different dynamic for the social production of space and the cultural construction of place through strategic struggles and narrative performances. Each boundary plant dynamic is a different territorialization process, based in part on a vegeculture that interweaves the biological properties of plants with social institutions and cosmology. Each shaped particular constructions of moral personhood and subjectivity. From this poststructuralist political ecology perspective, the vitality, peace, and protection contained in boundary plants show how cultural capital and structural power relate to specific contested histories of accumulation and legitimization. These cosmological meanings correspond to different experiences of personhood in societies at the margins of colonial capitalism and neoliberal globalization. This makes boundary plants sites of autonomy and cultural distinctiveness. This admittedly highly theoretical analysis suggests that the similarities of form, function, and meaning among the case studies is a relatively superficial issue compared to the particular histories of boundary plants and societies. The focus in this book has not been on the fact that cordyline leaves protect the edges of yards in Papua New Guinea, French Polynesia, and St. Vincent, but has instead examined the different histories, institutions, and cosmologies that are rooted below them.

This book has used each of these models to examine different aspects of the case studies. Each provides a lens for interpreting the data, but none of them fully explains these boundary plants. Dracaena and cordyline were technologies of power for making place, social relations, and meanings on the moving frontiers of the Bantu and Austronesian expansions, and the similarity of this process among tropical agrarian vegecultures may account for some of the similarity of these trajectories. Both mediated the disruptions of colonial rule, and both shape the messy contradictions of continuity and change in the postcolonial societies of tropical Africa and Oceania. These stories interact in the St. Vincent case study, where the creole creativity of the island's postslavery peasantry used cordyline to construct new institutions and meanings against the violence of the plantations. Each trajectory is like a ball set in motion by a different organization of social labor. Each then followed a different historical pathway of continuity and change and a different mix of determinism and originality. Their juxtaposition allows comparison. The contrast between the use of cordyline in Ma'ohi cultural revitalization and the continuity of dracaena in the Oku monarchy, for example, shows how boundaries of social rank transform and persist. Comparing Kilimanjaro with the Papua New Guinea highlands leads to the analysis of how flexible land boundaries relate to fluid social ones. Overall, understanding the past of boundary plants depends more on comparative social history than botanical similarities.

Boundary plants and the roots of power today

Each story in this book argues that lines of living plants left tracks and traces on society and culture. Each formed a different meshwork of polymarcation. In most of these case studies, people reported boundary plants becoming disentangled from collective relations, stripped of layers of meaning, and reduced to monomarcating economic institutions tied to private property. The evidence suggests that these changes are more nuanced than a generic simplification. Chagga farmers on Kilimanjaro report that disentangled *masale* means that peace and social order are in decline, while the Oku monarchy struggles to align demographic and political messes with the social order of *nkeng*. Papua New Guineans use *tanget* plants to assert claims on and against people beyond their *wantoks*, tourists, and transnational corporations. Missionary governance led French Polynesians to discard the connotations of *auti* for social rank and revitalize it as an emblem of indigenous autonomy. In St. Vincent, the Spiritual Baptists escape structural violence by flying to the edge of heaven with the red dragon. All of these demonstrate boundary plants interacting with the expansion of the scale of social relations from localities to regional and global networks. These polymarcating institutions have transformed in content and application, particularly because of colonialism, but none have become inert monomarcating property markers like concrete posts and cadastral maps. Dracaena and cordyline are still ways to negotiate appropriate property relations and perform moral personhood, but in changing ways. They are unlikely to become simplified to purely economic institutions as long as kinship, gender, ancestors, rank, indigeneity, and cultural autonomy matter in Africa, Oceania, and the Caribbean.

The disentanglement that people report instead indicates their experiences of the changing roots of power. It is difficult to explain and interpret these roots using classical social theory about how power creates continuity and order. Émile Durkheim would argue that boundary plants are mechanisms for mechanical solidarity in small-scale societies in Tanzania and Papua New Guinea, and elements of organic solidarity in Cameroonian and French Polynesian stratified societies. This functionalist perspective explains how vegecultures bind people together but does not address colonial conflict and contradiction. Max Weber would suggest that the worlds of boundary plants are becoming disenchanted as rational-legal institutions eclipse traditional ones. As title deeds replace boundary plants, they will dwindle to thin decorative stalks instead of robust clumps of vitality and order. This approach to reordered economics and meaning does not account for the ways that societies continue to re-enchant their worlds despite whatever iron cages state bureaucrats put them into.

Conflict-based approaches address power as a coercive relationship. A typical Marxian approach shows how the expansion of the capitalist world-system remade property relations on its frontier to support asymmetrical patterns of wealth accumulation. This illuminates the impacts of cash cropping, land speculation, and plantation slavery in our case studies, but cannot explain why particular social forms and cultural commitments grew around boundary plants, sometimes in ways counter to elite interests. Finally, a Foucauldian perspective would argue that modernity consists of "placeless power and powerless places" (Castells and Henderson 1987, 7), and that new global mechanisms of power and knowledge are imposing new sorts of governmentality and territorialization on local subjectivities. In contrast, boundary plants demonstrate that diversely disciplined subjects exist in different economic systems, and that the creative production of territory and construction of personhood can occur even in the most oppressive regimes. For the people who live with boundary plants, their power disciplines bodies, fills landscapes with vitality, and delights their senses. To avoid flattening these experiences of power, this book has proceeded like a sailboat moving upwind, by tacking a bit to starboard with Durkheim and correcting the course to port with Marx. Understanding the roots of power requires an eclectic toolkit of analytical concepts, not a theoretical orthodoxy. Poststructuralist political ecology has been this book's major tool for roughing out each case study's form, and I have relied on finer instruments like vegetative agency, social capital, and relational personhood to delineate their final shapes.

The significance of boundary plants today is that their meanings of economic, social, and moral order provide foils of continuity for people experiencing change. Their social entanglements are like roots knitting the soil together and preventing erosion. Dracaena links the people of Kilimanjaro to their ancestors and the people of Oku to their monarchy; cordyline ties together the bundles of powers that constitute access to land in Papua New Guinea and confer indigenous identity in French Polynesia. Vincentians both fix themselves to land in their provision grounds with cordyline and escape their limitations with the plant's leaves. None of these meanings are mere epiphenomena of the economics of landholding. These institutional and conceptual entanglements form landscapes, motivate social dynamics, and point toward peace and justice. Ultimately, boundary plants are important components of social-ecological systems because they relate one part of a system to another, allow people to accumulate and convert economic, social, and cultural capital, and establish fixed points of reference in shifting historical contexts.

Beyond the bounds

The future of boundary plants is likely to be persistence on the edges of global capitalism in the tropics, but these vegetative institutions need not be limited to a marginal role. The Evolutionary Theory of Land Rights (ETLR) argues that societies create property rights institutions as population pressure on resources and market incentives push and pull them toward new arrangements and moral norms (Platteau 2000, 327). Under conditions of abundant uncommodified land, there is no need for farmers to invest extra physical and social labor in the landscape, so access to land is based on group membership. The resulting horticultural common property system relies on family labor, hand tools, and shifting cultivation to produce sustainable livelihoods in multicropped gardens. Population growth eventually introduces new pressures and contradictions, and agrarian economic theory points to two possible pathways for systemic change. The mostly pessimistic Malthusian view insists that scarcity generates conflict and violence, allowing "positive checks" like war and famine to re-balance people and resources. Outmigration is one solution to this Malthusian dilemma and may help explain aspects of the Austronesian and Bantu expansions (respectively, Kirch 2000a; Kopytoff 1987). The more optimistic Boserupian approach notes that farmers can increase food production by investing more labor into the land to create long-lasting "landesque capital" like terracing and irrigation (Boserup 1990; Håkansson and Widgren 2014). The ETLR approach extends Ester Boserup's focus on innovation to include social institutions and cultural meanings as responses to population pressure.[2] Economic relations, social structure, cultural identities, and personhood are all co-emergent by-products of population pressure on resources in relatively closed social-ecological systems (Platteau 2000, 77). The problem is that there are no closed systems anywhere on the planet, so the question for policymakers is how to induce this evolutionary process artificially. The World Bank has long advocated ETLR as a strategy for modernist development programs that can spark a virtuous spiral of innovation, efficiency, and poverty reduction (Gardner and Lewis 2015).

Policies based on ETLR often fail because they mistake the formal maps and abstract equations of institutional economics for the lived-in territories of real lives. First, it assumes that societies somehow want to be more efficient, and that private property is the natural functional result. This modernist structural-functionalism is teleological, ethnocentric, and blithely innocent of the lively economic, social, and cultural entanglements described in this book. At worst, it views non-Western landholding as hopelessly affixed in "rigid tribal institutions" (Acemoglu and Robinson 2010, 35). Second, the ETLR model is simplistic in that it requires culturally similar people everywhere to climb the same evolutionary ladder. A substantivist approach to economics and culture rebuts this narrow view of human variation and insists that economies are systems of meaning as well as patterns of production, exchange, and consumption (Wilk and Cliggett 2019). Third, property rights institutions come into existence because of social goals, not just because of the self-interest demanded by neoclassical economic theory. Indigenous Tanzanian irrigation systems, for example, developed as property rights systems not just to allocate water, but to help cattle-poor patrons attract land-poor clients to form neighborhoods (Sheridan 2002; 2012). Fourth, ETLR tends to overlook the increasing inequality and gender differentiation that privatization usually causes and the institutional and structural power that support and naturalize them (Platteau 2000, 96). Without attention to long-term and colonial histories of inequality, development policy is unlikely to address the structural causes of poverty, as seen most acutely in the St. Vincent chapter. Finally, ETLR assumes that local social-ecological systems have some degree of autonomy from a generally beneficent state. As our five boundary plant stories have shown, states are often more intrusive and disruptive than supportive (e.g., Ferguson 1990; Scott 1998), and there is often more security in local social institutions (like boundary plants) than in national bureaucratic ones (Sheridan 2008).

These critiques have not led land tenure specialists to abandon the ETLR model, but to revise it. The crux of their argument is that if farming communities had more control over their own resource access institutions, these local systems would meet local needs instead of responding to central government demands. Local institutions could have a stronger bargaining position vis-à-vis transnational corporations, and communities would have the legal space to evolve their own property relations. Jean-Philippe Platteau summarizes this position for sub-Saharan Africa, but his prescription applies to Oceania and the Caribbean equally well;

What the region requires is a pragmatic and gradualist approach that reinstitutionalizes indigenous land tenure, promotes the adaptability of its existing arrangements, avoids a regimented tenure model, and relies as much as possible on informal procedures at [the] local level.

(2000, 182)

This strategy was widely discussed in African studies in the 1990s (Bassett and Crummey 1993; Bruce and Migot-Adholla 1994; Downs and Reyna 1988), started to be applied worldwide in the early 21st century, and was only recently recognized as a major policy shift (Alden Wily 2018a; 2018b; Batterbury and Fernando 2006). By legalizing collective forms of resource access, this change affects the approximately two billion people who operate family farms on about 40% of the Earth's land mass. Eight-four percent of these are smallholder farms like the ones described in this book (Lowder et al. 2016). State recognition of boundary plants would allow these people to make the law more consistent with their ways of experiencing economic security, social order, and culturally meaningful landscapes, instead of being disrupted by it. The policy choice here is not between creating private property regimes by fiat in the name of economic efficiency (which is often disastrous, Shipton 1988), or leaving local systems completely undisturbed in the name of social justice (as if states had such influence over transnational corporations and other states). Instead there is a pluralistic "third way" based on hybrid state-local systems with decentralized governance (Otto and Hoekema 2012). These systems and the boundary plants embedded in them would be cheap, socially relevant, accessible, and oriented toward culture-specific ideas of justice, fairness, and peace. The third way would create a more inclusive "people's law" oriented toward livelihood security instead of the top-down discipline of government law (Alden Wily 2012). An applied "anthropology of the good" (Ortner 2016) would, in turn, evaluate and promote local by-laws and inclusive new social-ecological systems (German et al. 2010).

This third way for the reinstitutionalization of boundary plants is not a panacea and would create new obstacles. New Institutional Economics critics would argue that devolution to localities would increase transaction costs as each community evolves different policies and practices. The more pluralistic world would require more translation of the contents and nuances of local law and demand new dispute settlement mechanisms. These new sources of friction would increase uncertainty instead of reducing it. Papua New Guinea would, for example, need new institutions to investigate and document the diverse forms and meanings of cordyline. The Oku Kwifon would have to start keeping records of dracaena placement. Land planners in St. Vincent would have to include transient living plants on their cadastral maps, and likely need to keep updating them.

Even advocates of decentralization recognize that states don't give up power easily. They tend to devolve authority to localities incompletely by appointing local administrators from the top down instead of permitting bottom-up administration. Messy and ambiguous legal pluralism could allow local elites to dominate negotiations over legitimate institutions and resource access (Hoekema 2012, 170). It is hard to imagine the Oku fon or Ma'ohi cultural experts not having the loudest voices in debates over what boundary plants mean and do. At the policy level, even well-meaning administrators would be strongly incentivized to smooth over the rough edges of social and cultural differences and ignore the radical implications of actually empowering local communities to seek their own paths toward sustainable livelihoods (Scoones 2009).

Finally, feminist scholars have long pointed out that in societies organized largely by kinship and gender, the validation of customary law tends to disempower women (Whitehead and Tsikata 2003). They charge that "custom" and "tradition" have been so thoroughly transformed by colonial law and postcolonial neoliberalism that women's institutional and structural power have been erased, which allows men to monopolize the accumulation of economic and social capital (Addison et al. 2021; Amadiume 1987). Simply empowering the Chagga, Oku, and Papua New Guinean boundary plants in this book would strengthen patriarchy along with local control, which then would substitute gender injustice and contradiction for state domination and governmentality. Solving this riddle requires that the new third way boundary plant institutions be attentive to women's strategies and narratives for asserting resource access and claiming care and support from other people (Van Allen 2015). Fortunately, the structural powers that flow through many boundary plants are already associated with feminine symbolic domains like fertility and motherhood. Building these meanings into the new local institutions would require that decentralization be a program for education about indigenous checks and balances on gendered power as well as law construction.

These worries about the complexities of "peoples' law" should not hinder reform. Boundary plants are logical starting points for a politics that redistributes power. A politics of boundary plant recognition would complement a redistribution of political power (Fraser 1995) and define development as the pursuit of human (and interspecies) dignity, not economic efficiency (Löfven 2015). This book has documented how boundary plants relate to culture-specific models of peace and order, and these can form the basis for creative new approaches that promote hope, vitality, and co-existence. African, Oceanic, and Caribbean governments could, for example, recognize dracaena and cordyline as important parts of their histories and push for global acknowledgement of boundary plant institutions under the 2003 UNESCO Convention for the Safeguarding of the Intangible Cultural Heritage. The tourists who climb Mt. Kilimanjaro could wear T-shirts emblazoned with images of *masale* leaves, and the Spiritual Baptists of St. Vincent could follow the red dragon to Africaland knowing they contributed to the emergence of a creole Afro-Caribbean culture from the structural violence of plantation slavery. These botanical manifestations of cultural landscapes, humane economics, social order, and moral personhood can serve as examples of how boundaries can tie social-ecological systems together instead of separating them into splintered domains – something that our planet needs desperately. At first glance, boundary plants may appear to be marginal spatial practices in peripheral places. This book brings them from the edges to the center to argue that these manifestations of the social production of space and the cultural construction of place in different societies can contribute to a more diverse and vital global social-ecological system.

Notes

1. For the general argument that musing about nonhuman agency is unproductive, see Martinez-Reyes (2017). The posthuman turn's rebuttal of anthropocentrism is a politics of representation but not one of redistribution (Fraser 1995) because it tells us how to think about plants instead of how to allocate material resources to them. Recognizing that plants want to live does not necessarily improve their chances for survival.
2. Demography and population pressure on resources are important themes in the chapters on Tanzania and Cameroon, but only tangential for Papua New Guinea, French Polynesia, and St. Vincent. The reasons for these different analyses is lack of evidence in regional scholarship, the pragmatics of field methods, and history. Both Kilimanjaro and Oku are relatively bounded social-ecological units for which robust historical demographic data exist. In contrast, my Papua New Guinean data is from a transect across multiple social-ecological systems that lack demographic data and easy spatial identification. French Polynesia was so thoroughly depopulated and its surviving population so radically transformed that only the rough outlines of precolonial demography and agricultural economics are visible through archaeological reconstruction. St. Vincent's indigenous population was exterminated and exiled, and the histories of slave plantations and outmigration are more matters of racist structural violence than population pressure on resources.

References

Abbott, Isabella. 1992. *Lā'au Hawai'i*. Honolulu: Bishop Museum Press.

Abbott, William. 1892. "Ethnological Collections in the U. S. National Museum from Kilima-Njaro, East Africa." *Report of the US National Museum* 46: 381–429.

Abernethy, Jane, Suelyn Ching Tune, and Julie Williams. 1983. *Made in Hawai'i*. Honolulu: University of Hawai'i Press.

Abrahams, Roger. 1983. *The Man-of-Words in the West Indies*. Baltimore: Johns Hopkins University Press.

Acemoglu, Daron, and James Robinson. 2010. "Why Is Africa Poor?" *Economic History of Developing Regions* 25 (1): 21–50.

Addison, Lincoln, Matthew Schnurr, Christopher Gore, Sylvia Bawa, and Sarah Mujabi-Mujuzi. 2021. "Women's Empowerment in Africa: Critical Reflections on the Abbreviated Women's Empowerment in Agriculture Index (A-WEAI)." *African Studies Review* 64 (2): 276–91.

Aiton, William. 1789. *Hortus Kewensis*. London: George Nicol.

Alden Wily, Liz. 2011. *Whose Land Is It? The Status of Customary Land Tenure in Cameroon*. Brussels: Centre for Environment and Development.

―――――. 2012. "From State to People's Law: Assessing Learning-by-Doing as a Basis of New Land Law." In *Fair Land Governance*, edited by J. Otto and A. Hoekema, 85–110. Leiden: Leiden University Press.

―――――. 2018a. "Customary Tenure: Remaking Property for the 21st Century." In *Comparative Property Law*, edited by M. Graziadei and L. Smith, 458–77. Cheltenham, UK: Edward Elgar Publishing.

―――――. 2018b. "Collective Land Ownership in the 21st Century: Overview of Global Trends." *Land* 7 (2): 68.

Allan, William. 1965. *The African Husbandman*. Edinburgh: Oliver Boyd.

Allen, Bryant. 2013. "Papua New Guinea: Indigenous Migrations in the Recent Past." In *Encyclopedia of Global Human Migration*, edited by I. Ness. Hoboken, NJ: Wiley-Blackwell.

Althaus, Georg. 1903. "Nachrichten aus Mamba." *Evangelische-Lutherisches Missionsblatt* 58 (3): 256–58.

Amadiume, Ifi. 1987. *Male Daughters, Female Husbands*. London: Zed Books.

Amanor, Kojo. 1994. *The New Frontier*. London: Zed Books.

Amin, Julius. 2021. "President Paul Biya and Cameroon's Anglophone Crisis: A Catalogue of Miscalculations." *Africa Today* 68 (1): 95–122.

Anderson, Alexander. 1983a. *Alexander Anderson's The St. Vincent Botanic Garden.* Edited by R. Howard and E. Howard. Cambridge, MA: Arnold Arboretum, Harvard University.

———. 1983b. *Alexander Anderson's Geography and History of St. Vincent, West Indies.* Edited by R. Howard and E. Howard. Cambridge, MA: Arnold Arboretum, Harvard University.

Anderson, Astrid. 2011. *Landscapes of Relations and Belonging.* Oxford: Berghahn Books.

Anderson, Edgar. 2020. *Plants, Man and Life,* 2e. Berkeley: University of California Press.

Anderson, John. 2001. *Between Slavery and Freedom.* Edited by Roderick McDonald. Philadelphia: University of Pennsylvania Press.

Anderson, Tim, and Gary Lee. 2010. *In Defence of Melanesian Customary Land.* Sydney: AID/WATCH.

Andrews, Edmund, and Irene Andrews. 1944. *A Comparative Dictionary of the Tahitian Language.* Chicago: Chicago Academy of Sciences.

Anthony, Ming, Angela Luhning, and Pierre Verger. 1995. "À la Recherche des Plantes Perdues, les Plantes Retrouvées par les Descendants Culturels des Yoruba au Brésil." *Revue d'Ethnolinguistique* 7: 113–40.

Anzaldúa, Gloria. 2012. *Borderlands,* 4e. San Francisco: Aunt Lute Books.

Argenti, Nicolas. 1998. "Air Youth: Performance, Violence and the State in Cameroon." *Journal of the Royal Anthropological Institute* 4 (4): 753–82.

———. 1999. "Ephemeral Monuments, Memory and Royal Sempiternity in a Grassfields Kingdom." In *The Art of Forgetting,* edited by A. Forty and S. Küchler, 21–52. New York: Berg.

———. 2002. "People of the Chisel: Apprenticeship, Youth, and Elites in Oku (Cameroon)." *American Ethnologist* 29 (3): 497–533.

———. 2006. "Remembering the Future: Slavery, Youth and Masking in the Cameroon Grassfields." *Social Anthropology* 14: 49–69.

———. 2007. *The Intestines of the State.* Chicago: University of Chicago Press.

———. 2011. "Things of the Ground: Children's Medicine, Motherhood, and Memory in the Cameroon Grassfields." *Africa* 81 (2): 269–94.

Armitage, Lynne. 2001. "Customary Land Tenure in Papua New Guinea: Status and Prospects." Proceedings of the International Association for the Study of Common Property Rights, 2001 Pacific Regional Conference, Queensland University of Technology, Brisbane, 12–23. Online document, http://hdl.handle. net/10535/589, accessed September 17, 2021.

Associated Press Archive. 2019. "Cameroon Women Protest PM's Visit to Bamenda." YouTube (video), May 14, 2019, https://www.youtube.com/watch?v=sQQhc-2Q6i_Q&ab_channel=APArchive, accessed August 29, 2021.

Atabong, Amindeh. 2018. "Cameroon Crisis Threatens Wildlife as Thousands Flee to Protected Areas." *African Arguments,* online document, https://africanarguments.org/2018/07/cameroon-crisis-threatens-wildlife-people-flee-protected-areas/, accessed September 1, 2021.

Aufenanger, Heinrich. 1961. "The *Cordyline* Plant in the Central Highlands of New Guinea." *Anthropos* 56 (3): 393–408.

Awafong, Francis. 2003. *The Impact of State Land Law on Customary Tenure in North West Cameroon.* Institute of Rural Development. Kiel, Germany: Wissenschaftsverlag Vauk.

Azong, Matilda. 2021. "Impact of Cultural Beliefs on Smallholders' Response to Climate Change: The Case of Bamenda Highlands, Cameroon." *International Journal of Environmental Studies* 78 (4): 663–78.

Azong, Matilda, Clare Kelso, and Kammila Naidoo. 2018. "Vulnerability and Resilience of Female Farmers in Oku, Cameroon, to Climate Change." *African Sociological Review* 22 (1): 31–53.

Azong, Matilda, and Clare Kelso. 2021. "Gender, Ethnicity and Vulnerability to Climate Change: The Case of Matrilineal and Patrilineal Societies in Bamenda Highlands Region, Cameroon." *Global Environmental Change* 67: 102241.

Bah, Njakoi John. 1996. *Oku Past and Present.* Jikijem, Cameroon. Manuscript copy in possession of the author.

———. 1998. "Marriage and Divorce in Oku." *Baessler-Archiv Neue Folge* 46 (1): 31–57.

———. 2000. *Some Oku Rituals (Western Grassfields, Cameroon).* Jikijem, Cameroon. Manuscript copy in possession of the author.

———. 2004. "Ntok Ebkuo: A Western Grassfields Palace (Cameroon)." *Anthropos* 99 (2): 435–50.

Bailey, P. 1968. "The Changing Economy of the Chagga Cultivators of Marangu, Kilimanjaro." *Geography* 53 (2): 163–69.

Bainkong, Godlove. 2014. "Honey Fetches FCFA 30-40 Million to Oku Annually." *Cameroon Tribune.* Online document, http://allafrica.com/stories/201409110620. html, accessed June 11, 2018.

Baland, Jean-Marie, François Bourguignon, Jean-Philippe Platteau, and Thierry Verdier, eds. 2020. *Handbook of Economic Development and Institutions.* Princeton: Princeton University Press.

Balick, Michael, ed. 2009. *Ethnobotany of Pohnpei.* Honolulu: University of Hawai'i Press.

Balick, Michael, and Paul Cox. 1996. *Plants, People, and Culture.* New York: Scientific American Library.

Ballard, Chris. 2013. "It's the Land, Stupid! The Moral Economy of Resource Ownership in Papua New Guinea." In *The Governance of Common Property in the Pacific Region*, edited by P. Larmour, 47–65. Canberra: ANU Press.

Banks, Joseph. 1896. *Journal of the Right Hon. Sir Joseph Banks [...].* Edited by J. Hooker. London: Macmillan.

Barker, Hugh. 2012. *Hedge Britannia.* London: Bloomsbury.

Barnes, Gerry, and Tom Williamson. 2006. *Hedgerow History.* Bollington, UK: Windgather.

Barnes, J. 1962. "African Models in the New Guinea Highlands." *Man* 62: 5–9.

Barr, Colin, and Sandrine Petit, eds. 2001. *Hedgerows of the World.* Aberdeen: IALE.

Barrett, Stanley. 2002. *Culture Meets Power.* Westport, CT: Praeger.

Barrau, Jacques. 1965. "Witnesses of the Past: Notes on Some Food Plants of Oceania." *Ethnology* 4 (3): 282–94.

Bartelt, Brian. 2006. *"Healers and Witches in Oku."* PhD diss., University of Southern California.

Barth, Fredrik. 2000. "Boundaries and Connections." In *Signifying Identities*, edited by A. Cohen, 17–36. London: Routledge.

Barth, Fredrik, ed. 1969. *Ethnic Groups and Boundaries.* Boston: Little, Brown, and Co.

Barton, Huw, and Tim Denham. 2018. "Vegecultures and the Social–biological Transformations of Plants and People." *Quaternary International* 489: 17–25.

Bassett, Thomas, and Donald Crummey, eds. 1993. *Land in African Agrarian Systems*. Madison: University of Wisconsin Press.

Bastin, Yvonne, A. Coupez, and Michael Mann. 1999. *Continuity and Divergence in the Bantu Languages*. Turvuren, Belgium: Musée royal de l'Afrique centrale.

Batterbury, Simon, and Jude Fernando. 2006. "Rescaling Governance and the Impacts of Political and Environmental Decentralization: An Introduction." *World Development* 34 (11): 1851–63.

Baudrillard, Jean. 1994. *Simulacra and Simulation*. Ann Arbor: University of Michigan Press.

Baye, Francis. 2008. "Changing Land Tenure Arrangements and Access to Primary Assets Under Globalization: A Case Study of Two Villages in Anglophone Cameroon." *African Development Review* 20 (1): 135–62.

Bayliss-Smith, Tim. 1997. "From Taro Garden to Golf Course? Alternative Futures for Agricultural Capital in the Pacific Islands." In *Environment and Development in the Pacific Islands*, edited by B. Burt and C. Clerk, 143–70. Canberra, Australia: National Centre for Development Studies, ANU.

Beaglehole, John, ed. 1967. *The Journals of Captain James Cook on His Voyages of Discovery*, vol. 3. Cambridge: Cambridge University Press.

Beckett, J. 1990. *The Agricultural Revolution*. Oxford: Basil Blackwell.

Behague, Gerard. 2006. "Regional and National Trends in Afro-Brazilian Religious Musics: A Case of Cultural Pluralism." *Latin American Music Review* 27 (1): 91–103.

Bellwood, Peter, and Eusebio Dizon. 2005. "The Batanes Archaeological Project and the 'Out of Taiwan' Hypothesis for Austronesian Dispersal." *Journal of Austronesian Studies* 1 (1): 1–33.

Bender, Matthew. 2019. *Water Brings No Harm*. Athens: Ohio University Press.

Benediktsson, Karl. 2002. *Harvesting Development*. Ann Arbor: University of Michigan Press.

Bennett, Jane. 2010. *Vibrant Matter*. Durham: Duke University Press.

Benoît, Catherine. 1990. "Outil Graphique et Analyse Anthropologique des Jardins de Case en Guadeloupe." *Histoire & Mesure* 5 (3–4): 315–42.

Bergl, Richard, John Coates, and Roger Fotso. 2007. "Distribution and Protected Area Coverage of Endemic Taxa in West Africa's Biafran Forests and Highlands." *Biological Conservation* 134: 195–208.

Berkes, Fikret, Johan Colding, and Carl Folke, eds. 2003. *Navigating Social-Ecological Systems*. Cambridge: Cambridge University Press.

Berleant-Schiller, Riva, and Lydia Pulsipher. 1986. "Subsistence Cultivation in the Caribbean." *New West Indian Guide* 60 (1–2): 1–40.

Berlin, Ira, and Philip Morgan. 1993. *Cultivation and Culture*. Charlottesville: University of Virginia Press.

Berrigan, Caitlin. 2014. "Life Cycle of a Common Weed." In *The Multispecies Salon*, edited by E. Kirksey, 165–80. Durham: Duke University Press.

Berry, Sara. 1989. "Social Institutions and Access to Resources." *Africa* 59 (1): 41–55.

———. 1993. *No Condition Is Permanent*. Madison: University of Wisconsin Press.

———. 2017. "Struggles Over Land and Authority in Africa." *African Studies Review* 60 (3): 105–25.

Besky, Sarah, and Jonathan Padwe. 2016. "Placing Plants in Territory." *Environment and Society* 7 (1): 9–28.

Bessire, Lucas, and David Bond. 2014. "Ontological Anthropology and the Deferral of Critique." *American Ethnologist* 41 (3): 440–56.

Besson, Jean. 1979. "Symbolic Aspects of Land in the Caribbean: The Tenure and Transmission of Land Rights among Caribbean Peasantries." In *Peasants, Plantations and Rural Communities in the Caribbean*, edited by M. Cross and A. Marks, 86–116. Surrey: University of Surrey.

———. 1984. "Land Tenure in the Free Villages of Trelawny, Jamaica: A Case Study in the Caribbean Peasant Response to Emancipation." *Slavery and Abolition* 5 (1): 3–23.

———. 1987. "A Paradox in Caribbean Attitudes to Land." In *Land and Development in the Caribbean*, edited by J. Besson and J. Momsen, 13–45. London: Macmillan.

———. 1993. "Reputation and Respectability Reconsidered: A New Perspective on Afro-Caribbean Peasant Women." In *Women and Change in the Caribbean*, edited by J. Momsen, 15–37. Kingston, Jamaica: Ian Randle.

———. 2000. "The Appropriation of Lands of Law by Lands of Myth in the Caribbean Region." In *Lands, Law and Environment*, edited by A. Abramson and D. Theodossopoulos, 116–35. London: Pluto Press.

———. 2002. *Martha Brae's Two Histories*. Chapel Hill: University of North Carolina Press.

———. 2016. *Transformations of Freedom in the Land of the Maroons*. Kingston, Jamaica: Ian Randle.

———. 2018. "Sidney W. Mintz's 'Peasantry' as a Critique of Capitalism: New Evidence from Jamaica." *Critique of Anthropology* 38 (4): 443–60.

Best, Eldson. 1976. *Maori Agriculture*. Wellington, New Zealand: A.R. Shearer.

Bhattacharya, Rajesh, and Ian Seda-Irizarry. 2017. "Primitive Accumulation." In *Routledge Handbook of Marxian Economics*, edited by D. Brennan, D. Kristjanson-Gural, C. Mulder and E. Olsen, 144–54. Abington, UK: Taylor & Francis.

Biersack, Aletta, and James Greenberg, eds. 2006. *Reimagining Political Ecology*. Durham: Duke University Press.

Bird, Michael, Scott Condie, Sue O'Connor, Damien O'Grady, Christian Reepmeyer, Sean Ulm, Mojca Zega, Frédérik Saltré, and Corey Bradshaw. 2019. "Early Human Settlement of Sahul Was Not an Accident." *Scientific Reports* 9 (1): 8220.

Bird-David, Nurit. 1999. "'Animism' Revisited: Personhood, Environment, and Relational Epistemology." *Current Anthropology* 40: S67–S91.

Blaikie, Piers, and Harold Brookfield. 1987. *Land Degradation and Society*. London: Methuen.

Bligh, William. 1792. *A Voyage to the South Sea Undertaken by Command of His Majesty [...]*. London: G. Nicol.

Bligh, William, and Edward Christian. 2001. *The Bounty Mutiny*. New York: Penguin Books.

Bloch, Maurice. 1998. "Why Trees, Too, Are Good to Think With: Towards an Anthropology of the Meaning of Life." In *The Social Life of Trees*, edited by L. Rival, 39–55. Oxford: Berg.

Blood, Cynthia, and Leslie Davis. 1999. *Oku-English Provisional Lexicon*. Yaoundé: SIL.

Boa, Sheena. 2001. "'Walking on the Highway to Heaven': Religious Influences and Attitudes Relating to the Freed Population of St. Vincent, 1834–1884." *Journal of Caribbean History* 35 (2): 179–207.

Boccagni, Paolo. 2019. "Multi-Sited Ethnography." In *SAGE Research Methods Foundations*, edited by P. Atkinson, S. Delamont, A. Cernat, J. Sakshaug, and R. Williams. United Kingdom: SAGE Publications Limited.

Bohannan, Paul. 1963. "'Land,' 'Tenure,' and Land-Tenure." In *African Agrarian Systems*, edited by D. Biebuyck, 101–11. London: Oxford University Press.

Bonnemere, Pascale. 1998. "Trees and People: Some Vital Links. Tree Products and Other Agents in the Life Cycle of the Ankave-Anga of Papua New Guinea." In *The Social Life of Trees*, edited by L. Rival, 113–31. Oxford: Berg.

Borofsky, Robert. 1997. "Cook, Lono, Obeyesekere, and Sahlins." *Current Anthropology* 38 (2): 255–82.

Borofsky, Robert, Laura Nader, Matei Candea, and Jonathan Friedman. 2019. "Where Have All the Comparisons Gone?" *Public Anthropology* blog, https://www.publicanthropology.org/where-have-all-the-comparisons-gone/.

Bos, J.J. 1984. *Dracaena in West Africa*. Wageningen: Agricultural University Wageningen.

Boserup, Ester. 1965. *The Conditions of Agricultural Growth*. Chicago: Aldine.

———. 1990. *Economic and Demographic Relationships in Development*. Baltimore: Johns Hopkins University Press.

Bouge, Louis-Joseph. 1952. "Première Législation Tahitienne, le Code Pomaré de 1819: Historique et Traduction." *Journal de la Société des Océanistes* 8: 5–26.

Bourdieu, Pierre. 1990. *The Logic of Practice*. Stanford: Stanford University Press.

Bowman, H., and J. Bowman. 1939. *Crusoe's Island in the Caribbean*. Indianapolis: Bobbs-Merrill Company.

Brassey, Annie. 1885. *In the Trades, the Tropics, and the Roaring Forties*. London: Longman, Green & Co.

Bretschneider, Emil. 2011. *History of European Botanical Discoveries in China*. Hamburg: Severus Verlag.

Brierley, John. 1976. "Kitchen Gardens in West-Indies, With a Contemporary Study from Grenada." *Journal of Tropical Geography* 43 (December): 30–40.

———. 1987. "Land Fragmentation and Land-Use Patterns in Grenada." In *Land and Development in the Caribbean*, edited by J. Besson and J. Momsen, 194–209. London: Macmillan.

Brighenti, Andrea, and Mattias Kärrholm. 2020. *Animated Lands*. Lincoln: University of Nebraska Press.

Brookfield, Harold. 1972. "Intensification and Disintensification in Pacific Agriculture: A Theoretical Approach." *Pacific Viewpoint* 13: 30–48.

Brookfield, Harold, and Paula Brown. 1963. *Struggle for Land*. Oxford: Oxford University Press.

Broughton, Daniel. 1794. *Hortus Eastensis*. St. Jago de la Vega, Jamaica: Alexander Aikman.

Brown, Paula. 1978. *Highland Peoples of New Guinea*. Cambridge: Cambridge University Press.

Brown, Paula, Harold Brookfield, and Robin Grau. 1990. "Land Tenure and Transfer in Chimbu, Papua New Guinea, 1958–1984: A Study in Continuity and Change, Accommodation and Opportunism." *Human Ecology* 18 (1): 21–50.

Brown, Vincent. 2008. *The Reaper's Garden*. Cambridge, MA: Harvard University Press.

Browne, Benarva. 2016. "The Experience of St. Vincent and the Grenadines in the Development of a National Land Policy for Sustainable Land Management in the Context of Climate Change." In *Island Systems Planning*, edited by A. Mohammed, S. Mahabir and N. Maingot, 48–67. St. Augustine, Trinidad and Tobago: Caribbean Network for Urban and Land Management..

Bruce, John, and Shem Migot-Adholla, eds. 1994. *Searching for Land Tenure Security in Africa*. Dubuque: Kendall/Hunt.

Bryans, Robin. 1967. *Trinidad and Tobago*. London: Faber and Faber.

Budji, Ivoline. 2020. "Utilizing Sounds of Mourning as Protest and Activism: The 2019 Northwestern Women's Lamentation March within the Anglophone Crisis in Cameroon." *Resonance: The Journal of Sound and Culture* 1 (4): 443–61.

Bunyan, John. 1987. *The Pilgrim's Progress*. Harmondsworth, UK: Penguin Books.

Burkhill, H.M. 1985. *The Useful Plants of West Tropical Africa*, 2e. Kew: Royal Botanic Gardens.

Busse, Mark. 2005. "Wandering Hero Stories in the Southern Lowlands of New Guinea." *Cultural Anthropology* 20 (4): 443–73.

Cameroon Gender and Environment Watch (CAMGEW). 2016. *CAMGEW 2016 Annual Report*. Oku, Cameroon: CAMGEW. Manuscript copy in possession of the author.

Candea, Matei. 2007. "Arbitrary Locations: In Defence of the Bounded Field-site." *Journal of the Royal Anthropological Institute* 13 (1): 167–84.

———. 2018. *Comparison in Anthropology*. Cambridge: Cambridge University Press.

Capone, Stefania. 2010. *Searching for Africa in Brazil*. Durham: Duke University Press.

Carmichael, Gertrude. 1833. *Domestic Manners and Social Condition of the White, Coloured, and Negro Population of the West Indies*. 2 vols. London: Whittaker, Treacher & Company.

Carney, Judith. 2001. *Black Rice*. Cambridge, MA: Harvard University Press.

———. 2020. "Subsistence in the Plantationocene: Dooryard Gardens, Agrobiodiversity, and the Subaltern Economies of Slavery." *Journal of Peasant Studies* 48 (5): 1075–99.

Carney, Judith, and Richard Rosomoff. 2009. *In the Shadow of Slavery*. Berkeley: University of California Press.

Carrier, James. 1998. "Property and Social Relations in Melanesian Anthropology." In *Property Relations*, edited by C.M. Hann, 85–103. Cambridge: Cambridge University Press.

Carson, Mike. 2002. "Ti Ovens in Polynesia: Ethnological and Archaeological Perspectives." *Journal of the Polynesian Society* 111 (4): 339–70.

Carsten, Janet, and Stephen Hugh-Jones. 1995. "Introduction: About the House – Lévi-Strauss and Beyond." In *About the House*, edited by J. Carsten and S. Hugh-Jones, 1–46. Cambridge: Cambridge University Press.

Cassidy, F.G., and R.B. Page. 2002. *Dictionary of Jamaican English*, 2e. Kingston: University of the West Indies Press.

Castells, Manuel, and Jeffrey Henderson. 1987. "Techno-Economic Restructuring, Socio-Political Processes and Spatial Transformation: A Global Perspective." In *Global Restructuring and Territorial Development*, edited by J. Henderson and M. Castells, 1–17. London: SAGE Publications.

Chand, Satish, and Charles Yala. 2009. "Land Tenure and Productivity: Farm-Level Evidence from Papua New Guinea." *Land Economics* 85 (3): 442–53.

Chanock, Martin. 1991. "Paradigms, Policies, and Property: A Review of the Customary Law of Land Tenure." In *Law in Colonial Africa*, edited by K. Mann and R. Roberts, 61–84. Portsmouth, NH: Heinemann.

Chao, Sophie. 2018. "In the Shadow of the Palm: Dispersed Ontologies among Marind, West Papua." *Cultural Anthropology* 33 (4): 621–49.

———. 2021. "The Beetle or the Bug? Multispecies Politics in a West Papuan Oil Palm Plantation." *American Anthropologist* 123: 476–489.

Cheek, Martin, Jean-Michel Onana, and Benedict John Pollard. 2000. *The Plants of Mount Oku and the Ijim Ridge, Cameroon*. Kew: Royal Botanical Gardens.

Cheka, Cosmas. 2008. "Traditional Authority at the Crossroads of Governance in Republican Cameroon." *Africa Development* 33 (2): 67–89.

Chem-Langhëë, Bongfen. 1995. "Slavery and Slave Marketing in Nso' in the Nineteenth Century." *Paideuma* 41: 177–190.

Chevannes, Barry. 1994. *Rastafari*. Syracuse: Syracuse University Press.

Chevannes, Barry, and Jean Besson. 1996. "The Continuity-Creativity Debate: The Case of Revival." *New West Indian Guide* 70 (3–4): 209–28.

Chilver, Elizabeth. 1990. "Thaumaturgy in Contemporary Traditional Religion: The Case of Nso' in Mid-Century." *Journal of Religion in Africa* 20 (3): 226–47.

Chilver, Elizabeth, and Phyllis Kaberry. 1967. *Traditional Bamenda*. Yaoundé: Cameroon Ministry of Primary Education and Social Welfare.

Clack, Timothy. 2009. "Infusing the Sacred: Syncretistic Landscapes, Ritual Performance and Religious Experience in Chaggaland." In *Culture, History, and Identity*, edited by T. Clack, 195–206. Oxford: Archaeopress.

Clarke, William. 1971. *Place and People*. Berkeley: University of California Press.

Closser, Svea. 2010. *Chasing Polio in Pakistan*. Nashville: Vanderbilt University Press.

Coe, Felix, and Gregory Anderson. 1996. "Ethnobotany of the Garífuna of Eastern Nicaragua." *Economic Botany* 50 (1): 71–107.

Cohen, Anthony. 1985. *The Symbolic Construction of Community*. London: Routledge.

Colding, Johan, and Stephan Barthel. 2019. "Exploring the Social-Ecological Systems Discourse 20 Years Later." *Ecology and Society* 24 (1): 2.

Coleman, Simon, and Pauline von Hellermann, eds. 2011. *Multi-Sited Ethnography*. London: Routledge.

Collar, N.J., and S.N. Stuart. 1988. *Key Forests for Threatened Birds in Africa*. Cambridge: International Council for Bird Preservation.

Collett, Jill, and Patrick Bowe. 1998. *Gardens of the Caribbean*. London: Macmillan Education.

Collins, David. 1803. *Practical Rules for the Management and Medical Treatment of Negro Slaves in the Sugar Colonies*. London: J. Barfield.

Colson, Elizabeth. 1997. "Places of Power and Shrines of the Land." *Paideuma* 43: 47–57.

Colthurst, John Bowen. 1977. *The Colthurst Journal*, edited by W.K. Marshall. Millwood, NY: KTO Press.

Comaroff, John. 1996. "Ethnicity, Nationalism, and the Politics of Difference in an Age of Revolution." In *The Politics of Difference*, edited by E. Wilmsen and P. McAllister, 162–84. Chicago: University of Chicago Press.

Comaroff, John, and Jean Comaroff. 2001. "On Personhood: An Anthropological Perspective from Africa." *Social Identities* 7 (2): 267–83.

———. 2009. *Ethnicity, Inc.* Chicago: University of Chicago Press.

Conklin, Beth. 2010. *Consuming Grief.* Austin: University of Texas Press.

Conklin, Harold. 1957. *Hanunóo Agriculture.* Rome: FAO.

———. 1967. "Ifugao Ethnobotany 1905–1965: The 1911 Beyer-Merrill Report in Perspective." *Economic Botany* 21 (3): 243–72.

Cooter, Robert. 1991. "Inventing Market Property: The Land Courts of Papua New Guinea." *Law & Society Review* 25 (4): 759–801.

Coppenrath, Gérald. 2003. *La Terre à Tahiti et dans les Îles.* Pape'ete: Haere Po.

Coulson, Andrew. 1982. *Tanzania: A Political Economy.* Oxford: Clarendon Press.

Cox, Edward. 1994. "Religious Intolerance and Persecution: The Shakers of St. Vincent, 1900–1934." *Journal of Caribbean History* 28 (2): 208–43.

Cox, Paul. 1982. "Cordyline Ovens (*Umu Ti*) in Samoa." *Economic Botany* 36 (4): 389–96.

Craparo, A.C.W., P.J.A. Van Asten, P. Läderach, L.T.P. Jassogne, and S.W. Grab. 2015. "*Coffea arabica* Yields Decline in Tanzania Due to Climate Change: Global Implications." *Agricultural and Forest Meteorology* 207: 1–10.

Craton, Michael. 1977. "Perceptions of Slavery: A Preliminary Excursion into the Possibilities of Oral History in Rural Jamaica." In *Old Roots in New Lands*, edited by A. Pescatello, 263–90. Westport, CT: Greenwood Press.

Crawford, Michael. 1983. "The Anthropological Genetics of the Black Caribs 'Garifuna' of Central America and the Caribbean." *American Journal of Physical Anthropology* 26: 161–92.

Crépeau, Robert, and Frédéric Laugrand. 2017. "Ontological Flows and Fluids: Rethinking Mana-Like Concepts." *Social Compass* 64 (3): 313–27.

Crocombe, Ron, and Robin Hide. 1987. "New Guinea." In *Land Tenure in the Pacific*, 3e, edited by R. Crocombe, 324–67. Suva, Fiji: University of the South Pacific.

Cronin, Mike, and Daryl Adair. 2002. *The Wearing of the Green.* New York: Routledge.

Crook, Tony. 1999. "Growing Knowledge in Bolivip, Papua New Guinea." *Oceania* 69 (4): 225–42.

Cunningham, Anthony. 2001. *Applied Ethnobotany.* London: Earthscan.

Cunningham, Anthony, Elias Ayuk, Steven Franzel, Bahiru Duguma, and Christian Asanga. 2002. *An Economic Evaluation of Medicinal Tree Cultivation: Prunus africana in Cameroon.* People and Plants Working Paper #10, UNESCO. Online document, http://unesdoc.unesco.org/images/0013/001330/133098e.pdf, accessed June 11, 2018.

Curry, George, and Gina Koczberski. 2009. "Finding Common Ground: Relational Concepts of Land Tenure and Economy in the Oil Palm Frontier of Papua New Guinea." *Geographical Journal* 175 (2): 98–111.

Curtin, Tim, and David Lea. 2006. "Land Titling and Socioeconomic Development in the South Pacific." *Pacific Economic Bulletin* 21 (1): 153–80.

Daly, Lewis, Katherine French, Theresa Miller, and Luíseach Nic Eoin. 2016. "Integrating Ontology into Ethnobotanical Research." *Journal of Ethnobiology* 36 (1): 1–9.

Dalziel, J.M. 1937. *The Useful Plants of West Tropical Africa.* London: Crown Agents.

Danowski, Déborah, and Eduardo Viveiros de Castro. 2017. *The Ends of the World*. Cambridge: Polity Press.

Darian-Smith, Eve. 1995. "Beating the Bounds: Law, Identity and Territory in the New Europe." *Political and Legal Anthropology Review* 18 (1): 63–73.

Darwin, Charles. 1915. *Works of Charles Darwin*. New York: D. Appleton and Company.

Davidson, George. 1787. *The Case of the Caribbs in St. Vincent's*. London: Thomas Coke.

Davies, John. 1851. *A Tahitian and English Dictionary*. Tahiti: London Missionary Society's Press.

Davy, John. 1854. *The West Indies, Before and Since Slave Emancipation*. London: W. & F.G. Cash.

de Deckker, Paul. 1990. "Les Îles de la Société 1767-1797." In *Encyclopedie de la Polynésie*, 2e, vol. 6, edited by P.-Y. Toullelan, 25–40. Pape'ete: Éditions de l'Alizé.

de Soto, Hernando. 2000. *The Mystery of Capital*. New York: Basic Books.

de Filippo, Cesare, Koen Bostoen, Mark Stoneking, and Brigitte Pakendorf. 2012. "Bringing Together Linguistic and Genetic Evidence to Test the Bantu Expansion." *Proceedings of the Royal Society of London B: Biological Sciences* 279 (1741): 3256–63.

DeFreitas, C.L. 1965. *The St. Vincent Botanic Gardens 1765–1965*. Kingstown, St. Vincent: Government Printer. St. Vincent National Archives, document #B1-1.

Delle, James. 2014. *The Colonial Caribbean*. New York: Cambridge University Press.

Denham, Tim. 2013. "Ancient and Historic Dispersals of Sweet Potato in Oceania." *Proceedings of the National Academy of Sciences* 110 (6): 1982–83.

Denham, Tim, S.B. Haberle, C. Lentfer, R. Fullagar, J. Field, M. Therin, N. Porch, and B. Winsborough. 2003. "Origins of Agriculture at Kuk Swamp in the Highlands of New Guinea." *Science* 301 (5630): 189–93.

Dening, Greg. 1992. *Mr. Bligh's Bad Language*. Cambridge: Cambridge University Press.

Depommier, D. 1983. *Aspects de la Foresterie Villageoise dans l'Ouest et le Nord (Cameroun)*. Nkolbisson, Cameroon: Institut de Recherche Agronomique and Centre Tropicale Forestier Technique.

Descola, Philippe. 2013. *Beyond Nature and Culture*. Chicago: University of Chicago Press.

Devenne, Francois, Odile Chapuis, and Francois Bart. 2002. "Chagga Farm Systems, Strengths and Stumbling Blocks." In *Kilimanjaro: Mountain, Memory, Modernity*, edited by F. Bart, M. Milline, F. Devenne and T. Martin, 177–99. Dar es Salaam: Mkuki na Nyota.

Diamond, Jared. 1999. *Guns, Germs, and Steel*. New York: W. W. Norton.

Diduk, Susan. 1993. "Twins, Ancestors and Socio-Economic Change in Kedjom Society." *Man* 28 (3): 551–71.

Dinnen, Sinclair. 2019. "Security Governance in Melanesia: Police, Prisons, and Crime." In *The Melanesian World*, edited by E. Hirsch and W. Rollason, 269–84. New York: Routledge.

Dirks, Nicholas, Geoff Eley, and Sherry Ortner, eds. 1994. *Culture/Power/History*. Princeton: Princeton University Press.

Donaldson, Emily. 2016. *"Living with Sacred Lands."* PhD diss., McGill University.

———. 2018. "Place, Destabilized: Ambivalent Heritage, Community and Colonialism in the Marquesas Islands." *Oceania* 88 (1): 69–89.

————. 2019. *Working with the Ancestors.* Seattle: University of Washington Press.

Douglas, Mary. 1966. *Purity and Danger.* London: Routledge and Kegan Paul.

————. 1973. *Natural Symbols.* New York: Vintage Books.

————. 1986. *How Institutions Think.* Syracuse: Syracuse University Press.

Dowdeswell, W.H. 1987. *Hedgerows and Verges.* London: Allen and Unwin.

Downs, R.E., and Stephen Reyna, eds. 1988. *Land and Society in Contemporary Africa.* Hanover: University Press of New England.

Duarte, Luiz. 2021. "The Vitality of Vitalism in Contemporary Anthropology: Longing for an Ever Green Tree of Life." *Anthropological Theory* 21 (2): 131–53.

Dumont d'Urville, Jules-Sébastian-César. 1832. "Sur les Îles du Grand Ocean." *Bulletin de la Société de Géographie* 17: 1–21.

Dundas, Charles. 1924. *Kilimanjaro and Its People.* London: Frank Cass and Co.

Durkheim, Émile. 1915. *The Elementary Forms of the Religious Life.* London: Allen and Unwin.

Duvall, Chris. 2019. *The African Roots of Marijuana.* Durham: Duke University Press.

Dwiartama, Angga, and Christopher Rosin. 2014. "Exploring Agency beyond Humans: The Compatibility of Actor-Network Theory (ANT) and Resilience Thinking." *Ecology and Society* 19 (3): 28.

Dze-Ngwa, Willibroad. 2011. "Boundary Dynamics and the Search for Geopolitical Space: The Case of the Mbororo in the North West Region of Cameroon." In *Boundaries and History in Africa*, edited by D. Abwa, A.-P. Temgoua, E.D.S. Fomin and W. Dze-Ngwa, 15–31. Yaoundé: University of Yaoundé I.

Earle, Timothy. 1997. *How Chiefs Come to Power.* Stanford: Stanford University Press.

Earle, Timothy, and Matthew Spriggs. 2015. "Political Economy in Prehistory: A Marxist Approach to Pacific Sequences." *Current Anthropology* 56 (4): 515–44.

Edmonds, Kevin. 2014. "Ganja and Globalization in St. Vincent." *Stabroek News* [Guyana]. https://www.stabroeknews.com/2014/12/15/features/ganja-globalization-st-vincent/

Edwards, Bryan. 1793. *The History, Civil and Commercial, of the British West Indies.* Dublin: Luke White.

————. 1819. *The History, Civil and Commercial, of the British West Indies*, 5e. London: G. and W.B. Whittaker.

Efird, Rob. 2017. "Perceiving Nature's Personhood: Anthropological Enhancements to Environmental Education." In *Routledge Handbook of Environmental Anthropology*, edited by H. Kopnina and E. Shoreman-Ouimet, 441–51. London: Routledge.

Ehret, Christopher. 1998. *An African Classical Age.* Charlottesville: University Press of Virginia.

Ehrlich, Celia. 1989. "Special Problems in an Ethnobotanical Literature Search: *Cordyline terminalis* (L.) Kunth., the 'Hawaiian Ti Plant'." *Journal of Ethnobiology* 9 (1): 51–63.

————. 2000. "'Inedible' to 'Edible': Firewalking and the Ti Plant [*Cordyline fruticosa* (L.) A. Chev.]." *Journal of the Polynesian Society* 109 (4): 371–400.

Elak-Oku Council. 2012. "Elak-Oku Council Development Plan." Oku, Cameroon: Elak-Oku Council and the National Community Driven Development Program. https://www.pndp.org/documents/26_CDP_Elak_Oku.pdf

Ellen, Roy. 2016. "Is There a Role for Ontologies in Understanding Plant Knowledge Systems?" *Journal of Ethnobiology* 36 (1): 10–28.

Ellen, Roy, and Réka Komáromi. 2013. "Social Exchange and Vegetative Propagation: An Untold Story of British Potted Plants." *Anthropology Today* 29 (1): 3–7.

Ellis, William. 1788. *An Authentic Narrative of a Voyage Performed by Captain Cook and Captain Clerke [...]*, vol. 2. Altenburg, Germany: Richter.

Ellis, William. 1829. *Polynesian Researches [...]*, vol. 2. London: Fisher, Son, & Jackson.

Elmhirst, Rebecca. 2015. "Feminist Political Ecology." In *Routledge Handbook of Political Ecology*, edited by T. Perreault, G. Bridge and J. McCarthy, 519–30. London: Routledge.

Eltis, David, and David Richardson. 2010. *Atlas of the Transatlantic Slave Trade.* New Haven: Yale University Press.

Emerson, Joseph. 1902. "Some Hawaiian Beliefs Regarding Spirits." *Hawaiian Historical Society Annual Report* 9: 10–17.

Emory, Kenneth. 1933. *Stone Remains in the Society Islands.* Bernice P. Bishop Museum Bulletin 119. Honolulu: Bishop Museum.

Ensminger, Jean, ed. 2002. *Theory in Economic Anthropology.* Lanham, MD: Rowman and Littlefield.

Ensminger, Jean, and Joseph Henrich, eds. 2014. *Experimenting With Social Norms.* New York: Russell Sage Foundation.

Erb, Maribeth, and Yosef Jelahut. 2007. "For the People or for the Trees? A Case Study of Violence and Conservation in Ruteng Nature Recreation Park." In *Biodiversity and Human Livelihoods in Protected Areas*, edited by N. Sodhi, 222–40. Cambridge: Cambridge University Press.

Escobar, Arturo. 1999. "After Nature: Steps to an Antiessentialist Political Ecology." *Current Anthropology* 40 (1): 1–30.

———. 2018. *Designs for the Pluriverse.* Durham: Duke University Press.

Evans, Julian. 1982. *Plantation Forestry in the Tropics.* Oxford: Clarendon Press.

Eyongetah, Tambi, and Robert Brain. 1974. *A History of the Cameroon.* Harlow, UK: Longman.

Falzon, Mark-Anthony. 2009. "Multi-Sited Ethnography: Theory, Praxis and Locality in Contemporary Research." In *Multi-Sited Ethnography*, edited by M.-A. Falzon, 1–23. Burlington: Ashgate.

Fankhauser, Barry. 1987. "A Beginner's Guide to *Umu Ti.*" *New Zealand Archaeological Association Newsletter* 30 (3): 144–57.

Fassmann, Robert. 1902. "Nachrichten aus Moschi." *Evangelisch-Lutherisches Missionsblatt* 1902: 345–349.

Feld, Steven, and Keith Basso. 1996. *Senses of Place.* Santa Fe: School of American Research Press.

Feldman-Savelsberg, Pamela. 1999. *Plundered Kitchens, Empty Wombs.* Ann Arbor: University of Michigan Press.

Ferguson, James. 1990. *The Anti-Politics Machine.* Cambridge: Cambridge University Press.

Ferme, Marianne. 2001. *The Underneath of Things.* Berkeley: University of California Press.

Fernandes, E.C.M., A. O'kting'ati, and J. Maghembe. 1985. "The Chagga Homegardens: A Multistoried Agroforestry Cropping System on Mt. Kilimanjaro (Northern Tanzania)." *Agroforestry Systems* 2 (2): 73–86.

Figgis, T.F. 1958. *Chagga Land Tenure Report 1957–1958.* Tanzania National Archives. Microfilm in Worldcat labeled "Tanzania Land Tenure Reports."

Filer, Colin. 2007. "Local Custom and the Art of Land Group Boundary Maintenance in Papua New Guinea." In *Customary Land Tenure & Registration in Australia and Papua New Guinea,* edited by J. Weiner and K. Glaskins, 135–73. Canberra: ANU Press.

———. 2011. "The New Land Grab in Papua New Guinea." State, Society, and Governance in Melanesia Discussion Paper 2011/2. Canberra: ANU.

Filer, Colin, David Henton, and Richard Jackson. 2000. *Landowner Compensation in Papua New Guinea's Mining and Petroleum Sectors.* Port Moresby: PNG Chamber of Mines and Petroleum.

Filer, Colin, and John Numapo. 2017. "The Political Ramifications of Papua New Guinea's Commission of Inquiry." In *Kastom, Property, and Ideology,* edited by S. McDonnell, M. Allen and C. Filer, 251–82. Canberra: ANU Press.

Finney, Ben. 1979. *Hokule'a.* New York: Dodd, Mead and Company.

———. 1994. *Voyage of Rediscovery.* Berkeley: University of California Press.

Firth, Raymond. 2012. *Tikopia Ritual and Belief.* New York: Routledge.

Fisiy, Cyprian. 1992. *Power and Privilege in the Administration of Law.* Leiden: African Studies Centre. Open access document, https://openaccess.leidenuniv.nl/handle/1887/446, accessed June 11, 2018.

———. 1994. "The Death of a Myth System: Land Colonization on the Slopes of Mount Oku, Cameroon." In *Land Tenure and Sustainable Land Use,* edited by R. Bakema, 12–20. Amsterdam: Royal Tropical Institute.

Fisiy, Cyprian, and Peter Geschiere. 1996. "Witchcraft, Violence and Identity: Different Trajectories in Postcolonial Cameroon." In *Postcolonial Identities in Africa,* edited by R. Werbner and T. Ranger, 193–221. London: Zed Books.

Fison, Lorimer. 1885. "The Nanga, or Sacred Stone Enclosure, of Wainimala, Fiji." *Journal of the Anthropological Institute of Great Britain and Ireland* 14: 14–31.

Fjellman, Stephen, and Miriam Goheen. 1984. "A Prince by Any Other Name? Identity and Politics in Highland Cameroon." *American Ethnologist* 11 (3): 473–86.

Fomin, E.D.S., and Michael Ndobegang. 2006. "African Slavery Artifacts and European Colonialism: The Cameroon Grassfields from 1600 to 1950." *The European Legacy* 11 (6): 633–46.

Fon, Dorothy. 2011. "Access to Arable Land by Rural Women in Cameroon." *Tropicultura* 29 (2): 65–69.

Foncha, Jicenta, Azinwie Asongwe, and Che Fuh. 2019. "Using Indigenous Knowledge in Agro-Forestry Practices: A Strategy for Livelihood Sustainability, in the Mount Oku Region of Cameroon." *International Journal of Food Science and Agriculture* 3 (4): 299–307.

Foncha, Jicenta, and Dora Ewule. 2020. "Community Forest Management: A Strategy for Rehabilitation, Conservation and Livelihood Sustainability: The Case of Mount Oku, Cameroon." *Journal of Geoscience and Environment Protection* 8: 1–14.

Fontein, Joost. 2015. *Remaking Mutirikwi.* Oxford: James Currey.

Forde, Maarit. 2019. "The Spiritual Baptist Religion." *Caribbean Quarterly* 65 (2): 212–40.

Forster, Georg. 2000. *A Voyage Round the World,* edited by N. Thomas and O. Berghof. Honolulu: University of Hawai'i Press.

Fortune, Reo. 1932. *Sorcerers of Dobu.* London: George Routledge and Sons.

Fosberg, Francis. 1985. "*Cordyline fruticosa* (L.) Chevalier (Agavaceae)." *Baileya* 22 (4): 180–81.

Foster, Robert. 1985. "Production and Value in the Enga Tee." *Oceania* 55 (3): 182–96.

Foucault, Michel. 1979. *Discipline and Punish*. New York: Vintage Books.

———. 1997. "The Subject and Power." In *Power*, edited by J. Faubion, 326–98. New York: New Press.

———. 2012. *The History of Sexuality, Vol. 1: An Introduction*. New York: Knopf Doubleday.

Fowler, Ian. 1995. "The Oku Iron Industry in Its Regional Setting: A Descriptive Account." *Baessler-Archiv Neue Folge* 43: 89–126.

———. 2011. "Kingdoms of the Cameroon Grassfields." *Reviews in Anthropology* 40 (4): 292–311.

Fowler, Ian., and David Zeitlyn, eds. 1996. *African Crossroads*. Providence: Berghahn.

Fox, James. 2006. "Comparative Perspectives on Austronesian Houses: An Introductory Essay." In *Inside Austronesian Houses*, edited by J. Fox, 1–28. Canberra: ANU Press.

Fraser, Adrian. 1986. "*Peasants and Agricultural Labourers in St. Vincent and the Grenadines 1899–1951*," PhD diss., University of Western Ontario.

———. 2011. *From Shakers to Spiritual Baptists*. Kingstown, St. Vincent: Kings-SVG Pub.

Fraser, Nancy. 1995. "From Redistribution to Recognition? Dilemmas of Justice in a 'Post-socialist' Age." *New Left Review* 212: 68–93.

Frederick, Nfor, and Balgah Nguh. 2020. "Constraints to the Development of Ecotourism Potentials along the Babessi-Oku Axis, North West Region of Cameroon." *Asian Journal of Geographical Research* 3 (4): 1–16.

Friends of the Earth. 2015. "Legal Action for Land Rights Papua New Guinea." Online document, https://www.foei.org/no-category/legal-action-land-rights-papua-new-guinea.

Galiani, Sebastian, and Itai Sened. 2014. *Institutions, Property Rights, and Economic Growth*. New York: Cambridge University Press.

Galvin, Shaila. 2018. "Interspecies Relations and Agrarian Worlds." *Annual Review of Anthropology* 47 (1): 233–49.

Garanger, José. 1964. "Recherches Archéologiques dans le District de Tautira (Tahiti, Polynésie Française)." *Journal de la Société des Océanistes* 20 (20): 5–22.

Gardner, Anne, John DeMarco, and Christian Asanga. 2001. *A Conservation Partnership: Community Forestry at Kilum-Ijim, Cameroon*. Rural Development Forestry Network, Network Paper 25h. Online document, https://odi.org/en/publications/a-conservation-partnership-community-forestry-at-kilum-ijim-cameroon/, accessed June 11, 2018.

Gardner, Katy, and David Lewis. 2015. *Anthropology and Development*, 2e. London: Pluto Press.

Gargiullo, Margaret. 2008. *A Field Guide to Plants of Costa Rica*. Oxford: Oxford University Press.

Garson, John, and Charles Read, eds. 1892. *Notes and Queries on Anthropology*. London: Royal Anthropological Institute of Great Britain and Ireland.

Gautier, Denis. 1996. "Ficus (*Moraceae*) as Part of Agrarian Systems in the Bamileke Region (Cameroon)." *Economic Botany* 50 (3): 318–26.

Gearing, Margaret. 1988. *"The Reproduction of Labor in a Migration Society."* PhD diss., University of Florida - Gainesville.

Gebauer, Paul. 1964. *Spider Divination in the Cameroons.* Milwaukee: Milwaukee Public Museum.

German, Laura, Waga Mazengia, Hailemichael Taye, Mesfin Tsegaye, Shenkut Ayele, Sarah Charamila, and Juma Wickama. 2010. "Minimizing the Livelihood Trade-Offs of Natural Resource Management in the Eastern African Highlands: Policy Implications of a Project in 'Creative Governance'." *Human Ecology* 38 (1): 31–47.

Gershon, Ilana. 2010. "Bruno Latour (1947–)." In *From Agamben to Žižek*, edited by J. Simons, 161–76. Edinburgh: Edinburgh University Press.

Geschiere, Peter. 1993. "Chiefs and Colonial Rule in Cameroon: Inventing Chieftaincy, French and British Style." *Africa* 63 (2): 151–75.

———. 2004. "Ecology, Belonging and Xenophobia: The 1994 Forest Law in Cameroon and the Issue of 'Community'." In *Rights and the Politics of Recognition in Africa*, edited by H. Englund and F. Nyamnjoh, 237–59. London: Zed Books.

———. 2013. *Witchcraft, Intimacy, and Trust.* Chicago: University of Chicago Press.

Gibbs, B., ed. 1947. *A Plan for the Development of the Colony of St. Vincent, Windward Islands, British West Indies.* Port-of-Spain, Trinidad: Guardian Commercial Printery.

Gibbs, Philip. 2016. "Grave Business in Enga." *Journal of the Polynesian Society* 125 (2): 115–32.

Giddens, Anthony. 1993. *The Giddens Reader*, edited by P. Cassell. Stanford: Stanford University Press.

Gillespie, Rosemary, Elin Claridge, and Sara Goodacre. 2008. "Biogeography of the Fauna of French Polynesia: Diversification Within and between a Series of Hot Spot Archipelagos." *Philosophical Transactions of the Royal Society of London. Series B, Biological Sciences* 363 (1508): 3335–46.

Gillespie, Susan. 2000. "Lévi-Strauss: *Maison* and *Société à Maisons*." In *Beyond Kinship*, edited by R. Joyce and S. Gillespie, 22–52. Philadelphia: University of Pennsylvania Press.

Gillson, Lindsey, Michael Sheridan, and Dan Brockington. 2003. "Representing Environments in Flux: Case Studies from East Africa." *Area* 35 (4): 371–89.

Glauning, Hans. 1906. "Bericht des Hauptmanns Glauning über seine Reise in den Nordbezirk." *Deutsches Kolonialblatt* 17 (1906): 235–41.

Glazier, Stephen. 1983. *Marchin' the Pilgrims Home.* Westport, CT: Greenwood Press.

Glesne, Corrine. 1985. *"Strugglin', but no Slavin'."* PhD diss., University of Illinois, Urbana-Champaign.

Gluckman, Max. 1965. *The Ideas in Barotse Jurisprudence.* New Haven: Yale University Press.

Godelier, Maurice. 1986. *The Making of Great Men.* Cambridge: Cambridge University Press.

Goheen, Miriam. 1996. *Men Own the Fields, Women Own the Crops.* Madison: University of Wisconsin Press.

González, Nancie. 1988. *Sojourners of the Caribbean.* Urbana: University of Illinois Press.

Government of Tanzania. 2013. *National Bureau of Statistics 2012 Population and Housing Census: Population Distribution by Administrative Areas.* Project Report. National Bureau of Statistics. URL http://repository.out.ac.tz/362/

Graeber, David. 2015. "Radical Alterity Is Just Another Way of Saying 'Reality': A Reply to Eduardo Viveiros de Castro." *HAU: Journal of Ethnographic Theory* 5 (2): 1–41.

Graham, Barbara. 2012. *Profile of the Small-Scale Farming in the Caribbean.* Rome: FAO. http://www.fao.org/3/a-au343e.pdf.

Grainger, James. 1764. *The Sugar-Cane: A Poem.* London: R. and J. Dodsley.

Green, Roger. 1991. "Near and Remote Oceania: Disestablishing 'Melanesia' in Culture History." In *Man and a Half,* edited by A. Pawley, 491–502. Auckland: Polynesian Society.

———. 2000. "Lapita and the Cultural Model for Intrusion, Integration and Innovation." In *Australian Archaeologist,* edited by A. Anderson and T. Murray, 372–92. Canberra: Coombs Academic.

———. 2005. "Sweet Potato Transfers in Polynesian Prehistory." In *The Sweet Potato in Oceania,* edited by C. Ballard, P. Brown and M. Bourke, 43–62. Sydney: Oceania Monographs.

Grisebach, August. 1864. *Flora of the British West Indian Islands.* London: Lovell Reeve & Company.

Grossman, Lawrence. 1997. "Soil Conservation, Political Ecology, and Technological Change on Saint Vincent." *GERE Geographical Review* 87 (3): 353–74.

———. 1998. *The Political Ecology of Bananas.* Durham: University of North Carolina Press.

Grove, Richard. 1995. *Green Imperialism.* Cambridge: Cambridge University Press.

———. 2006. "The British Empire and the Origins of Forest Conservation in the Eastern Caribbean 1700-1800." In *Islands, Forests and Gardens in the Caribbean,* edited by R. Grove, K. Hiebert and R. Anderson, 132–73. Oxford: Macmillan Caribbean.

Griffiths, A.W.M. 1930. *Land Tenure Moshi District.* Microfilm in Worldcat labeled "Tanzania Land Tenure Reports."

Griffiths, Mark, ed. 1992. "Cordyline." In *New Royal Horticultural Society Dictionary of Gardening,* vol. 2, 718–719. London: Macmillan.

Gudeman, Stephen. 2016. *Anthropology and Economy.* Cambridge: Cambridge University Press.

Gufler, Hermann. 2009. "Reenactment of a Myth: The Fon of Oku Visits Lake Mawes (Cameroon)." *Anthropos* 104 (2): 347–57.

———. 2009-2010. "'The Land Has Become Bad': Finding a Solution to the Troubled Relationship between the Ntul and the Past Fons of Oku (Cameroon)." *Archiv für Völkerkunde* 59–60: 107–24.

Gufler, Hermann, and Nkajoi John Bah. 2006. "The Establishment of the Princes' Society in Oku, Cameroon: An Enhancement of Traditional Culture or Its Adulteration?" *Anthropos* 101 (1): 55–80.

Guilding, Lansdown. 1825. *An Account of the Botanic Garden in the Island of St Vincent from Its Establishment to the Present Time.* Glasgow: Richard Griffin and Co.

Gullick, Charles. 1985. *Myths of a Minority.* Assen, Netherlands: Van Gorcum.

Gunson, Niel. 1962. "An Account of the Mamaia or Visionary Heresy of Tahiti, 1826–1841." *Journal of the Polynesian Society* 71 (2): 209–43.

———. 1969. "Pomare II of Tahiti and Polynesian Imperialism." *Journal of Pacific History* 4: 65–82.

Gunson, Niel, and Paul de Deckker. 1990. "Crises Religieuses et Politiques 1827–1842." In *Encyclopedie de la Polynésie,* 2e, vol. 6, edited by P.-Y. Toullelan, 121–36. Pape'ete: Éditions de l'Alizé.

Gupta, Akhil, and James Ferguson, eds. 1997a. *Culture, Power, Place.* Durham: Duke University Press.

Gupta, Akhil, and James Ferguson. 1997b. "Discipline and Practice: 'The Field' as Site, Method, and Location in Anthropology." In *Anthropological Locations*, edited by A. Gupta and J. Ferguson, 1–46. Berkeley: University of California Press.

Gutmann, Bruno. 1926. *The Law of the Chagga.* Munich: Beck. Human Relations Area Files. New Haven: Yale University Press.

———. 1932. *The Tribal Teachings of the Chagga.* Munich: Beck. Human Relations Area Files. New Haven: Yale University Press.

———. 2017. *Poetry and Thinking of the Chagga.* Oxford: Signal Books Limited.

Guyer, Jane. 1981. "Household and Community in African Studies." *African Studies Review* 24 (2–3): 87–137.

———. 2016. *Legacies, Logics, Logistics.* Chicago: University of Chicago Press.

Guyer, Jane, and Samuel Belinga. 1995. "Wealth in People as Wealth in Knowledge: Accumulation and Composition in Equatorial Africa." *Journal of African History* 36: 91–120.

Haberle, Simon, Carol Lentfer, Shawn O'Donnell, and Tim Denham. 2012. "The Paleoenvironments of Kuk Swamp from the Beginnings of Agriculture in the Highlands of Papua New Guinea." *Quaternary International* 249: 129–39.

Håkansson, N. Thomas. 2003. "Rain and Cattle: Gendered Structures and Political Economy in Precolonial Pare, Tanzania." In *Gender at Work in Economic Life*, edited by G. Clark, 19–40. Walnut Creek, CA: AltaMira Press.

———. 2009. "Politics, Cattle, and Ivory: Regional Interaction and Changing Land-Use Prior to Colonialism." In *Culture, History, and Identity*, edited by T. Clack, 141–154. Oxford: Archaeopress.

———. 2019. "Criticizing Resilience Thinking: A Political Ecology Analysis of Droughts in Nineteenth-century East Africa." *Economic Anthropology* 6 (1): 7–20.

Håkansson, N. Thomas, and Mats Widgren, eds. 2014. *Landesque Capital.* Walnut Creek, CA: Left Coast Press.

Hall, Douglas. 1978. "The Flight from the Estates Reconsidered: The British West Indies 1838–42." *Journal of Caribbean History* 10: 7–24.

Hall, Douglas, and Thomas Thistlewood. 1999. *In Miserable Slavery.* Kingston, Jamaica: University of the West Indies Press.

Haller, Tobias. 2010. *Disputing the Floodplains.* Leiden: Brill.

Handler, Jerome. 1971. "The History of Arrowroot and the Origin of Peasantries in the British West Indies." *Journal of Caribbean History* 2: 46–93.

———. 2002. "Plantation Slave Settlements in Barbados, 1650s to 1834." In *In the Shadow of the Plantation*, edited by A.O. Thompson, 123–161. Kingston, Jamaica: Ian Randle.

Handy, E.S. 1930. *History and Culture in the Society Islands.* Honolulu: Bishop Museum.

Handy, E.S., E. Handy, and Mary Pukui. 1972. *Native Planters in Old Hawaii.* Honolulu: Bishop Museum Press.

Hann, C.M. 1998. "Introduction: The Embeddedness of Property." In *Property Relations*, edited by C.M. Hann, 1–47. Cambridge: Cambridge University Press.

Hannerz, Ulf. 2009. "Afterword: The Long March of Anthropology." In *Multi-Sited Ethnography*, edited by M.-A. Falzon, 271–81. Burlington: Ashgate.

Haraway, Donna. 2003. *The Companion Species Manifesto*. Chicago: Prickly Paradigm Press.

———. 2016. *Staying With the Trouble*. Durham: Duke University Press.

Haraway, Donna, Noboru Ishikawa, Scott Gilbert, Kenneth Olwig, Anna Tsing, and Nils Bubandt. 2016. "Anthropologists Are Talking – About the Anthropocene." *Ethnos* 81 (3): 535–64.

Haraway, Donna, Anna Tsing, and Gregg Mitman. 2019. "Reflections on the Plantationocene." *Edge Effects*, University of Madison-Wisconsin, https://edgeeffects.net/haraway-tsing-plantationocene/.

Hardin, Garrett. 1968. "The Tragedy of the Commons." *Science* 162: 1243–48.

Hargreaves, Dorothy, and Bob Hargreaves. 1960. *Tropical Blossoms*. Portland: Hargreaves Industrial.

Hart, Gillian. 2002. *Disabling Globalization*. Berkeley: University of California Press.

Hartigan, John Jr. 2017. *Care of the Species*. Minneapolis: University of Minnesota Press.

Harvey, David. 2009. *Cosmopolitanism and the Geographies of Freedom*. New York: Columbia University Press.

Hasu, Päivi. 1999. *Desire and Death*. Saarijärvi, Finland: Finnish Anthropological Society.

———. 2009a. "People of the Banana Garden: Placing the Dead at the Ultimate Home in Kilimanjaro." In *Culture, History, and Identity*, edited by T. Clack, 77–87. Oxford: Archaeopress.

———. 2009b. "For Ancestors and God: Rituals of Sacrifice among the Chagga of Tanzania." *Ethnology* 48 (3): 195–213.

Hather, Jon. 2013. "The Identification of Charred Root and Tuber Crops from Archaeological Sites in the Pacific." In *Tropical Archaeobotany*, edited by J. Hather, 51–65. London: Routledge.

Hauser, Mark. 2017. "A Political Ecology of Water and Enslavement: Water Ways in Eighteenth-Century Caribbean Plantations." *Current Anthropology* 58 (2): 227–56.

Hawkesworth, John, ed. 1789. *An Account of the Voyages Undertaken by the Order of His Present Majesty [...]*, 4th ed. Perth, Scotland: R. Morison and Son.

Healey, Christopher. 1990. *Maring Hunters and Traders*. Berkeley: University of California Press.

Heckenberger, Michael, Afukaka Kuikuro, Urissapá Tabata Kuikuro, Christian Russell, Morgan Schmidt, Carlos Fausto, and Bruna Franchetto. 2003. "Amazonia 1492: Pristine Forest or Cultural Parkland?" *Science* 301: 1710–14.

Heider, Karl. 1970. *The Dugum Dani*. Chicago: Aldine Publishing Company.

Hemp, Andreas. 2005. "Climate Change-Driven Forest Fires Marginalize the Impact of Ice Cap Wasting on Kilimanjaro." *Global Change Biology* 11 (7): 1013–23.

———. 2006a. "The Banana Forests of Kilimanjaro: Biodiversity and Conservation of the Chagga Homegardens." *Biodiversity and Conservation* 15 (4): 1193–1217.

———. 2006b. "Continuum or Zonation? Altitudinal Gradients in the Forest Vegetation of Mt. Kilimanjaro." *Plant Ecology* 184 (1): 27–42.

———. 2006c. "Vegetation of Kilimanjaro: Hidden Endemics and Missing Bamboo." *African Journal of Ecology* 44 (3): 305–28.

———. 2009. "Climate Change and Its Impact on the Forests of Kilimanjaro." *African Journal of Ecology* 47 (s1): 3–10.

Hemp, Andreas, and Claudia Hemp. 2009. "Environment and Worldview: The Chagga Homegardens Part I: Ethnobotany and Ethnozoology." In *Culture, History, and Identity*, edited by T. Clack, 235–71. Oxford: Archaeopress.

Henney, Jeannette. 1973. "The Shakers of St. Vincent: A Stable Religion." In *Religion, Altered States of Consciousness, and Social Change*, edited by E. Bourguignon, 219–63. Columbus: Ohio State University Press.

———. 1974. "Spirit-Possession Belief and Trance Behavior in Two Fundamentalist Groups in St. Vincent." In *Trance, Healing, and Hallucination*, edited by F.D. Goodman, J. Henney and E. Pressel, 1–111. New York: John Wiley and Sons.

Henry, Teuira. 1893. "*Te Umu-Ti*, a Raiatean Ceremony." *Journal of the Polynesian Society* 2 (2): 105–8.

———. 1928. *Ancient Tahiti*. Honolulu: Bishop Museum.

Herdt, Gilbert. 1987. *The Sambia*. New York: Holt, Rinehart, and Winston.

Herskovits, Melville, and Frances Herskovits. 1947. *Trinidad Village*. New York: A. A. Knopf.

Heywood, Paolo. 2018. "The Ontological Turn: School or Style?" In *Schools and Styles of Anthropological Theory*, edited by M. Candea, 224–35. London: Routledge.

Higman, B.W. 1984. *Slave Populations of the British Caribbean, 1807–1834*. Baltimore: Johns Hopkins University Press.

———. 2001. *Jamaica Surveyed*. Kingston, Jamaica: University of the West Indies Press.

Hills, Theo. 1988. "The Caribbean Peasant Food Forest, Ecological Artistry or Random Chaos." In *Small Farming and Peasant Resources in the Caribbean*, edited by J. Brierley and H. Rubenstein, 1–28. Winnipeg: University of Manitoba Dept. of Geography.

Hinkle, Anya. 2004. "The Distribution of a Male Sterile Form of Ti (*Cordyline fruticosa*) in Polynesia: A Case of Human Selection?" *Journal of the Polynesian Society* 113 (3): 263–90.

———. 2005. "*Population Structure and Reproductive Biology of the Ti Plant (Cordyline fruticosa) with Implications for Polynesian Prehistory*." PhD diss., University of California, Berkeley.

———. 2007. "Population Structure of Pacific *Cordyline fruticosa* (Laxmanniaceae) With Implications for Human Settlement of Polynesia." *American Journal of Botany* 94 (5): 828–39.

Hirsch, Eric. 1995. "Landscape: Between Place and Space." In *The Anthropology of Landscape*, edited by E. Hirsch and M. O'Hanlon, 1–30. Oxford: Clarendon Press.

Hitchings, Russell. 2003. "People, Plants and Performance: On Actor Network Theory and the Material Pleasures of the Private Garden." *Social & Cultural Geography* 4 (1): 99–114.

Hoekema, André. 2012. "If Not Private Property, Then What? Legalising Extra-Legal Rural Land Tenure via a Third Road." In *Fair Land Governance*, edited by J. Otto and A. Hoekema, 135–79. Leiden: Leiden University Press.

Hogbin, Ian, and Camilla Wedgwood. 1953. "Local Grouping in Melanesia." *Oceania* 23 (4): 241–76.

Holand, Ivar. 1996. "More People, More Trees: Population Growth, The Chagga Irrigation System, and the Expansion of a Sustainable Agroforestry System on Mount Kilimanjaro." MSc, Norwegian University of Science and Technology. https://www.researchgate.net/publication/235218855_More_people_more_trees_population_growth_the_Chagga_irrigation_system_and_the_expansion_of_a_sustainable_agroforestry_system_on_Mount_Kilimanjaro

Holbraad, Martin, and Morten Pedersen. 2017. *The Ontological Turn*. Cambridge: Cambridge University Press.

Holcomb, Bobby, and Claire Leimbach. 1992. *Bobby: Visions Polynésiennes*. Avalon, Australia: Pacific Bridge.

Holmes, Seth. 2013. *Fresh Fruit, Broken Bodies*. Berkeley: University of California Press.

Home, Robert, and Hilary Lim, eds. 2004. *Demystifying the Mystery of Capital*. London: Cavendish Publishing Ltd.

Hornborg, Alf. 2001. *The Power of the Machine*. Walnut Creek, CA: AltaMira Press.

———. 2009. "Zero-Sum World: Challenges in Conceptualizing Environmental Load Displacement and Ecologically Unequal Exchange in the World-System." *International Journal of Comparative Sociology* 50 (3–4): 237–62.

———. 2013. "Revelations of Resilience: From the Ideological Disarmament of Disaster to the Revolutionary Implications of (P)Anarchy." *Resilience* 1 (2): 116–29.

———. 2021. "Objects don't Have Desires: Toward an Anthropology of Technology Beyond Anthropomorphism." *American Anthropologist* 123 (4): 753–66.

Houseman, Michael. 1998. "Painful Places: Ritual Encounters with one's Homelands." *Journal of the Royal Anthropological Institute* 4 (3): 447–67.

Howard, Richard. 1954. "A History of the Botanic Garden of St. Vincent, British West Indies." *Geographical Review* 44 (3): 381–93.

Huguenin, Paul. 1902. *Raiatea La Sacrée*. Neuchatel, Switzerland: P. Attinger.

Ingold, Tim. 2015. *The Life of Lines*. London: Routledge.

———. 2016. *Lines: A Brief History*. London: Routledge Classics.

Ingram, Verina. 2014. *"Win-wins in Forest Product Value Chains?"* PhD diss., Amsterdam Institute for Social Science Research.

Ingram, Verina, Marjam Ros-Tonen, and Tom Dietz. 2015. "A Fine Mess: Bricolaged Forest Governance in Cameroon." *International Journal of the Commons* 9 (1): 41–64.

Institute für Länderkunde (IFL). 1907. Leibniz-Institute für Länderkunde photographic archive. Spiritan Mission, Kibosho. Photo by Dr. Förster, item # Af045-1029, http://ifl.wissensbank.com/starweb/IFL/DEU/OPACG/servlet.starweb

International Mission Photography Archive (IMP). n.d. Photo of Moshi Market, ca. 1909, by Ernst Hohfeld. International Mission Photography Archive, ca.1860–1960, Leipzig Mission, Album 10 photo #132.

Isaacs, Philemon. 2014. *St. Vincent and the Grenadines National Land Policy*. St. Vincent: Government of St. Vincent and the Grenadines. http://centreforapplied-developmentstudies.com/wp-content/uploads/2019/03/National-Land-Policy.pdf

Isager, Lotte, and Søren Ivarsson. 2002. "Contesting Landscapes in Thailand: Tree Ordination as Counter-Territorialization." *Critical Asian Studies* 34 (3): 395–417.

Jacka, Jerry. 2009. "Global Averages, Local Extremes: The Subtleties and Complexities of Climate Change in Papua New Guinea." In *Anthropology and Climate Change*, edited by S. Crate and M. Nuttall, 197–208. Walnut Creek, CA: Left Coast Press.

———. 2015. *Alchemy in the Rain Forest*. Durham: Duke University Press.

Jackson, Michael, and Ivan Karp, eds. 1990. *Personhood and Agency*. Uppsala: Uppsala University.

James, R.W., and G.M. Fimbo. 1973. *Customary Land Law of Tanzania*. Nairobi: East African Literature Bureau.

Jeffreys, M. 1961. "Oku Blacksmiths." *The Nigerian Field* 26 (3): 137–44.

John, Karl. 2006. *Land Reform in Small Island Developing States*. College Station: Virtualbookworm.com Publishing.

Johnston, Harry. 1886. *The Kilima-Njaro Expedition*. London: Kegan Paul.

Jordan, Fiona, Russell Gray, Simon Greenhill, and Ruth Mace. 2009. "Matrilocal Residence Is Ancestral in Austronesian Societies." *Proceedings of the Royal Society B: Biological Sciences* 276 (1664): 1957–64.

Juul, Kristine, and Christian Lund, eds. 2002. *Negotiating Property in Africa*. Portsmouth, NH: Heinemann.

Kaberry, Phyllis. 1950. "Land Tenure among the Nsaw of the British Cameroons." *Africa* 20 (4): 307–23.

Kairi Consultants. 2008. *Final Report: St Vincent and the Grenadines Country Poverty Assessment 2007/2008*. Tunapuna, Trinidad: Kairi Consultants Ltd.

Kahn, Jennifer. 2007. "Power and Precedence in Ancient House Societies: A Case Study from the Society Island Chiefdoms." In *The Durable House*, edited by R. Beck, 198–223. Carbondale: Center for Archaeological Investigations, Southern Illinois University.

Kahn, Jennifer, and Patrick Kirch. 2013. "Residential Landscapes and House Societies of the Late Prehistoric Society Islands." *Journal of Pacific Archaeology* 4 (1): 50–72.

Kahn, Miriam. 2011. *Tahiti Beyond the Postcard*. Seattle: University of Washington Press.

Kayser, Manfred. 2010. "The Human Genetic History of Oceania: Near and Remote Views of Dispersal." *Current Biology* 20 (4): R194–R201.

Keesing, Roger. 1982. "Kastom in Melanesia: An Overview." *Mankind* 13 (4): 297–301.

———. 1984. "Rethinking 'Mana'." *Journal of Anthropological Research* 40 (1): 137–156.

Keleman Saxena, Alder, Deepti Chatti, Katy Overstreet, and Michael Dove. 2018. "From Moral Ecology to Diverse Ontologies: Relational Values in Human Ecological Research, Past and Present." *Current Opinion in Environmental Sustainability* 35: 54–60.

Kelly, Tara. 2012. *"Plants, power, possibility."* D. Phil diss., University of Oxford.

Keming, David. 2015. "SCNC Rebellion in Oku, Bui Division, North West Region of Cameroon." *Scholars Journal of Arts, Humanities and Social Sciences* 3 (4a): 869–80.

Kenn, Charles. 1949. *Fire-Walking from the Inside*. Los Angeles: Franklin Thomas.

Kennedy, Geoff. 2008. *Diggers, Levellers, and Agrarian Capitalism*. Lanham, MD: Lexington Books.

Kenyatta, Jomo. 1965. *Facing Mt. Kenya*. New York: Vintage Books.

Kepler, Angela. 1998. *Hawaiian Heritage Plants*. Honolulu: University of Hawai'i Press.

Kim, Julie. 2013. "The Caribs of St. Vincent and Indigenous Resistance During the Age of Revolutions." *Early American Studies* 11 (1): 117–32.

Kimambo, Isaria. 1996. "Environmental Control and Hunger in the Mountains and Plains of Northeastern Tanzania." In *Custodians of the Land*, edited by G. Maddox, J. Giblin and I. Kimambo, 71–95. Athens: Ohio University Press.

Kingsley, Charles. 1871. *At Last: A Christmas in the West Indies*. London: Harper & Bros.

Kirch, Patrick. 1994. *The Wet and the Dry*. Chicago: University of Chicago Press.

———. 1996. *Legacy of the Landscape*. Honolulu: University of Hawai'i Press.

———. 1997a. *The Lapita Peoples*. Blackwell: Wiley.

———. 1997b. "Microcosmic Histories: Island Perspectives on 'Global' Change." *American Anthropologist* 99 (1): 30–42.

———. 2000a. *On the Road of the Winds*. Berkeley: University of California Press.

———. 2000b. "Temples as 'Holy Houses': The Transformation of Ritual Architecture in Traditional Polynesian Societies." In *Beyond Kinship*, edited by R. Joyce and S. Gillespie, 103–14. Philadelphia: University of Pennsylvania Press.

———. 2010. "Peopling of the Pacific: A Holistic Anthropological Perspective." *Annual Review of Anthropology* 39: 131–48.

Kirch, Patrick, and Jennifer Kahn. 2007. "Advances in Polynesian Prehistory: A Review and Assessment of the Past Decade (1993–2004)." *Journal of Archaeological Research* 15 (3): 191–238.

Kirksey, Eben, and Stefan Helmreich. 2010. "The Emergence of Multispecies Ethnography." *Cultural Anthropology* 25 (4): 545–76.

Kirsch, Stuart. 2008. "Social Relations and the Green Critique of Capitalism in Melanesia." *American Anthropologist* 110 (3): 288–98.

Kitalyi, Aichi, Robert Otsyina, Charles Wambugu, and Deborah Kimaro. 2013. *FAO Characterisation of Global Heritage Agroforestry Systems in Tanzania and Kenya*. Rome, FAO. http://www.fao.org/3/bp876e/bp876e.pdf

Kocher Schmid, Christin. 1991. *Of People and Plants*. Basel: Ethnologisches Seminar der Universität und Museum für Völkerkunde.

Koenig, Dolores. 2016. "The Year 2015 in Sociocultural Anthropology: Material Life and Emergent Cultures." *American Anthropologist* 118 (2): 346–58.

Kohn, Eduardo. 2013. *How Forests Think*. Berkeley: University of California Press.

———. 2015. "Anthropology of Ontologies." *Annual Review of Anthropology* 44: 311–27.

Koloss, Hans-Joachim. 1992. "Kwifon and Fon in Oku: On Kingship in the Cameroon Grasslands." In *Kings of Africa*, edited by E. Beumers and H.-J. Koloss, 33–42. Berlin: Foundation Kings of Africa.

———. 2000. *World-view and Society in Oku (Cameroon)*. Berlin: Dietrich Reimer.

———. 2008. "Court Art and Masks in the Kingdoms of the North West Province of Cameroon." In *Cameroon: Art and Kings*, edited by L. Homberger, 68–110. Zürich: Museum Rietberg.

———. 2012. *Cameroon: Thoughts and Memories*. Berlin: Dietrich Reimer.

Kopytoff, Igor. 1987. "The Internal African Frontier: The Making of African Political Culture." In *The African Frontier*, edited by I. Kopytoff, 3–85. Bloomington: Indiana University Press.

Krapf, Johann. 1860. *Travels, Researches, and Missionary Labours during an Eighteen Years' Residence in Eastern Africa*. London and Boston: Ticknor and Fields.

Krause, Arno. 1905. "News from Nkoaranga at Meru." In *Evangelisch-Lutherisches Missionsblatt: Extracts on Arusha and Meru, 1897–1914*, edited by T. Spear, 58–59. Madison: University of Wisconsin-Madison African Studies Program.

Kumar, B., and P. Nair. 2004. "The Enigma of Tropical Homegardens." *Agroforestry Systems* 61 (1): 135–52.

Kuper, Hilda. 2003. "The Language of Sites in the Politics of Space." In *The Anthropology of Space and Place*, edited by S. Low and D. Lawrence-Zúñiga, 247–63. Malden: Blackwell.

Kuwahara, Makiko. 2005. *Tattoo*. Oxford: Berg.

Lamont, Michèle, and Virág Molnár. 2002. "The Study of Boundaries in the Social Sciences." *Annual Review of Sociology* 28 (1): 167–95.

Landweer, Lynn, and Peter Unseth. 2012. "An Introduction to Language Use in Melanesia." *International Journal of the Sociology of Language* 214: 1–3.

Lang, Andrew. 1899. "The Fire Walk by Europeans." *The Athenaeum* 3755 (October 14): 528.

Langley, S., and Andrew Lang. 1901. "The Fire Walk Ceremony in Tahiti." *Folklore* 12 (4): 446–55.

Lans, Cheryl, and Karla Georges. 2011. "Women's Knowledge of Herbs Used in Reproduction in Trinidad and Tobago." In *Ethnomedicinal Plants*, edited by M. Rai, D. Acharya and J. Ríos, 115–34. Boca Raton, FL: CRC Press.

Larrue, Sebastien, Jean-Yves Meyer, and Thomas Chiron. 2010. "Anthropogenic Vegetation Contributions to Polynesia's Social Heritage: The Legacy of Candlenut Tree (*Aleurites moluccana*) Forests and Bamboo (*Schizostachyum glaucifolium*) Groves on the Island of Tahiti." *Economic Botany* 64 (4): 329–39.

Latour, Bruno. 1988. *The Pasteurization of France*. Cambridge, MA: Harvard University Press.

———. 2005. *Reassembling the Social*. Oxford: Oxford University Press.

———. 2012. *We Have Never Been Modern*. Cambridge, MA: Harvard University Press.

Lavondès, Anne. 1990. "Ressources, Échanges, et Consummation." In *Encyclopédie de la Polynésie*, 2e, vol. 5, edited by A. Lavondès, 57–72. Pape'ete: Éditions de L'Alizé.

Lavondès, Anne, and M. Charleux. 1990. "Le Corps, le Vêtement et la Parure." In *Encyclopedie de la Polynésie*, 2e, vol. 5, edited by A. Lavondès, 73–88. Pape'ete: Éditions de l'Alizé.

Lawry, Steven, Cyrus Samii, Ruth Hall, Aaron Leopold, Donna Hornby, and Farai Mtero. 2017. "The Impact of Land Property Rights Interventions on Investment and Agricultural Productivity in Developing Countries: A Systematic Review."*Journal of Development Effectiveness* 9 (1): 61–81.

Lea, David, and Timothy Curtin. 2011. *Land Law and Economic Development in Papua New Guinea*. Newcastle upon Tyne: Cambridge Scholars.

Leach, James. 2004. *Creative Land*. New York: Berghahn Books.

Leach, Melissa, Robin Mearns, and Ian Scoones. 1999. "Environmental Entitlements: Dynamics and Institutions in CBRNM." *World Development* 27 (2): 225–47.

Leenhardt, Maurice. 1946. "Le Ti en Nouvelle Caledonie." *Journal de la Société des Océanistes* 2 (2): 192–93.

Lefebvre, Henri. 1991. *The Production of Space*. Cambridge, MA: Blackwell.

Lema, Anderson. 1995a. "Gender Roles in Lyamungo Villages, Tanzania." In *Towards Common Ground*, edited by A. Sigot, L. A. Thrupp and J. Green, 65–80. Nairobi: ACTS Press.

———. 1995b. "Land Degradation in Local Communities on the Southern Slopes of Lake Kilimanjaro [*sic*]: Towards a Social Science Explanation and Prospects for Sustainable Development." In *The Tanzanian Peasantry*, edited by P. Forster and S. Maghimbi, 95–110. Aldershot, UK: Avebury.

Lemonnier, Pierre. 2012. *Mundane Objects*. Walnut Creek, CA: Left Coast Press.

Lepofsky, Dana. 1994. *"Prehistoric Agricultural Intensification in the Society Islands, French Polynesia."* PhD diss., University of California – Berkeley.

———. 1999. "Gardens of Eden? An Ethnohistoric Reconstruction of Maohi (Tahitian) Cultivation." *Ethnohistory* 46 (1): 1–29.

———. 2003. "The Ethnobotany of Cultivated Plants of the Maohi of the Society Islands." *Economic Botany* 57 (1): 73–92.

Le Roy, Alexandre. 1893. *Au Kilima-ndjaro*. Paris: L. de Soye et Fils.

Lévi-Strauss, Claude. 1982. *The Way of the Masks*. Seattle: University of Washington Press.

Levy, Claude. 1980. *Emancipation, Sugar, and Federalism*. Gainesville: University Press of Florida.

Lewellen, Ted. 2002. *The Anthropology of Globalization*. Westport, CT: Bergin and Garvey.

Lewis, Ferd. 2019. "Ti Leaves can't Overcome University of Hawaii's Turnovers." *Honolulu Star-Advertiser*, Sept. 15, 2019.

Lim, T.K. 2015. *Edible Medicinal and Non-Medicinal Plants*, vol. 9. Dordrecht: Springer.

Lindsey, N.A. 1901. *Cruising in the Madiana*. Boston: N.A. Lindsey Company.

LiPuma, Edward. 1988. *The Gift of Kinship*. Cambridge: Cambridge University Press.

Lloyd, Geoffrey. 2019. "The Clash of Ontologies and the Problems of Translation and Mutual Intelligibility." *HAU: Journal of Ethnographic Theory* 9 (1): 36–43.

Löfven, Stephan. 2015. "In Pursuit of Dignity," Human Development Reports. New York: United Nations Development Programme. http://hdr.undp.org/en/content/pursuit-dignity

Lohmann, Roger. 2005. "The Afterlife of Asabano Corpses: Relationships with the Deceased in Papua New Guinea." *Ethnology* 44 (2): 189.

Long, Edward. 1774. *The History of Jamaica*. London: T. Lowndes.

Low, Setha. 2016. *Spatializing Culture*. New York: Routledge.

Low, Setha, ed. 1999. *Theorizing the City*. New Brunswick: Rutgers University Press.

Low, Setha, and Denise Lawrence-Zúñiga. 2003. "Locating Culture." In *The Anthropology of Space and Place*, edited by S. Low and D. Lawrence-Zúñiga, 1–47. Malden, MA: Blackwell.

Lowder, Sarah, Jakob Skoet, and Terri Raney. 2016. "The Number, Size, and Distribution of Farms, Smallholder Farms, and Family Farms Worldwide." *World Development* 87: 16–29.

Lum, Kenneth. 2000. *Praising His Name in the Dance*. Amsterdam: Harwood Academic Publishers.

Lutkehaus, Nancy. 1995. *Zaria's Fire*. Durham: Carolina Academic Press.

MacLeod, Heather. 1986. *The Conservation of Oku Mountain Forest, Cameroon*. Cambridge: International Council for Bird Preservation.

Maisels, F., E. Keming, M. Kemei, and C. Toh. 2001. "The Extirpation of Large Mammals and Implications for Montane Forest Conservation: The Case of the Kilum-Ijim Forest, North-West Province, Cameroon." *Oryx* 35 (4): 322–31.

Malinowski, Bronislaw. 1944. *A Scientific Theory of Culture and Other Essays*. Chapel Hill: University of North Carolina Press.

———. 1961. *Argonauts of the Western Pacific*. New York: EP Dutton and Company.

———. 2013. *Coral Gardens and Their Magic*. Hoboken, NJ: Taylor and Francis.

Mamdani, Mahmood. 1996. *Citizen and Subject*. Princeton: Princeton University Press.

Marck, Jeff. 2008. "Proto Oceanic Society Was Matrilineal." *Journal of the Polynesian Society* 117 (4): 345–82.

Marcus, Anthony. 2003. "Imaginary Worlds: The Last Years of Eric Wolf." *Social Anthropology* 11 (1): 113–27.

Marcus, George. 1995. "Ethnography in/of the World System: The Emergence of Multi-Sited Ethnography." *Annual Review of Anthropology* 24: 95–117.

———. 2016. "Multi-Sited Ethnography: Notes and Queries." In *Multi-Sited Ethnography*, edited by M.-A. Falzon, 181–96. Abington, UK: Routledge.

Marealle, Petro. 1965. "Chagga Customs, Beliefs and Traditions." *Tanganyika Notes and Records* 64: 56–61.

———. 2002. *Maisha ya Mchagga Hapa Duniani na Ahera*. Dar es Salaam: Mkuki na Nyota Publishers.

Maro, Paul. 2009. "The Impact of Human Population on Land Management." In *Culture, History, and Identity*, edited by T. Clack, 207–12. Oxford: Archaeopress.

Marshall, E.J.P., and A.C. Moonen. 2002. "Field Margins in Northern Europe: Their Functions and Interactions with Agriculture." *Agriculture, Ecosystems and Environment* 89 (1–2): 5–21.

Marshall, Woodville. 1983. "'Vox Populi': The St. Vincent Riots and Disturbances of 1862." In *Trade, Government, and Society in Caribbean History, 1700–1920*, edited by B. Higman, 85–115. Kingston, Jamaica: Heinemann.

———. 1985. "Apprenticeship and Labour Relations in Four Windward Islands." In *Abolition and Its Aftermath*, edited by D. Richardson, 203–23. New York: Frank Cass and Company.

———. 1993. "Provision Ground and Plantation Labor in Four Windward Islands: Competition for Resources during Slavery." In *Cultivation and Culture*, edited by I. Berlin and P. Morgan, 203–20. Charlottesville, VA: University of Virginia Press.

———. 2003. "The Post-Slavery Labour Problem Revisited." In *Slavery, Freedom and Gender*, edited by B. Moore, B. Higman, C. Campbell and P. Bryan, 115–32. Kingston, Jamaica: University of West Indies Press.

———. 2011. "The Emergence and Survival of the Peasantry." In *General History of the Caribbean*, vol. 4, edited by K. Laurence, 149–90. London: Macmillan.

Martinez-Reyes, José. 2017. "Enviromateriality: Exploring the Links between Political Ecology and Material Culture Studies." In *Routledge Handbook of Environmental Anthropology*, edited by H. Kopnina and E. Shoreman-Ouimet, 71–80. London: Routledge.

Marx, Karl. 1915. *Capital*, vol. 1. Chicago: Charles H. Kerr and Company.

Mauss, Marcel. 1938. "Une Catégorie de l'Esprit Humain: La Notion de Personne Celle de 'Moi'." *Journal of the Royal Anthropological Institute of Great Britain and Ireland* 68: 263–81.

_____. 1967. *The Gift*. New York: W.W. Norton and Co.

Maynard, Kent. 2004. *Making Kedjom Medicine*. Westport, CT: Praeger.

Mbunda-Samba, Patrick, Paul Mzeka, Mathias Niba, and Clare Wirmum, eds. 1993. *Rites of Passage and Incorporation in the Western Grassfields of Cameroon*. Bamenda, Cameroon: Kaberry Research Centre.

McAlvay, Alex, Chelsey Armstrong, Janelle Baker, Linda Black Elk, Samantha Bosco, Natalia Hanazaki, Leigh Joseph, Tania Eulalia Martínez-Cruz, Mark Nesbitt, Meredith Palmer, Walderes Cocta Priprá de Almeida, Jane Anderson, Zemede Asfaw, Israel Borokini, Eréndira Juanita Cano-Contreras, Simon Hoyte, Maui Hudson, Ana Ladio, Guillaume Odonne, Sonia Peter, John Rashford, Jeffrey Wall, Steve Wolverton, and Ina Vandebroek. 2021. "Ethnobiology Phase VI: Decolonizing Institutions, Projects, and Scholarship."*Journal of Ethnobiology* 41 (2): 170–91.

McCay, Bonnie, and James Acheson, eds. 1987. *The Question of the Commons*. Tucson: University of Arizona Press.

McDaniel, Lorna. 1990. "The Flying Africans: Extent and Strength of the Myth in the Americas." *New West Indian Guide* 64 (1–2): 28–40.

McGinnis, Michael, and Elinor Ostrom. 2014. "Social-Ecological System Framework: Initial Changes and Continuing Challenges." *Ecology and Society* 19 (2): 30.

Mcintosh, Janet. 2018. "Personhood, Self, and Individual." In *International Encyclopedia of Anthropology*, edited by H. Callan. Hoboken, NJ: Wiley Blackwell.

Menzies, Archibald. 1920. *Hawaii Nei 128 Years Ago*. Honolulu: New Freedom.

Merceron, François, ed. 1988. *Dictionnaire Illustré de la Polynésie*. 4 vols. Pape'ete: Les Editions de l'Alizé.

Merlin, Mark. 1989. "The Traditional Geographical Range and Ethnobotanical Diversity of *Cordyline fruticosa* (L.) Chevalier." *Ethnobotany* 1: 25–39.

_____. 2000. "A History of Ethnobotany in Remote Oceania." *Pacific Science* 54 (3): 275–87.

Merrifield, Andrew. 2013. *Henri Lefebvre*. Hoboken, NJ: Taylor and Francis.

Merrill, Elmer. 1917. *An Interpretation of Rumphius's Herbarium Amboinense*. Manila, Philippines: Dept. of Agriculture and Natural Resources.

Meyer, Birgit, and Peter Geschiere. 1999. "Introduction." In *Globalization and Identity*, edited by B. Meyer and P. Geschiere, 1–15. Oxford: Blackwell.

Miller, Teresa. 2019. *Plant Kin*. Austin: University of Texas Press.

Mintz, Sidney. 1962. "Living Fences in the Fond-des-Negres Region, Haiti." *Economic Botany* 16 (2): 101–5.

_____. 1965. "The Caribbean as a Socio-Cultural Area." *Journal of World History* 9 (1): 912–37.

_____. 1974. *Caribbean Transformations*. Chicago: Aldine.

_____. 1985. "From Plantations to Peasantries in the Caribbean." In *Caribbean Contours*, edited by S. Mintz and S. Price, 127–53. Baltimore: Johns Hopkins University Press.

Mintz, Sidney, and Douglas Hall. 1960. *The Origins of the Jamaican Internal Marketing System*. Yale University Publications in Anthropology 57. New Haven: Yale University Press.

Mintz, Sidney, and Richard Price. 1976. *An Anthropological Approach to the Afro-American Past*. Philadelphia: Institute for the Study of Human Issues.

_____. 1992. *The Birth of African-American Culture*. Boston: Beacon Press.

Moerenhout, J.A. 1993. *Travels to the Islands of the Pacific Ocean*. Lanham, MD: University Press of America.

Molina-Venegas, Rafael, Markus Fischer, Neduvoto Mollel, and Andreas Hemp. 2020. "Connecting Plant Evolutionary History and Human Well-being at Mt. Kilimanjaro, Tanzania." *Botanical Journal of the Linnean Society* 194 (4): 397–409.

Momsen, Janet. 1987. "Land Settlement as an Imposed Solution." In *Land and Development in the Caribbean*, edited by J. Momsen and J. Momsen, 46–69. London: Macmillan.

Monroe, Cameron. 2013. "Power and Agency in Precolonial African States." *Annual Review of Anthropology* 42 (1): 17–35.

Montlahuc, Marie-Laure, and Gerard Philippson. 2002. "Before Coffee, Pre-Colonial Agricultural Glossary." In *Kilimanjaro: Mountain, Memory, Modernity*, edited by F. Bart, M. Milline, F. Devenne and T. Martin, 59–69. Dar es Salaam: Mkuki na Nyota.

Moore, Donald. 1993. "Contesting Terrain in Zimbabwe's Eastern Highlands: Political Ecology, Ethnography, and Peasant Resource Struggles." *Economic Geography* 69 (4): 380–401.

———. 2005. *Suffering for Territory*. Durham: Duke University Press.

Moore, Jason, ed. 2016. *Anthropocene or Capitalocene?* Oakland: PM Press.

Moore, Sally Falk. 1976. "The Secret of the Men: A Fiction of Chagga Initiation and Its Relation to the Logic of Chagga Symbolism."*Africa* 46 (4): 357–70.

———. 1977. "The Chagga of Kilimanjaro." In *The Chagga and Meru of Tanzania*, edited by W. O'Barr, 1–85. London: International African Institute.

———. 1986. *Social Facts and Fabrications*. Cambridge: Cambridge University Press.

———. 2009. "Past in the Present: Tradition, Land and 'Customary' Law on Kilimanjaro 1880–1980." In *Culture, History, and Identity*, edited by T. Clack, 39–76. Oxford: Archaeopress.

———. 2016. *Comparing Impossibilities*. Chicago: Hau Books.

Morgan, Lewis Henry. 1878. *Ancient Society*. New York: Henry Holt and Company.

Morrison, James. 2010. *After the Bounty*, edited by D. Maxton. Washington, DC: Potomac Books.

Müller, E. 1897. "Die Station Madschame." *Evangelisch-Lutherisches Missionsblatt* 14: 266–70.

Mullin, Michael. 1994. *Africa in America*. Urbana: University of Illinois Press.

Munson, Robert. 2013. *The Nature of Christianity in Northern Tanzania*. Lanham, Maryland: Lexington Books.

Murdock, George. 1959. *Africa: Its Peoples and Their Culture History*. New York: McGraw-Hill.

Myers, Fred. 1991. *Pintupi Country, Pintupi Self.* Berkeley: University of California Press.

Myers, Natasha. 2015. "Conversations on Plant Sensing: Notes from the Field." *NatureCulture* 1 (3): 35–66.

Myhre, Knut. 2006. "Divination and Experience: Explorations of a Chagga Epistemology." *Journal of the Royal Anthropological Institute* 12 (2): 313–30.

———. 2016. "Membering and Dismembering: The Poetry and Relationality of Animal Bodies in Kilimanjaro." In *Cutting and Connecting*, edited by K. Myhre, 114–31. New York: Berghahn Books.

———. 2017. *Returning Life*. New York: Berghahn Books.

———. 2019a. "Life and Its Inflections in Kilimanjaro: Becoming and Being Beyond the Metaphoric." *HAU: Journal of Ethnographic Theory* 9 (2): 405–20.

———. 2019b. "Tales of a Stitched Anus: Fictions, Analytics, and Personhood in Kilimanjaro, Tanzania." *Journal of the Royal Anthropological Institute* 25 (1): 9–28.

Myhre, Knut, ed. 2016a. *Cutting and Connecting*. New York: Berghahn Books.

Nadasdy, Paul. 2021. "How Many Worlds Are There? Ontology, Practice, and Indeterminacy." *American Ethnologist* 48 (4): 357–69.

Nanau, Gordon. 2011. "The Wantok System as a Socio-Economic and Political Network in Melanesia." *Omnes* 2 (1): 31–55.

Nanton, Philip. 1983. "The Changing Pattern of State Control in St. Vincent and the Grenadines." In *Crisis in the Caribbean*, edited by R. Cohen and F. Ambursley, 223–46. New York: Monthly Review Press.

National Institute of Statistics, Cameroon (NIS). 2015. Oku population data. http://cameroon.opendataforafrica.org/, accessed 9/26/2017.

Ndaluka, Thomas, and Frans Wijsen, eds. 2014. *Religion and State in Tanzania Revisited*. Zürich: LIT Verlag.

Ndenecho, Emmanuel. 2011. *Ethnobotanic Resources of Tropical Montane Forests*. Bamenda, Cameroon: Langaa Research and Publishing.

Ndishangong, Thaddeus. 1984. *A Historical Study of Self-Reliant Development in Rural Societies*. Yaoundé: University of Yaoundé. Manuscript copy in possession of the author.

Netting, Robert. 1993. *Smallholders, Householders*. Stanford: Stanford University Press.

Neumann, Katharina, and Elisabeth Hildebrand. 2009. "Early Bananas in Africa: The State of the Art." *Ethnobotany Research and Applications* 7: 353–62.

Newbury, Colin. 1980. *Tahiti Nui*. Honolulu: University Press of Hawai'i.

Newell, Jennifer. 2010. *Trading Nature*. Honolulu: University of Hawai'i Press.

Newell, Sasha. 2018. "The Affectiveness of Symbols: Materiality, Magicality, and the Limits of the Antisemiotic Turn." *Current Anthropology* 59 (1): 1–22.

Ngum III, Fon. 2001. "Oku Forest – Our Life and Our Future." *Forests, Trees, and People Newsletter* 45: 19–21.

Ngute, Alain, Rob Marchant, and Aida Cuni-Sanchez. 2021. "Climate Change, Perceptions, and Adaptation Responses among Farmers and Pastoralists in the Cameroon Highlands." In *Handbook of Climate Change Management*, edited by W. Filho, J. Luetz, and D. Ayal. Basel: Springer International Publishing.

Nicholson, Sharon. 2016. "An Analysis of Recent Rainfall Conditions in Eastern Africa." *International Journal of Climatology* 36 (1): 526–32.

Njoh, Ambe. 1992. "Institutional Impediments to Private Residential Development in Cameroon." *Third World Planning Review* 14 (1): 21–37.

Nkwi, Paul. 1976. *Traditional Government and Social Change*. Fribourg, Switzerland: University Press.

———. 1995. "Slavery and Slave Trade in the Kom Kingdom of the 19th Century." *Paideuma* 41: 239–49.

Nkwi, Paul, and Jean-Pierre Warnier. 1982. *Elements for a History of the Western Grassfields*. Yaoundé: University of Yaoundé, Department of Sociology.

Nkwi, Walter. 2010. *Voicing the Voiceless*. Bamenda, Cameroon: Langaa Research and Publishing.

Nolan, Justin, and Nancy Turner. 2011. "Ethnobotany: The Study of People – Plant Relationships." In *Ethnobiology*, edited by E. Anderson, D. Pearsall, E. Hunn and N. Turner, 133–47. Hoboken, NJ: John Wiley and Sons.

North, Douglass. 1990. *Institutions, Institutional Change and Economic Performance.* Cambridge: Cambridge University Press.

Notué, Jean-Paul, and Bianca Triaca. 2005. *Mankon.* Milan: 5 Continents Editions.

Nurse, Derek, and Gérard Philippson. 2006. "Toward a Historical Classification of the Bantu Languages." In *The Bantu Languages*, edited by D. Nurse and G. Philippson, 164–81. London: Routledge.

Nurse, M.C., C.R. McKay, J.T. Young, and C.A. Asanga. 1994. "Biodiversity Conservation through Community Forestry, in the Mountain Forests of Cameroon." XXI World Conference: Global Partnership for Bird Conservation. Rosenheim, Germany: Global Partnership for Bird Conservation Rosenheim. ODI Rural Development Forestry Network Paper No. 18d. Online document at https://odi.org/documents/3274/1106.pdf, accessed August 20, 2022.

Nwokeji, G.U., and David Eltis. 2002. "Characteristics of Captives Leaving the Cameroons for the Americas, 1822–37." *Journal of African History* 43 (2): 191–210.

Nyanchi, Godwill, Kiming Ngala, Nyuyki Bodzemo, Akoni Ngwainbi, and Geroge Diom. 2020. "Challenges of Rural Landscape Mosaic and Beautification in Oku, North West Region of Cameroon." *International Journal of Science and Qualitative Analysis* 6 (1): 1–7.

Obeyesekere, Gananath. 1992. *The Apotheosis of Captain Cook.* Princeton: Princeton University Press.

Odner, Knut. 1971. "A Preliminary Report on an Archaeological Survey on the Slopes of Kilimanjaro." *Azania* 6: 131–49.

Ogden, Laura, Billy Hall, and Kimiko Tanita. 2013. "Animals, Plants, People, and Things." *Environment and Society* 4 (1): 5–24.

O'kting'ati, A.K.U., and Hussein Mongi. 1986. "Agroforestry and the Small Farmer: A Case Study of Kilema and Kirua Vunjo in Kilimanjaro." *International Tree Crops Journal* 3 (4): 257–65.

Oldaker, Alan. 1957. "Tribal Customary Land Tenure in Tanganyika." *Tanganyika Notes and Records* 47: 117–44.

Oliver, Douglas. 1974. *Ancient Tahitian Society.* Honolulu: University Press of Hawai'i.

Olwig, Karen. 1997. "Caribbean Family Land: A Modern Commons." *Plantation Society in the Americas* 4 (2–3): 135–58.

Oreszczyn, S., and A. Lane. 2000. "The Meaning of Hedgerows in the English Landscape: Different Stakeholder Perspectives and the Implications for Future Hedge Management." *Journal of Environmental Management* 60 (1): 101–18.

Orliac, Catherine. 1990. "Les Habitations de Polynésie." In *Encyclopedie de la Polynésie*, 2e, vol. 5, edited by A. Lavondès, 25–40. Pape'ete: Éditions de L'Alizé.

Orr, Yancey, Stephen Lansing, and Michael Dove. 2015. "Environmental Anthropology: Systemic Perspectives." *Annual Review of Anthropology* 44: 153–68.

Ortner, Sherry. 1973. "On Key Symbols." *American Anthropologist* 75 (5): 1338–46.

———. 2016. "Dark Anthropology and Its Others: Theory since the Eighties." *HAU: Journal of Ethnographic Theory* 6 (1): 47–73.

Oslisly, Richard, Geoffroy de Saulieu, and Pascal Nlend Nlend. 2015. "Archaeological Data and Oral Traditions from the Oku and Kovifem Sites in Northwest Cameroon." *Azania* 50 (1): 76–91.

Osterhoudt, Sarah. 2017. *Vanilla Landscapes*. Bronx: New York Botanical Garden.

Ostrom, Elinor. 1990. *Governing the Commons*. Cambridge: Cambridge University Press.

———. 2009. "A General Framework for Analyzing Sustainability of Social-Ecological Systems." *Science* 325 (5939): 419–22.

Otto, Jan Michiel, and André Hoekema, eds. 2012. *Fair Land Governance*. Leiden: Leiden University Press.

Page, Ben. 2003. "The Political Ecology of 'Prunus Africana' in Cameroon." *Area* 35 (4): 357–70.

Paiement, Jason. 2007. *"The Tiger and the Turbine."* PhD diss., McGill University.

Panoff, Francoise, and Francis Panoff. 1972. "Maenge Taro and Cordyline: Elements of a Melanesian Key." *Journal of the Polynesian Society* 81 (3): 375–90.

Parkes, Annette. 1997. "Environmental Change and the Impact of Polynesian Colonization: Sedimentary Records from Central Polynesia." In *Historical Ecology in the Pacific Islands*, edited by P. Kirch and T. Hunt, 166–99. New Haven: Yale University Press.

Parkinson, Sydney. 1794. *A Journal of a Voyage to the South Seas in Her Majesty's Ship the* Endeavor. London: Stanfield Parkinson.

Parsons, Talcott. 1954. *Essays in Sociological Theory*, revised ed. Glencoe, IL: Free Press.

Paulson, Susan, and Lisa Gezon, eds. 2005. *Political Ecology across Spaces Scales and Social Groups*. New Brunswick: Rutgers University Press.

Pearsall, Deborah. 2015. *Paleoethnobotany*, 3e. Walnut Creek, CA: Left Coast Press.

Peet, Richard, and Michael Watts, eds. 2004. *Liberation Ecologies*, 2e. London: Routledge.

Peluso, Nancy Lee. 2003. "Nature in the Global South: Environmental Projects in South and Southeast Asia." In *Territorializing Local Struggles for Resource Control*, edited by P. Greenough and A. Tsing, 231–52. Durham: Duke University Press.

Peluso, Nancy Lee, and Christian Lund. 2011. "New Frontiers of Land Control: Introduction." *Journal of Peasant Studies* 38 (4): 667–81.

Pétard, Paul. 1986. *Quelques Plantes Utiles de Polynésie Française et Raau Tahtiti*. Pape'ete: Editions Haere Po No Tahiti.

Pétard, Paul, Maurice Leenhardt, and André Guillaumin. 1946. "Le Ti." *Journal de la Société des Océanistes* 2: 191–208.

Peters, Pauline. 2004. "Inequality and Social Conflict over Land in Africa." *Journal of Agrarian Change* 4 (3): 269–314.

———. 2009. "Challenges in Land Tenure and Land Reform in Africa: Anthropological Contributions." *World Development* 37 (8): 1317–25.

———. 2013. "Conflicts over Land and Threats to Customary Tenure in Africa." *African Affairs* 112 (449): 543–62.

Petitt, Andrea. 2022. "Conceptualizing the Multispecies Triad: Toward a Multispecies Intersectionality." *Feminist Anthropology*. https://doi.org/10.1002/fea2.12099

Philippson, Gérard, and Marie-Laure Montlahuc. 2006. "Kilimanjaro Bantu (E60 and E74)." In *The Bantu Languages*, edited by D. Nurse and G. Philippson, 475–500. London: Routledge.

Pieroni, Andrea, and Ina Vandebroek, eds. 2009. *Traveling Cultures and Plants*. New York: Berghahn Books.

Pigliasco, Guido. 2010. "We Branded Ourselves Long Ago: Intangible Cultural Property and Commodification of Fijian Firewalking." *Oceania* 80 (2): 161–81.

Platteau, Jean-Philippe. 2000. *Institutions, Social Norms and Economic Development*. London: Routledge.

Podolefsky, Aaron. 1987. "Population Density, Land Tenure, and Law in the New Guinea Highlands: Reflections on Legal Evolution." *American Anthropologist* 89 (3): 581–95.

Pollan, Michael. 2001. *The Botany of Desire*. New York: Random House.

Pollard, Ernest, D. Hooper, and N.W. Moore. 1977. *Hedges*. London: Collins.

Pollock, Nancy. 1984. "Breadfruit Fermentation Practices in Oceania." *Journal de la Société des Océanistes* 40 (79): 151–64.

Portlock, Nathaniel. 1789. *A Voyage Round the World*. London: John Stockdale.

Powell, Dulcie. 1972. *The Botanic Garden, Liguanea*. Kingston, Jamaica: Institute of Jamaica.

———. 1977. "The Voyage of the Plant Nursery, H.M.S. *Providence*, 1791–1793." *Economic Botany* 31 (4): 387–431.

Powell, J. 1976. "Ethnobotany." In *New Guinea Vegetation*, edited by K. Paijmans, 106–83. Canberra: ANU Press.

Prance, Ghillean. 1995. "Ethnobotany Today and in the Future." In *Ethnobotany*, edited by R. Schultes and S. von Reis, 60–8. Portland, Oregon: Dioscorides Press.

Prebble, Matthew. 2008. "No Fruit on That Beautiful Shore: What Plants Were Introduced to the Subtropical Polynesian Islands Prior to European Contact?" In *Islands of Inquiry*, edited by G. Clark, F. Leach and S. O'Connor, 227–52. Canberra: ANU Press.

Price, David. 1985. "The Palace and Its Institutions in the Chiefdom of Ngambe." *Paideuma* 31: 85–103.

Price, Richard. 2001. "The Miracle of Creolization: A Retrospective." *New West Indian Guide* 75 (1–2): 35–64.

———. 2002. *First-Time*, 2e. Chicago: University of Chicago Press.

Pukui, Mary, E.W. Haertig, and C.A. Lee. 2002. *Nana I Ke Kumu*. Honolulu: Hui Hanai.

Pulsipher, Lydia. 1994. "The Landscapes and Ideational Roles of Caribbean Slave Gardens." In *The Archaeology of Garden and Field*, edited by N. Miller and K. Gleason, 202–21. Philadelphia: University of Pennsylvania Press.

Quaranta, Ivo. 2010. "Politics of Blame: Clashing Moralities and the AIDS Epidemic in Nso' (North-West Province, Cameroon)." In *Morality, Hope and Grief*, edited by H. Dilger and U. Luig, 173–91. New York: Berghahn Books.

Quinlan, Robert, and Edward Hagen. 2008. "New Genealogy: It's Not Just for Kinship Anymore." *Field Methods* 20 (2): 129–54.

Rackham, Oliver. 1986. *The History of the Countryside*. London: J. M. Dent.

Ragone, Diane. 1991. "Ethnobotany of Breadfruit in Polynesia." In *Islands, Plants, and Polynesians*, edited by P. Cox and S. Banack, 203–20. Portland: Dioscorides Press.

Rapaport, Moshe. 1996. "Between Two Laws: Tenure Regimes in the Pearl Islands." *Contemporary Pacific* 8 (1): 33–49.

Rappaport, Roy. 1968. *Pigs for the Ancestors*. New Haven: Yale University Press.

———. 1984. *Pigs for the Ancestors*, 2e. New Haven: Yale University Press.

Rashford, John. 1988. "Packy Tree, Spirits and Duppy Birds." *Jamaica Journal* 21 (3): 2–10.

_____. 1989. "Leaves of Fire: Jamaica's Crotons." *Jamaica Journal* 21 (4): 19–25.

_____. 1994. "Jamaica's Settlement Vegetation, Agroecology, and the Origin of Agriculture." *Caribbean Geography* 5 (1): 32–50.

Raum, Otto. 1940. *Chaga Childhood*. Hamburg: LIT Verlag.

Ravault, François. 1982. "Land Problems in French Polynesia." *Pacific Perspective* 10 (2): 31–65.

Ribot, Jesse, and Nancy Lee Peluso. 2003. "A Theory of Access." *Rural Sociology* 68 (2): 153–81.

Richardson, Bonham. 1989. "Catastrophes and Change on St-Vincent." *National Geographic Research* 5 (1): 111–25.

_____. 1997. *Economy and Environment in the Caribbean*. Barbados: University of the West Indies Press.

Riesenfeld, Alphonse. 1950. *The Megalithic Culture of Melanesia*. Leiden: Brill.

Riesman, Paul. 1986. "The Person and the Life Cycle in African Social Life and Thought." *African Studies Review* 29 (2): 71–138.

Robbins, Joel. 1995. "Dispossessing the Spirits: Christian Transformations of Desire and Ecology among the Urapmin of Papua New Guinea." *Ethnology* 34 (3): 211–24.

Robbins, Paul. 2020. *Political Ecology*, 3e. Oxford: Wiley.

Robertson, A. F. 1984. *People and the State*. Cambridge: Cambridge University Press.

Robineau, Claude, Jean-François Baré, Alain Babadzan, and François Ravault. 1990. "Le Royaume Chrétien des Pomare 1815–1827." In *Encyclopedie de la Polynésie*, 2e, vol. 6, edited by P.-Y. Toullelan, 57–72. Pape'ete: Éditions de l'Alizé.

Robineau, Claude, and José Garanger. 1990. "Les Marae." In *Encyclopedie de la Polynésie*, 2e, vol. 4, edited by J. Garanger, 57–72. Pape'ete: Éditions de L'Alizé.

Robineau, Claude, and Jean-Louis Rallu. 1990. "Une Société Nouvelle 1815–1827." In *Encyclopédie de la Polynésie*, 2e, vol. 6, edited by P.-Y. Toullelan, 73–88. Pape'ete: Éditions de l'Alizé.

Robotham, Don. 2018. "Introduction: Interrogating Postplantation Caribbean Society." *Journal of Latin American and Caribbean Anthropology* 23 (3): 409–15.

Roseberry, William. 1989. *Anthropologies and Histories*. New Brunswick: Rutgers University Press.

Ross, Eric. 1998. *The Malthus Factor*. London: Zed Books.

Rowlands, Michael. 2005a. "The Embodiment of Sacred Power in the Cameroon Grassfields." In *Social Transformations in Archaeology*, edited by K. Kristiansen and M. Rowlands, 398–415. London: Routledge.

_____. 2005b. "The Archaeology of Colonialism and Constituting the African Peasantry." In *Social Transformations in Archaeology*, edited by K. Kristiansen and M. Rowlands, 375–96. London: Routledge.

Rubenstein, Hymie. 1987. *Coping with Poverty*. Boulder: Westview Press.

_____. 2006. "'Bush,' 'Garden' and 'Mountain' on the Leeward Coast of St. Vincent and the Grenadines, 1719–1995." In *Islands, Forests and Gardens in the Caribbean*, edited by R. Anderson, R. Grove and K. Hiebert, 194–213. Oxford: Macmillan Caribbean.

Rudin, Harry. 1938. *Germans in the Cameroons 1884–1914*. New Haven: Yale University Press.

Rumphius, Georgius. 1743. *Herbarium Amboinense*, vol. 4. Amsterdam: Apud Fransicum Changuion, Joannem Catuffe, Hermannum Uytwerf.

Ryan, John. 2012. "Passive Flora? Reconsidering Nature's Agency through Human-Plant Studies (HPS)." *Societies* 2 (3): 101–21.

Sahlins, Marshall. 1963. "Poor Man, Rich Man, Big-Man, Chief: Political Types in Melanesia and Polynesia." *Comparative Studies in Society and History* 5 (3): 285–303.

———. 1985. *Islands of History*. Chicago: University of Chicago Press.

———. 1995. *How "Natives" Think*. Chicago: University of Chicago Press.

———. 2018. *What the Foucault?* Chicago: Prickly Paradigm Press.

Salmond, Anne. 2009. *Aphrodite's Island*. Berkeley: University of California Press.

———. 2011. *Bligh*. Berkeley: University of California Press.

Sauer, Carl. 1969. *Agricultural Origins and Dispersals*, 2e. Cambridge, MA: MIT Press.

Saulieu, Geoffroy de, Yannick Garcin, David Sebag, Pascal Nlend Nlend, David Zeitlyn, Pierre Deschamps, Guillemette Ménot, Pierpaolo Di Carlo, and Richard Oslisly. 2021. "Archaeological Evidence for Population Rise and Collapse between ~2500 and ~500 cal. yr BP in Western Central Africa." *Afrique: Archéologie & Arts* 17: 11-32.

Saura, Bruno. 1990. "Culture et Renouveau Culturel." In *Encyclopedie de la Polynésie*, 2e, vol. 7, edited by F. Ravault, 57–69. Pape'ete: Editions de l'Alizé.

———. 2013. *Tahiti Ma'ohi*. Pirae, French Polynesia: Au Vent des Îles.

Saura, Bruno, and Dorothy Levy. 2013. *Bobby, l'Enchanteur du Pacifique*. Pirae, French Polynesia: Au vent des Îles.

Schiebinger, Londa. 2004. *Plants and Empire*. Cambridge, MA: Harvard University Press.

Schneider, Katharina. 2013. "Pigs, Fish, and Birds: Toward Multispecies Ethnography in Melanesia." *Environment and Society* 4 (1): 25–40.

Schoenbrun, David. 1998. *A Green Place, a Good Place*. Portsmouth, NH: Heinemann.

Schoenbrun, David, and Jennifer Johnson. 2018. "Introduction: Ethnic Formation with Other-Than-human Beings." *History in Africa* 45: 307–45.

Schomburgk, R.H. 1848. *The History of Barbados*. London: Longman, Brown, Green and Longmans.

Schram, Ryan. 2015. "Notes on the Sociology of Wantoks in Papua New Guinea." *Anthropological Forum* 25 (1): 3–20.

Schroeder, Richard. 2012. *Africa after Apartheid*. Bloomington: Indiana University Press.

Schultes, Richard, and Albert Hofmann. 1979. *Plants of the Gods*. New York: McGraw-Hill.

Scoones, Ian. 2009. "Livelihoods Perspectives and Rural Development." *Journal of Peasant Studies* 36 (1): 171–96.

Scott, James. 1998. *Seeing Like a State*. New Haven: Yale University Press.

Serra, Narcis, and Joseph Stiglitz, eds. 2008. *The Washington Consensus Reconsidered*. Oxford: Oxford University Press.

Setel, Philip. 1999. *A Plague of Paradoxes*. Chicago: University of Chicago Press.

Shanklin, Eugenia. 1990. "Installation Rites in Kom Royal Court Compounds." *Paideuma* 36: 291–302.

———. 2007. "Exploding Lakes in Myth and Reality: An African Case Study." *Geological Society, London, Special Publications* 273 (1): 165–76.

Shannon, J.L. 1956. "Care and Management of Dairy Goats in Trinidad and Tobago." *Journal of the Agricultural Society of Trinidad and Tobago* 56: 118–34.

Sheller, Mimi. 2007. "Arboreal Landscapes of Power and Resistance." In *Caribbean Land and Development Revisited*, edited by J. Besson and J. Momsen, 207–18. New York: Palgrave Macmillan.

Shephard, C.Y. 1945. *Peasant Agriculture in the Leeward and Windward Islands.* Trinidad: Imperial College of Tropical Agriculture.

———. 1947. "Peasant Agriculture in the Leeward and Windward Islands." *Tropical Agriculture* 24: 61–7.

———. 1948. "Problems of Peasant Agriculture in the British West Indies." Sixth International Conference of Agricultural Economists, Aug-Sept 1947, Dartington Hall, UK. London: Oxford University Press.

Sheridan, Michael. 2002."An Irrigation Intake Is Like a Uterus': Culture and Agriculture in Precolonial North Pare, Tanzania." *American Anthropologist* 104 (1): 79–92.

———. 2008. "Tanzanian Ritual Perimetrics and African Landscapes: The Case of Dracaena." *International Journal of African Historical Studies* 41 (3): 491–521.

———. 2012. "Water: Irrigation and Resilience in the Tanzanian Highlands." In *Ecology and Power*, edited by A. Hornborg, B. Clark and K. Hermele, 168–81. London: Routledge.

———. 2014. "The Social Life of Landesque Capital and a Tanzanian Case Study." In *Landesque Capital*, edited by N.T. Håkansson and M. Widgren, 155–71. New York: Routledge.

———. 2016. "Boundary Plants, the Social Production of Space, and Vegetative Agency in Agrarian Societies." *Environment and Society* 7 (1): 29–49.

———. Forthcoming. "When Rain Is a Person: Rainmaking, Relational Persons, and Posthuman Ontologies in Sub-Saharan Africa." In *Climate Change Epistemologies in Southern Africa: Social and Cultural Dimensions*, edited by J. Ahrens and E. Halbmayer. London: Routledge.

Sheridan, Michael, and Celia Nyamweru, eds. 2008. *African Sacred Groves*. Oxford: James Currey.

Shipton, Parker. 1988. "The Kenyan Land Tenure Reform: Misunderstandings in the Public Creation of Private Property." In *Land and Society in Contemporary Africa*, edited by R. Downs and S. Reyna, 91–135. Hanover, NH: University Press of New England.

———. 1994. "Land and Culture in Tropical Africa: Soils, Symbols, and the Metaphysics of the Mundane." *Annual Review of Anthropology* 23: 347–77.

Shivji, Issa. 1994. *A Legal Quagmire*. London: IIED.

———. 2000. "Contradictory Perspectives on Rights and Justice in the Context of Land Tenure Reform in Tanzania." In *Beyond Rights Talk and Culture Talk*, edited by M. Mamdani, 37–60. New York: St. Martin's Press.

Silber, Ilana. 1995. "Space, Fields, Boundaries: The Rise of Spatial Metaphors in Contemporary Sociological Theory." *Social Research* 62 (2): 323–55.

Sillitoe, Paul. 1983. *Roots of the Earth*. Manchester: Manchester University Press.

———. 1988. "From Head-dresses to Head-messages: The Art of Self-Decoration in the Highlands of Papua New Guinea." *Man* 23 (2): 298–318.

———. 1999. "Beating the Boundaries: Land Tenure and Identity in the Papua New Guinea Highlands." *Journal of Anthropological Research* 55 (3): 331–60.

Simon, Scott. 2015. "Real People, Real Dogs, and Pigs for the Ancestors: The Moral Universe of 'Domestication' in Indigenous Taiwan." *American Anthropologist* 117 (4): 693–709.

Simpson, George. 1962a. "Folk Medicine in Trinidad." *Journal of American Folklore* 75 (298): 326–40.

⸻. 1962b. "The Shango Cult in Nigeria and in Trinidad." *American Anthropologist* 64 (6): 1204–19.

Simpson, Philip. 2000. *Dancing Leaves.* Christchurch, New Zealand: Canterbury University Press.

⸻. 2012. "*Cordyline fruticosa* (Ti Plant)." *Invasive Species Compendium.* https://www.cabi.org/isc/datasheet/11866.

Sinoto, Yosihiko. 1996. "Mata'ire'a Hill, Huahine: A Unique Prehistoric Settlement, and a Hypothetical Sequence of *Marae* Development in the Society Islands." In *Oceanic Culture History*, edited by J. Davidson, G. Irwin, F. Leach, A. Pawley and D. Brown, 541–53. Dunedin, New Zealand: New Zealand Journal of Archaeology.

Siragusa, Laura, Clinton Westman, and Sarah Moritz. 2020. "Shared Breath: Human and Nonhuman Copresence through Ritualized Words and Beyond." *Current Anthropology* 61 (4): 471–94.

Slone, Thomas, ed. 2001. *One Thousand One Papua New Guinean Nights.* Oakland: Masalai Press.

Smith, Grafton Elliot. 1915. *The Migrations of Early Culture.* London: Longmans Green and Company.

Smith, Joseph. 1915. "St. Vincent: Report on the Administration of the Roads and Land Settlement Fund from 1st January, 1911, to 31st March, 1914." *Colonial Reports – Miscellaneous #90.* London: HM Stationery Office.

Soini, Eija. 2005a. "Land Use Change Patterns and Livelihood Dynamics on the Slopes of Mt. Kilimanjaro, Tanzania." *Agricultural Systems* 85 (3): 306–23.

⸻. 2005b. "Changing Livelihoods on the Slopes of Mt. Kilimanjaro, Tanzania: Challenges and Opportunities in the Chagga Homegarden System." *Agroforestry Systems* 64 (2): 157–67.

Soja, Edward. 1996. *Thirdspace.* Cambridge, MA: Blackwell.

⸻. 2013. *Seeking Spatial Justice.* Minneapolis: University of Minnesota Press.

Solefack, M.C.M. 2017. "Plant Assemblages along an Altitudinal Gradient of Mount Oku Forests (Cameroon)." *Journal of Agriculture and Ecology Research International* 11 (2): 1–10.

Spinelli, Joseph. 1973. "*Land Use and Population in St. Vincent, 1763–1960.*" PhD diss., University of Florida.

Spiro, Melford. 1965. "A Typology of Social Structure and the Patterning of Social Institutions: A Cross-cultural Study." *American Anthropologist* 67 (5): 1097–1119.

Spriggs, Matthew. 2015. "Oceania: Lapita Migration." In *The Global Prehistory of Human Migration*, edited by P. Bellwood, 767–80. Chichester, UK: Wiley.

St. Vincent, Government of. 1927. *The Laws of St. Vincent Containing the Ordinances of the Colony in Force on the 4th Day of May, 1926*, revised ed., edited by J. Rae. London: Waterlow and Sons.

Stahl, Kathleen. 1964. *History of the Chagga People of Kilimanjaro.* London: Mouton and Company.

Steiner, Franz. 1954. "Chagga Truth: A Note on Gutmann's Account of the Chagga Concept of Truth in *Das Recht der Dschagga*." *Africa* 24 (4): 364–69.

Stephen, James. 1830. *The Slavery of the British West India Colonies Delineated*. London: Saunders and Benning.

Stevenson, Karen. 1990. "'*Heiva*': Continuity and Change of a Tahitian Celebration." *Contemporary Pacific* 2 (2): 255–78.

Stewart, Kristine. 2009. "Effects of Bark Harvest and Other Human Activity on Populations of the African Cherry (*Prunus africana*) on Mount Oku, Cameroon." *Forest Ecology and Management* 258 (7): 1121–28.

Stewart, Pamela, and Andrew Strathern. 1998. "Shifting Places, Contested Spaces: Land and Identity Politics in the Pacific." *Australian Journal of Anthropology* 9 (2): 209–24.

———. 2001. *Humors and Substances*. Westport, CT: Bergin & Garvey.

Stone, Glenn. 1994. "Agricultural Intensification and Perimetrics: Ethnoarchaeological Evidence from Nigeria." *Current Anthropology* 35 (3): 317–24.

Strathern, Andrew. 1971. *The Rope of Moka*. London: Cambridge University Press.

Strathern, Andrew, and Pamela Stewart. 2001. "Rappaport's Maring: The Challenge of Ethnography." In *Ecology and the Sacred*, edited by E. Messer and M. Lambek, 277–90. Ann Arbor: University of Michigan Press.

Strathern, Marilyn. 1988. *The Gender of the Gift*. Berkeley: University of California Press.

———. 2006. "Divided Origins and the Arithmetic of Ownership." In *Accelerating Possession*, edited by B. Maurer and G. Schwab, 135–73. New York: Columbia University Press.

———. 2018. "Persons and Partible Persons." In *Schools and Styles of Anthropological Theory*, edited by M. Candea, 236–46. London: Routledge.

———. 2020. "A Clash of Ontologies? Time, Law, and Science in Papua New Guinea." In *Science in the Forest, Science in the Past*, edited by G. Lloyd and A. Vilaça, 43–74. Chicago: HAU Books.

Stump, Daryl, and Mattias Tagseth. 2009. "The History of Pre-colonial and Early Colonial Agriculture on Mount Kilimanjaro: A Review." In *Culture, History, and Identity*, edited by T. Clack, 107–24. Oxford: Archaeopress.

Sullivan, Sian. 2017. "What's Ontology Got to Do With It? On the Knowledge of Nature and the Nature of Knowledge in Environmental Anthropology." In *Routledge Handbook of Environmental Anthropology*, edited by H. Kopnina and E. Shoreman-Ouimet, 155–69. London: Routledge.

Sunjo, Tata. 2015. "Double Decades of Existence of the Kilum-Ijim Community Forest in Cameroon: What Conservation Lessons?" *Journal of International Wildlife Law & Policy* 18 (3): 223–43.

Suzman, James. 2021. *Work*. New York: Penguin Press.

Swanson, Tod. 2009. "Singing to Estranged Lovers: Runa Relations to Plants in the Ecuadorian Amazon." *Journal for the Study of Religion, Nature and Culture* 3 (1): 36–65.

Tanzania National Archives (TNA). n.d. *Kilimanjaro District Book*. CAMP project. Microfilm 977, reels 3 and 4.

Tardits, Claude. 1980. *Le Royaume Bamoum*. Paris: A. Colin.

Tatham, Maeburn. 1911. *St. Vincent: Report on the Administration of the Roads and Land Settlement Fund*. Colonial Reports, Miscellaneous #77. London: HM Stationery Office.

Taylor, Christopher. 2012. *The Black Carib Wars*. Jackson: University Press of Mississippi.

Te'ehuteatuaonoa (a.k.a. 'Jenny'). 1829. "Pitcairn's Island: The *Bounty*'s Crew." *United Service Journal and Naval and Military Magazine* II: 589–93. Pitcairn Islands Study Center, Pacific Union College, https://library.puc.edu/pitcairn/pitcairn/jenny.shtml.

Tetiarahi, Gabriel. 1987. "The Society Islands: Squeezing Out the Polynesians." In *Land Tenure in the Pacific*, 3e, edited by R. Crocombe, 45–58. Suva, Fiji: University of the South Pacific.

Thomas, Clive. 1984. *Plantations, Peasants, and State*. Los Angeles: Center for Afro-American Studies, University of California.

Thomas, David, Susan Anders, and Nkengla Penn. 2000. "Conservation in the Community: The Kilum-Ijim Forest Project, Cameroon." *Ostrich* 71 (1–2): 157–61.

Thomas, Nicholas. 1989. "The Force of Ethnology: Origins and Significance of the Melanesia/Polynesia Division." *Current Anthropology* 30 (1): 27–41.

Thomas-Hope, Elizabeth. 1995. "Island Systems and the Paradox of Freedom: Migration in the Post-Emancipation Leeward Islands." In *Small Islands, Large Questions*, edited by K. Olwig, 161–75. London: Frank Cass.

Thompson, James, Jens Gebauer, and Andreas Buerkert. 2010. "Fences in Urban and Peri-Urban Gardens of Khartoum, Sudan." *Forests, Trees and Livelihoods* 19 (4): 379–91.

Tomich, Dale. 1993. "*Une Petite Guinée*: Provision Ground and Plantation in Martinique, 1830–1848." In *Cultivation and Culture*, edited by I. Berlin and P. Morgan, 221–42. Charlottesville: University of Virginia Press.

Tomlinson, Matt, and Ty Kāwika Tengan. 2016. *New Mana*. Acton, Australia: ANU Press.

Tony, Christina. 2014. "Message from the WATUR." Bismarck Ramu Group, Land is Life – Land Justice for Papua New Guinea Project. https://www.globalgiving.org/projects/land-is-life-bismark-ramu-group/updates/?subid=52171

Toppin-Allahar, Christine. 2013. *Land Law and Agricultural Production in the Eastern Caribbean*. Rome: FAO. http://www.fao.org/3/a-i3204e.pdf.

Toullelan, Pierre-Yves. 1990. "Des Baleiniers aux Planteurs 1797–1842." In *Encyclopedie de la Polynésie*, 2e, vol. 6, edited by P.-Y. Toullelan, 89–104. Pape'ete: Éditions de l'Alizé.

Toullelan, Pierre-Yves, P. O'Reilly, and H. Vernier. 1990. "La Société Polynésienne 1842–1940." In *Encyclopedie de la Polynésie*, 2e, vol. 7, edited by P.-Y. Toullelan, 73–88. Pape'ete: Éditions de l'Alizé.

Trouillot, Michel-Rolph. 1988. *Peasants and Capital*. Baltimore: Johns Hopkins University Press.

———. 1992. "The Caribbean Region: An Open Frontier in Anthropological Theory." *Annual Review of Anthropology* 21: 19–42.

———. 2002. "Culture on the Edges: Caribbean Creolization in Historical Context." In *From the Margins*, edited by B. Axel, 189–210. Durham: Duke University Press.

Tsing, Anna. 2015. *The Mushroom at the End of the World*. Princeton: Princeton University Press.

Turner, Jonathan. 2003. *Human Institutions*. Lanham, MD: Rowman and Littlefield.

Turner, Victor. 1967. *The Forest of Symbols*. Ithaca: Cornell University Press.

_____. 1969. *The Ritual Process*. Ithaca: Cornell University Press.

_____. 1977. "Process, System, and Symbol: A New Anthropological Synthesis." *Daedalus* 106: 61–80.

Tyerman, Daniel, and George Bennet. 1831. *Journal of Voyages and Travels by the Rev. Daniel Tyerman and George Bennet, Esq.* [...], vol. 1, edited by J. Montgomery. Boston: Crocker and Brewster.

United Kingdom House of Commons. 1842. *Report from the Select Committee on West India Colonies*. London: House of Commons.

University of the West Indies Development Mission (UWI). 1969. *The Development Problem in St. Vincent*. Kingston, Jamaica: Institute of Social and Economic Research.

Van Allen, Judith. 2015. "What Are Women's Rights Good for? Contesting and Negotiating Gender Cultures in Southern Africa." *African Studies Review* 58 (3): 97–128.

van Andel, Tinde, Sofie Ruysschaert, Kobeke van de Putte, and Sara Groenendjik. 2013. "What Makes a Plant Magical? Symbolism and Sacred Herbs in Afro-Surinamese Winti Rituals." In *African Ethnobotany in the Americas*, edited by R. Voeks and J. Rashford, 247–84. New York: Springer.

van der Veen, Margreet. 2014. "The Materiality of Plants: Plant–people Entanglements." *World Archaeology* 46 (5): 799–812.

Vandergeest, Peter, and Nancy Lee Peluso. 1995. "Territorialization and State Power in Thailand." *Theory and Society* 24 (3): 385–426.

Vansina, Jan. 1992a. "Kings in Tropical Africa." In *Kings of Africa*, edited by E. Beumers and H.-J. Koloss, 19–26. Berlin: Foundation Kings of Africa.

_____. 1992b. "History of Central African Civilization." In *Kings of Africa*, edited by E. Beumers and H.-J. Koloss, 13–18. Berlin: Foundation Kings of Africa.

_____. 1995. "New Linguistic Evidence and 'the Bantu Expansion'." *Journal of African History* 36 (2): 173–95.

Vaughn, Sarah, Bridget Guarasci, and Amelia Moore. 2021. "Intersectional Ecologies: Reimagining Anthropology and Environment." *Annual Review of Anthropology* 50: 275–90.

Vayda, Andrew, and Bradley Walters. 1999. "Against Political Ecology." *Human Ecology* 27 (1): 167–79.

Viveiros de Castro, Eduardo. 2014. *Cannibal Metaphysics*. Minneapolis: Univocal.

Voeks, Robert. 1990. "Sacred Leaves of Brazilian Candomblé." *Geographical Review* 80 (2): 118–31.

_____. 2012. "Spiritual Flora of Brazil's African Diaspora: Ethnobotanical Conversations in the Black Atlantic." *Journal for the Study of Religion, Nature and Culture* 6 (4): 501–22.

Volkens, Georg. 1897. *Der Kilimandscharo*. Berlin: Geographische Verlagshandlung Dietrich Reimer.

von Clemm, Michael. 1964. "Agricultural Productivity and Sentiment on Kilimanjaro." *Economic Botany* 18 (2): 99–121.

von den Steinen, Karl. 1925. *Die Marquesaner und ihre Kunst*. Berlin: Dietrich Reimer.

_____. 2007. *L'art du Tatouage aux Îles Marquises*. Pape'ete: Haere po.

von der Decken, Karl Klaus, and Otto Kersten. 1978. *Reisen in Ost-Afrika in den Jahren 1859 bis 1865*. Graz, Austria: Akademische Druck und Verlagsanstalt.

von Freyhold, Micaela. 1979. *Ujamaa Villages in Tanzania*. London: Heineman.

Waddell, Eric. 1972. *The Mound Builders*. Seattle: University of Washington Press.

Wagner, Roy. 1986. *Symbols That Stand for Themselves*. Chicago: University of Chicago Press.

Walker, Frederick. 1937. "Economic Progress of St. Vincent, BWI, Since 1927." *Economic Geography* 13 (3): 217–34.

Walker, Peter, and Pauline Peters. 2001. "Maps, Metaphors, and Meanings: Boundary Struggles and Village Forest Use on Private and State Land in Malawi." *Society and Natural Resources* 14 (5): 411–24.

Wallin, Paul. 1993. *"Ceremonial Stone Structures."* PhD diss., Uppsala University.

Walter, Annie, and Vincent Labot. 2007. *Gardens of Oceania*. Canberra: Australian Center for International Agricultural Research.

Wambeng, Sam. 1993. "Delivery and Naming in Oku." In *Rites of Passage and Incorporation in the Western Grassfields of Cameroon*, edited by P. Mbunda-Samba, P. Mzeka, M. Niba and C. Wirmum, 110–21. Bamenda: Kaberry Research Centre.

Ward, Gerard, and Elizabeth Kingdon. 1995a. "Land Use and Tenure: Some Comparisons." In *Land, Custom and Practice in the South Pacific*, edited by R. Ward and E. Kingdon, 6–35. Cambridge: Cambridge University Press.

———. 1995b. "Land Tenure in the Pacific Islands." In *Land, Custom and Practice in the South Pacific*, edited by R. Ward and E. Kingdon, 36–64. Cambridge: Cambridge University Press.

Wardle, Huon. 1999. "Jamaican Adventures: Simmel, Subjectivity and Extraterritoriality in the Caribbean." *Journal of the Royal Anthropological Institute* 5 (4): 523–39.

Warf, Barney, and Santa Arias, eds. 2009. *The Spatial Turn*. New York: Routledge.

Warner, Ashton. 1831. *Negro Slavery Described by a Negro*. London: S. Maunder.

Warnier, Jean-Pierre. 1975. *"Pre-colonial Mankon."* PhD diss., University of Pennsylvania.

———. 2007. *The Pot-King*. Leiden: Brill.

———. 2012. *Cameroon Grassfields Civilization*. Bamenda, Cameroon: Langaa Research & Publishing.

———. 2014. "Review of *Cameroon: Thoughts and Memories*, by HJ Koloss." *American Anthropologist* 116 (4): 875–76.

Watts, Michael. 2015. "Now and Then: The Origins of Political Ecology and the Rebirth of Adaptation as a Form of Thought." In *Routledge Handbook of Political Ecology*, edited by T. Perreault, G. Bridge and J. McCarthy, 19–50. London: Routledge.

Weiner, Annette. 1976. *Women of Value, Men of Renown*. Austin: University of Texas Press.

———. 1992. *Inalienable Possessions*. Berkeley: University of California Press.

West Indian Bulletin. 1910. "Distribution of Economic Plants from West Indian Botanic Stations." *West Indian Bulletin* 11: 146–52.

West Indies Royal Commission (WIRC). 1897. *Report of the West India Royal Commission*. London: HM Stationery Office.

———. 1945. *West India Royal Commission Report* [The Moyne Report]. London: HM Stationery Office.

West, Paige. 2005. "Translation, Value, and Space: Theorizing an Ethnographic and Engaged Environmental Anthropology." *American Anthropologist* 107 (4): 632–42.

_____. 2006. *Conservation Is Our Government Now*. Durham: Duke University Press.

_____. 2012. *From Modern Production to Imagined Primitive*. Durham: Duke University Press.

Westermark, George. 1997. "Clan Claims: Land, Law and Violence in the Papua New Guinea Eastern Highlands." *Oceania* 67 (3): 218–33.

Whistler, Arthur. 2000. *Tropical Ornamentals*. Portland, OR: Timber Press.

_____. 2009. *Plants of the Canoe People*. Kaua'i, HI: National Tropical Botanical Garden.

Whitehead, Ann, and Dzodzi Tsikata. 2003. "Policy Discourses on Women's Land Rights in Sub-Saharan Africa: The Implications of the Re-turn to the Customary." *Journal of Agrarian Change* 3 (1–2): 67–112.

Whyte, Nicola. 2007. "Landscape, Memory and Custom: Parish Identities c. 1550–1700." *Social History* 32 (2): 166–86.

Widenmann, August. 1899. "Die Kilimandscharo-Bevölkerung: Anthropologisches und Ethnographisches aus dem Dschaggalande." *Petermanns Geographische Mitteilungen* 129: 1–101.

Widgren, Mats. 2012. "Resilience Thinking Versus Political Ecology: Understanding the Dynamics of Small-Scale Labour-Intensive Farming Landscapes." In *Resilience and the Cultural Landscape*, edited by T. Plieninger and C. Bieling, 95–110. Cambridge: Cambridge University Press.

Widgren, Mats, and John Sutton, eds. 2004. *Islands of Intensive Agriculture in Eastern Africa*. Oxford: James Currey.

Wilk, Richard, and Lisa Cliggett. 2019. *Economies and Cultures*, 2e. London: Routledge.

Williams, Park, and Chris Funk. 2011. "A Westward Extension of the Warm Pool Leads to a Westward Extension of the Walker Circulation, Drying Eastern Africa." *Climate Dynamics* 37 (11): 2417–35.

Williams, R.O., and R.O. Williams Jr. 1951. *The Useful and Ornamental Plants in Trinidad and Tobago*, 4e. Port-of-Spain, Trinidad: Guardian Commercial Printery.

Wilson, James, and William Wilson. 1799. *A Missionary Voyage to the Southern Pacific Ocean*. London: T. Chapman.

Wilson, Mark. 1947. *Report of the Arusha-Moshi Lands Commission*. Dar es Salaam, Government Printer. Michigan State University library, microfilm #13918.

Wilson, Peter. 1973. *Crab Antics*. New Haven: Yale University Press.

Wimmer, Andreas. 2008. "The Making and Unmaking of Ethnic Boundaries: A Multilevel Process Theory." *American Journal of Sociology* 113 (4): 970–1022.

Winer, Lise. 2008. *Dictionary of the English/Creole of Trinidad & Tobago*. Montreal: McGill-Queen's University Press.

Winter, Christoph. 2009. "The Social History of the Chagga in Outline with Special Reference to the Evolution of the Homegardens." In *Culture, History, and Identity*, edited by T. Clack, 272–305. Oxford: Archaeopress.

Wolf, Eric. 1972. "Ownership and Political Ecology." *Anthropological Quarterly* 45 (3): 201–5.

_____. 1982. *Europe and the People without History*. Berkeley: University of California Press.

_____. 1999. *Envisioning Power*. Berkeley: University of California Press.

_____. 2001. *Pathways of Power*. Berkeley: University of California Press.

World Bank. 2020. *Understanding Poverty: Land.* Online document, accessed Oct. 26, 2021, https://www.worldbank.org/en/topic/land#1

World Intellectual Property Organization (WIPO). n.d. "Bees, Geographic Indications, and Development." Online document, http://www.wipo.int/ipadvantage/en/details.jsp?id=5554, accessed Sept. 26, 2017.

Wright, G. 1929. "Economic Conditions in St. Vincent, BWI." *Economic Geography* 5 (3): 236–59.

Yen, Douglas. 1973. "Agriculture in Anutan Subsistence." In *Anuta*, edited by D. Yen and J. Gordon, 113–48. Honolulu: Bishop Museum.

———. 1987. "The Hawaiian Ti Plant (*Cordyline fruticosa* L.): Some Ethnobotanical Notes." *Notes from the Waimea Arboretum and Botanical Garden* 14 (1): 8–11.

Young, J. 1925. "The Umu-Ti. Ceremonial Fire Walking as Practised in the Eastern Pacific." *Journal of the Polynesian Society* 34 (3): 214–22.

Young, Virginia. 1993. *Becoming West Indian.* Washington, DC: Smithsonian Institution Press.

Young, William. 1806. "A Tour through the Several Islands of Barbadoes, St. Vincent, Antigua, Tobago, and Grenada, in the Years 1791 and 1792." In *History, Civil and Commercial, of the British Colonies in the West Indies*, vol. 4, edited by B. Edwards, 243–76. London: G. and W.B. Whittaker.

Zahorka, Herwig. 2007. "The Shamanic Belian Sentiu Rituals of the Benuaq Ohookng, With Special Attention to the Ritual Use of Plants." *Borneo Research Bulletin* 38: 127–48.

Zama, Isaac. 2001. *"Land Tenure Rights, Community Participation and Conflict Management in Community Forestry in Cameroon."* PhD diss., University of Wisconsin – Madison.

Zane, Wallace. 1999. *Journeys to the Spiritual Lands.* Oxford: Oxford University Press.

Zeitlyn, David. 1993. "Spiders in and Out of Court, or, 'The Long Legs of the Law': Styles of Spider Divination in Their Sociological Contexts." *Africa* 63 (2): 219–40.

Ziegler, Alan. 2002. *Hawaiian Natural History, Ecology, and Evolution.* Honolulu: University of Hawai'i Press.

Zuria, Iriana, and Edward Gates. 2006. "Vegetated Field Margins in Mexico: Their History, Structure and Function, and Management." *Human Ecology* 34 (1): 53–77.

Index

Note: *Italic* page numbers refer to figures and page numbers followed by "n" refer to end notes.

Printed in the United States
by Baker & Taylor Publisher Services